# Fundamentals of Biofuel Production Processes

# Fundamentals of Biofuel Production Processes

Debabrata Das

Jhansi L. Varanasi

## CRC Press
Taylor & Francis Group
Boca Raton London New York

CRC Press is an imprint of the
Taylor & Francis Group, an **informa** business

CRC Press
Taylor & Francis Group
6000 Broken Sound Parkway NW, Suite 300
Boca Raton, FL 33487-2742

First issued in paperback 2020

ISBN-13: 978-1-138-08661-6 (hbk)
ISBN-13: 978-0-367-77994-8 (pbk)

### Library of Congress Cataloging-in-Publication Data

Names: Das, Debabrata, 1953- author. | Varanasi, Jhansi L., author.
Title: Fundamentals of biofuel production processes / Debabrata Das and Jhansi L. Varanasi.
Description: Boca Raton : Taylor & Francis, CRC Press, [2019] | Includes bibliographical references and index. |
Identifiers: LCCN 2018058272 (print) | LCCN 2018060129 (ebook) | ISBN 9781351617512 (Adobe PDF) | ISBN 9781351617499 (Mobipocket) | ISBN 9781351617505 (ePub) | ISBN 9781138086616 (hardback) | ISBN 9781315110929 (ebook)
Subjects: LCSH: Biomass energy.
Classification: LCC TP339 (ebook) | LCC TP339 .D374 2019 (print) | DDC 662.6/6--dc23
LC record available at https://lccn.loc.gov/2018058272

**Visit the Taylor & Francis Web site at**
**http://www.taylorandfrancis.com**

**and the CRC Press Web site at**
**http://www.crcpress.com**

# Contents

# Foreword

The participating 196 countries in the United Nations Climate Change Conference 2015 held in Paris agreed to reduce the greenhouse gas emissions to maintain the global atmospheric temperature. The accelerating rate of world population causes the global energy demands, which are mostly fulfilled by the fossil fuels. These fossil fuels are responsible for most greenhouse gas emissions. With the increasing concerns over climate change and the world $CO_2$ emissions, most countries have now shifted their focus towards carbon neutral renewables as alternative sources of energy.

In this respect, biomass energy appears to be the most abundant renewable form of energy and has tremendous potential to substitute for the conventional fossil fuels. It is estimated that biomass contributes 14 percent of the total 18 percent renewables in the energy mix, accounting for 10 percent of the total global energy supply. Traditionally, biomass has been used for local energy demands such as direct combustion, cooking, drying, and charcoal production in developing countries. However, with energy independency and environmental concerns, the modern use of biomass has made it a global trade commodity. Biomass feedstocks can be used readily as energy carriers for electricity generation sectors as well as for transportation sectors.

Biofuels represent a generic term for the transportation fuels obtained from the biomass feedstock. Different varieties of fuels can be obtained from biomass feedstock, such as ethanol, methanol, butanol, biodiesel, hydrogen, and methane. Biofuel combustion will produce significantly less greenhouse gases, little acid rain ingredients, no oxygen depletion, and less environmental pollution. Biofuels are manufactured by various methods such as direct thermal, thermochemical, electrochemical, and biological methods.

A book on biofuel production processes would certainly help in compiling all the necessary information related to it. I congratulate the authors, Debabrata Das and Jhansi L. Varanasi, for seeing the need for such a book and producing it. This book, *Fundamentals of Biofuel Production Processes*, comprehensively covers all aspects of the biofuel production processes. Microbiology, biochemistry, molecular biology, various feedstock, various biofuels production processes, influence of reactor configurations on the biofuels production, physicochemical parameters on biofuels production, scale-up and energy analysis, process economics, policy, and environmental impact are all included in this book. I strongly recommend this excellent book to energy scientists, environmentalists, engineers, policy makers, and students who are interested in biofuel production from organic wastes.

**T. Nejat Veziroglu,**
*International Association for Hydrogen Energy*

# Preface

*Nothing in life is to be feared, it is only to be understood. Now is the time to understand more, so that we may fear less.*

—**Marie Curie**

Energy plays an important role in human society because it contributes to the technological developments and social progress of a country. This contribution has great impact on the quality of our life. Due to rapid industrialization and urbanization, energy consumption has been increased extensively. According to the International Energy Agency (IEA), the world's primary energy need is expected to grow by 55 percent between 2005 and 2030, at an average annual rate of 1.8 percent. At present, the world energy needs are supplied mostly through fossil fuels such as coal, natural gases, and petrochemical sources that will be exhausted in less than 100 years as predicted by the World Energy Forum. Fossil fuels also are responsible for the greenhouse effect resulting from the generation of carbon dioxide.

Renewable energy sources as alternatives to fossil fuel energy sources may play a vital role in overcoming the energy shortage problem in future. Renewable energy sources not only help to meet our energy demand but also safe guard our environment. Biomass is considered an important renewable energy source. The emergence of biofuel production from biomass has endured since the dawn of early civilization. Solid biofuels such as wood and cow dung have been used for cooking and heating purposes for ages. Similarly, liquid biofuels have been used in the automotive industries since its inception. The first internal combustion engine was designed to run on a blend of ethanol and turpentine.

The severe oil crisis in the 1970s prompted renewed interest in biofuels as alternate sources of fuel. Countries like the United States and Brazil started large-scale production of bioethanol and biodiesel for use as transportation fuels. Most of these fuels were produced using food-based crops (first-generation biofuels) such as sugar cane, corn, and oil palm. However, the shift in arable land usage for fuel crops led to a shortage in food resources and an increase in the price of food crops. Over the past few decades, various efforts are being made to use alternate feedstock to improve the biofuel technologies, which can provide sustainable solution to world energy issues, climate change, and high oil prices. Biomethanation processes have been successfully operated in industrial sectors and rural places. The biohythane process comprised of biohydrogen production followed by biomethanation has been successfully developed for the maximization of the gaseous energy recovery.

This book covers all the fundamentals of the state-of-the-art biofuel production technology, which will be helpful for the research community, entrepreneurs, academicians, and industrialists. It is a comprehensive collection of chapters related to microbiology, biochemistry, molecular biology, feedstock requirements, bioreactors development, and scale-up studies of biofuel production processes. Moreover, the book includes the benefits of using the biofuels with respect to not only to meet energy demands but also to safeguard our environment. A comprehensive energy analysis of the biofuel production processes is also a unique feature of this book. This book can be a perfect handbook for young researchers involved in bioenergy, scientists of process industries, policymakers, research faculty, and others who wish to know the fundamentals of biofuel production technology. This book is written in a manner so that it can be useful to both senior students and graduates of energy engineering. It should also be useful in environmental engineering courses. Each chapter begins with a fundamental explanation for general readers and ends with in-depth scientific details suitable for expert readers. Various bioengineering and bioenergy laboratories may find this book a ready reference for their routine use.

We hope this book will be useful to our readers!

**Debabrata Das**

**Jhansi L. Varanasi**

# Authors

Dr. Debabrata Das pursued his doctoral studies at the Indian Institute of Technology (IIT) Delhi. He is a Senior Professor at IIT Kharagpur. He also was associated as the MNRE Renewable Energy Chair Professor. He pioneered the promising research and development of bioenergy production processes by applying fermentation technology. He has been involved actively in the research of hydrogen biotechnology for the last 20 years. His commendable contributions towards development of a commercially competitive and environmentally benign bioprocess began with the isolation and characterization of the high-yielding bacterial strain *Enterobacter cloacae* IIT-BT 08, which as of today is known to be the highest producer of hydrogen by fermentation. He conducted basic scientific research on the standardization of physicochemical parameters for the maximum productivity of hydrogen by fermentation and made significant contribution towards enhancement of hydrogen yield by redirection of biochemical pathways. Besides pure substrates, use of several other industrial wastewater such as distillery effluent, starchy wastewater, deoiled cake of several agricultural seeds like, groundnut, coconut, and cheese whey also were explored successfully as feedstock for hydrogen fermentation. He also has established thermophilic hydrogen production process for the utilization of the higher temperature industrial wastes such as distillery effluent. His recent work on the biohythane process for the maximization of gaseous energy recovery from organic wastes is worth mentioning. Prof. Das also has been involved in $CO_2$ sequestration, biohydrogen, biodiesel, and phycobillin protein production from microalgae research work for the last 10 years. He has been leading the Indian Group in the Indo-Danish-sponsored research project of Department of Biotechnology, Government of India on "High rate algal biomass production for food, biochemicals and biofuels." He organized one International Conference on "Algal Biorefinery: a Potential Source of Food, Feed, Biochemicals, Biofuels and Biofertilizers (ICAB 2013)" at IIT Kharagpur from January 10 to 12, 2013. He also explored the potentiality of microalgae in the operation of Microbial Fuel Cells (MFCs). The major aim of the research work was to synchronize the bioremediation of wastewater with clean energy generation. Presently, he has a Google h-index of 46 for his research work. He has 145 research publications in peer-reviewed journals and has contributed more than 35 chapters in the books published by international publishers. He is the author of *Biohydrogen Production: Fundamentals and Technology Advances* published by CRC Press, New York and *Biohythane: Fuel for the future* published by Pan Stanford Publishing, Singapore. He is the editor of *Algal Biorefinery: An Integrated Approach* and *Microbial Fuel Cell: A Bioelectrochemical System That Converts Waste to Watts*

published separately by Springer, Switzerland, and Capital Publishing Company, India. He has two Indian patents. He is the Editor-in-Chief of *American Journal of Biomass and Bioenergy*. He is the member of the editorial board of *International Journal of Hydrogen Energy, Biotechnology for Biofuels, Indian Journal of Biotechnology*, and *INAE Letters*. He has successfully completed six pilot plant studies in different locations in India. Recently, he successfully demonstrated a 10 $m^3$ pilot plant for the biohydrogen production process under the Technology Mission Project of Ministry of New and Renewable Energy (MNRE). He is involved in several national and international sponsored research projects such as NSF, USA; DAAD, Germany; MNRE, DBT, and DRDO, Government of India. He has been awarded the zIAHE Akira Mitsue Award 2008 and the Malaviya Memorial Award 2013 for his contribution to hydrogen research. He is a Fellow of the International Association for Hydrogen Energy 2016, Indian National Academy of Engineering 2015, Institute of Engineers (India) 2012, Biotechnology Research Society of India 2011, and West Bengal Academy of Science and Technology, 2004. He offered two 12 weeks web-based courses on Industrial Biotechnology and Aspects of Biochemical Engineering under NPTEL. He delivered a series of lectures in the GIAN program on Biotechnology and process engineering for biofuels production 2016 at the National Institute of Technology, Jalandhar.

**Ms. Jhansi L. Varanasi** earned her Masters on Technology degree in Biotechnology from Jawaharlal Nehuru Technological University Kakinada, Vizianagram campus, Andhra Pradesh, India in 2012. She then moved to the Indian Institute of Technology (IIT) Kharagpur, West Bengal, India, where she is currently pursuing her PhD under the guidance of Prof. Debabrata Das. Her research interests focus on the design and development of microbial electrochemical technologies for renewable energy generation and biochemical production.

# List of Symbols

| | |
|---|---|
| \$ | dollar |
| € | euro |
| A | Arrhenius constant, feed rate of substrate $(g\,L^{-1}\,h^{-1})$ |
| D | dilution rate, ti me$^{-1}$ |
| $e^-$ | electron |
| E | redox potential (Volt) |
| $E_0$ | standard redox potential |
| $E_a$ | activation energy $(kJ\,mol^{-1})$ |
| EJ | Exajoule $(10^{18}$ joules$)$ |
| F | volumetric flow rate, volume/time |
| Fe | iron |
| $F_S$ | feed rate of the substrate, mass/time |
| $gL^{-1}$ | gram per liter |
| gal | gallon |
| h | hour |
| $H^+$ | proton |
| ha | hectare |
| K | Kelvin |
| kg | kilogram |
| kJ | kilojoule |
| $K_S$ | saturation constant, mass/volume |
| kW | kilowatt |
| $kW_{el}$ | kilowatt electric |
| mg $L^{-1}$ | milligram per litre |
| min | minute |
| mL | milliliter |
| mol | mole |
| mV | millivolt |
| mW | milliwatt |
| Ni | nickel |
| °C | degree centigrade |
| Pa | Pascal |
| ppm | parts per million |
| Pt | platinum |
| R | universal gas constant $(J\,mol^{-1}\,K^{-1})$ |
| $R^2$ | coefficient of determination |
| $r_S$ | rate of substrate degradation, mass/volume time |
| S | substrate concentration, mass/volume |
| $S_0$ | initial substrate concentration, mass/volume |
| T | absolute temperature (K) |
| t | tonne, time |
| $t_{batch}$ | time required in batch process |

| | |
|---|---|
| $t_{\text{down time}}$ | down time or idle time |
| $V$ | volume of the reactor |
| $X$, $X_t$ | cell mass concentration at time t, mass/vol |
| $X_0$ | initial cell mass concentration, mass/vol |
| $Y_{X/S}$ | yield coefficient, mass of cell/mass of substrate consumed |
| $\mu E$ | microeinsteins |
| $\tau_{\text{CSTR}}$ | space time of CSTR |
| $\tau_{\text{PFR}}$ | space time of PFR |
| $\mu$ | specific growth rate of the microorganism, $\text{time}^{-1}$ |
| $\mu_{\text{max}}$ | maximum specific growth rate, $\text{time}^{-1}$ |

# List of Abbreviations

| | |
|---|---|
| **ABE** | acetone-butanol-ethanol |
| **ACC** | acetyl CoA carboxylase |
| **ACP** | acyl carrier protein |
| **AD** | anaerobic digestion |
| **AEM** | anion exchange membrane |
| **APHA** | American Public Health Association |
| **ASTM** | American Society for Testing and Materials |
| **ATCC** | American Type Culture Collection |
| **ATP** | adenosine triphosphate |
| **BES** | bioelectrochemical system |
| **BOD** | biochemical oxygen demand |
| **BP** | British Petroleum |
| **BTU** | British thermal unit |
| **CBP** | consolidated bioprocessing |
| **CEM** | cation exchange membrane |
| **CFU** | colony forming units |
| **$CH_4$** | methane |
| **CHP** | combined heat and power |
| **CIBE** | International Confederation of European Beet Growers |
| **CM** | cane molasses |
| **CNG** | compressed natural gas |
| **CO** | carbon monoxide |
| **$CO_2$** | carbon dioxide |
| **CoA** | coenzyme A |
| **COD** | chemical oxygen demand |
| **CSTR** | continuous stirred-tank reactor |
| **DAG** | diacylglycerol |
| **DCW** | dry cell weight |
| **DF** | dark fermentation |
| **DGAT** | diacylglycerol acyltransferase |
| **DGGE** | density gradient gel electrophoresis |
| **DNA** | deoxyribonucleic Acid |
| **EAB** | electrochemically active bacteria |
| **EIA** | Energy Information and Administration |
| **ETBR** | ethidium bromide |
| **FAME** | fatty acid methyl ester |
| **FFA** | free fatty acid |
| **G3P** | glycerol-3-phosphate |
| **GDOC** | groundnut deoiled cake |
| **GHGs** | greenhouse gasses |
| **GPAT** | glycerol-3-phosphate acyltransferase |
| **GTL** | gas-to-liquid |

| | |
|---|---|
| $H_2$ | hydrogen |
| H/D | height-to-diameter |
| $H_2SO_4$ | sulphuric acid |
| HCl | hydrochloric acid |
| HER | hydrogen evolution reaction |
| HRT | hydraulic retention time |
| HTL | hydrothermal liquefaction |
| IEA | International Energy Agency |
| KOH | potassium hydroxide |
| LCA | life cycle assessment |
| MBR | membrane bioreactor |
| MEC | microbial electrolysis cell |
| METs | microbial electrochemical technologies |
| MFC | microbial fuel cell |
| MLVSS | mixed liquor volatile suspended solids |
| MTCC | microbial type culture collection |
| $N_2$ | nitrogen |
| NADH | nicotinamide adenine dinucleotide |
| NADPH | nicotinamide adenine dinucleotide phosphate |
| NaOH | sodium hydroxide |
| NER | net energy ratio |
| $O_2$ | Oxygen |
| OLR | organic loading rate |
| ORR | oxidation reduction reaction |
| PA | phosphatidic acid |
| PAP | phosphatidic acid phosphatase |
| PBR | packed-bed reactor |
| PEM | proton exchange membrane |
| PFL | pyruvate formate lyase |
| PFOR | pyruvate ferredoxin oxidoreductase |
| PFR | plug flow reactor |
| PHA | polyhydroxyalkanoate |
| PUFA | polyunsaturated fatty acids |
| R&D | research and development |
| RFA | Renewable Fuel Association |
| RNA | ribonucleic acid |
| rRNA | ribosomal ribonucleic acid |
| RuBisCO | ribulose-1,5-bisphosphate carboxylase oxygenase |
| SBR | sequential batch reactor |
| SEM | scanning electron microscopy |
| SHF | separate hydrolysis and fermentation |
| SRC | short rotation crops |
| SRT | solid retention time |
| SSCF | simultaneous scarification and co-fermentation |
| SSF | simultaneous scarification and fermentation |
| TAG | triacylglycerol |

| | |
|---|---|
| **TDS** | total dissolved solids |
| **TEA** | terminal electron acceptor |
| **TS** | total solids |
| **TSS** | total suspended solids |
| **TVS** | total volatile solids |
| **UASB** | upflow anaerobic sludge blanket |
| **UNFCCC** | United Nations Framework Convention on Climate Change |
| **USR** | upflow solids reactor |
| **UV** | ultraviolet |
| **VFAs** | volatile fatty acids |
| **VS** | volatile solids |
| **VSS** | volatile suspended solids |

# 1 Introduction to Biofuels

## 1.1 INTRODUCTION

Energy has an inescapable role in the human society because it dictates the technological developments and social progress of a country and, in turn, thus improves the quality of life. With the rapid economic progress and growing world population, the demand for energy is expected to increase exponentially. According to the International Energy Agency (IEA), the world's primary energy demand is expected to grow by 55% between 2005 and 2030, at an average annual rate of 1.8%. At present, the world energy demands are predominantly supplied through fossil fuels such as coal, natural gas, and petrochemical sources. However, these sources are non-renewable and will be exhausted in less than 100 years as predicted by the World Energy Forum (Coelho 2012). The increasing energy demands and shortage of fossil fuel production have led to an increase in the costs of the petroleum fuels that jeopardize the economic progress of nations. Apart from the fuel crisis, accelerated global warming is another major crisis that the world faces today. The increasing $CO_2$ concentrations and other greenhouse gases (GHGs) in the atmosphere are primarily due to the combustion of fossil fuels. In view of the rising world population, depleting fossil fuels resources, and the global warming crisis, the utmost challenge for the international community today is to attain a sustainable economic development as well as environmental security. The growing concerns of significant global climate change and national energy security have triggered the quest for renewable sources of energy.

Several renewable energy sources have been developed that could replace the conventional fossil fuels. Figure 1.1 shows the global renewable energy consumption for 2017. Among the various renewables, biomass appears to be a most abundant and promising source of energy since it is easily available, environmentally friendly, and is non-toxic in nature. By the application of different conversion technologies, biomass can be readily converted to various forms of liquid and gaseous biofuels such as ethanol, butanol, hydrogen, methane, and so on, which can serve as potential domestic and transportation fuels. This chapter aims to acquaint the readers with the state-of-the-art of biofuels and the related technologies. A detailed evaluation of the current status and concerns is presented with an outlook on a biofuel-based economy. In addition, a brief overview of the focus of the book and the subsequent chapters has been provided for the readers' reference.

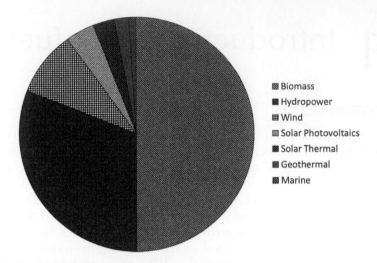

**FIGURE 1.1**    Global renewable energy consumption by fuel type.

## 1.2   HISTORY OF BIOFUELS

The emergence of biofuels has endured since the dawn of early civilization. Solid biofuels such as wood, charcoal, cow dung, and so on have been used for cooking and heating purposes for ages. Similarly, liquid biofuels have been used in the automotive industries since its inception (Nigam and Singh 2011). The first internal combustion engine was designed to run on a blend of ethanol and turpentine. The first ever diesel engine, invented by the German scientist Rudolph Diesel, was intended to run using vegetable oil (Webb 2013). During the industrial revolution from the mid-1700s through the 1800s, coal began to dominate over biomass as a primary source of energy (McLamb 2010). Coal-based fuel sources were much cheaper and efficient compared to the biofuels that suffered from serious disadvantages such as tedious conversion technology and high price. Consequently, the increased supply of fossil fuel-based energy sources led to a steep decrease in biofuel production and its usage. Incidentally, during World War I, the shortage in petroleum fuels brought back the demand for bioethanol (Webb 2013). In addition, the severe oil crisis in the 1970s prompted renewed interest in biofuels as alternate sources of fuel. Countries like the United States and Brazil started large-scale production of bioethanol and biodiesel for use as transportation fuels. Most of these fuels were produced using food-based crops (first-generation biofuels) such as sugarcane, corn, oil palm, and so on. However, the shift in arable land usage for fuel crops led to a shortage in food resources and increase in food crop prices. Over the past few decades, various efforts have utilized alternate feedstocks to improve the biofuel production technologies to provide a sustainable solution to world energy issues, climate change, and high oil prices.

## 1.3  BIOFUEL TYPES

Biofuels are developed and utilized in three major forms: solid biofuels such as wood, charcoal, and so on; liquid biofuels such as bioethanol, biodiesel, and so on; and gaseous biofuels such as biogas, biohydrogen, biohythane, and so on (Guo et al. 2015). Feedstocks, depending on their nature, can be classified into first-, second-, third-, and fourth-generation biofuels. The first-generation biofuels are produced using sugar, starch, or vegetable oil (Chandel et al. 2011). They predominately include food crops such as sugarcane, corn, soybean, oil palm, and so on. If used in large quantities, these feedstocks can have a detrimental impact on food supply. At present, the first-generation biofuels are the only fuels that are produced commercially. The second-generation biofuels are considered greener as compared to the first-generation biofuels because they are produced using non-food crops such as agricultural and forestry wastes, industrial wastes, municipal wastes, and so on (Nigam and Singh 2011). Currently, these fuels are tested only in the laboratory and their availability and complex conversion technologies limit their practical feasibility. The third-generation feedstock refers to the biofuels derived from microbial sources (Fischer et al. 2008). The most commonly used third-generation feedstock are algae, which have shown immense potential for biofuel production with higher yields and lower resource inputs (Demirbas 2010). However, despite numerous advantages, there are several technical barriers for implementation of this technology on a commercial scale. Similarly, the fourth-generation biofuels correspond to the genetically modified feedstocks and hosts that are currently being developed using metabolic engineering and synthetic biology tools to enhance the overall biofuel yields (Peralta-Yahya et al. 2012). The details of different generation of biofuels are discussed in Chapter 2.

## 1.4  RECENT TRENDS IN GLOBAL BIOFUEL PRODUCTION

The deterioration of the environment through harmful carbon emissions has provided the impetus for shifting from fossil fuels to renewables. In addition, the 2015 Paris Agreement made by the UNFCCC (United Nations Framework Convention on Climate Change) to combat climate change has intensified investments in bioenergy (World Energy Council 2016). Table 1.1 lists the total biofuel production across different countries in 2017. It can be observed that the world biofuel production has increased by 11.4% since 2007, thereby contributing towards the lower carbon energy systems. The global ethanol production has increased by 3.3% while the global biodiesel production increased by 4% (British Petroleum 2018). It is estimated that biofuels account for 10.3% of the total global energy supply of which 87% is from the forestry sector, 10% from the agricultural sector, and the remaining 3% comes from the waste energy generated by municipal solid wastes and landfill gas (World Bioenergy Association 2017). Apart from the energy sector, biofuels are readily used in industrial combined heat and power (CPH) processes and domestic purposes such as powering cooking stoves. With the increasing use of biorefineries, various other value-added biochemicals are being generated from biomass along with biofuels (Cherubini 2010).

**TABLE 1.1**

**Biofuel Production Across Different Countries in 2017**

| Country | Production (Thousand Tons Oil Equivalent) | Annual Production Rate (2007–2017) |
|---|---|---|
| United States | 36,936 | 12.9% |
| Canada | 1,239 | 21.4% |
| Brazil | 18,465 | 6.6% |
| Germany | 3,293 | 2.2% |
| United Kingdom | 617 | 9.3% |
| France | 2,224 | 13.4% |
| Belgium | 471 | 35.8% |
| Portugal | 315 | 15.5% |
| Spain | 1,541 | 16% |
| Australia | 144 | 13.4% |
| China | 2,147 | 6.9% |
| India | 435 | 22% |
| Thailand | 1,846 | 33.9% |
| Indonesia | 2,326 | 48.1% |
| World | 84,121 | 11.4% |

*Source:* British Petroleum. BP Statistical Review of World Energy, 2018.

Various initiatives have been undertaken by different countries to expand the biofuel industries. In the United States, General Motors started a project to produce E85 fuel using cellulose ethanol with a projected cost of $1 a gallon (Lamonica 2008). Theoretically, this process is claimed to be five times more energy efficient compared to corn-based ethanol. In Brazil, the government has taken the necessary steps to make the use of blended ethanol and biodiesel mandatory throughout the country (André Cremonez et al. 2015). The United States and Brazil together produce 70% of the total world bioethanol (Ciais et al. 2013). The European Union, in its biofuel directive, has targeted 10% biofuel usage for transportation in all the member states by 2020. Similarly, in Canada, the government aims to blend 45% of the country's gasoline consumption with 10% ethanol. In China and Southeast Asia, the palm oil industries have initiated contributions towards diesel fuel requirements. In India, the National Policy on Biofuel was set up to achieve energy independence by replacing gasoline with alternatives such as 20% bioethanol (E20) and 20% biodiesel (B20) (Singh et al. 2017). The continuing development in biofuels along with supportive governmental policies projects a positive future for the biofuel-based economy. It is expected that bioenergy will supply 30% of the total world energy demand by the year 2050 (Guo et al. 2015).

## 1.5 ENVIRONMENTAL AND SOCIAL-ECONOMIC IMPACT

Compared to the conventional fossil fuels, biofuels are considered environment friendly fuels since they aid in reducing the net greenhouse gas emissions. In addition, the use of biomass contributes towards waste management and mitigates the energy security issues. However, the overall environmental benefit largely depends upon the type of feedstock used for the biofuel production (Naik et al. 2010). Before cane ethanol production, the sugarcane fields were burned after the harvest to remove leaves and fertilize the land with ash, which created significant carbon pollution (Sandhu et al. 2013). Public health authorities have encouraged the elimination of this practice owing to its negative environmental impact. The advancements in fertilizers and natural pesticides have eliminated the need to burn fields, but they give rise to potential chemical pollution problems. Moreover, the increasing land usage to produce biofuel crops poses a severe threat to the natural flora and fauna. It is estimated that replacing 5% of conventional fuels with biofuels has a relatively minor impact on carbon emissions across Canada compared to the United States or Brazil (Nigam and Singh 2011). It is observed that the fertilizer inputs and the transportation of biomass across large distances are the major contributors affecting the net energy savings of the biofuel plant (Larson 2008). Thus, to have a real impact on the country's total greenhouse gas emissions, specific types of biofuels must be targeted based upon the availability of biomass. Second-generation feedstocks such as industrial, agricultural, and municipal solid wastes have shown tremendous potential to reduce the net carbon emissions. However, the technology is still far from being economically viable (Zhang and Zhang 2012).

It is estimated that the biofuels, in the present scenario, are not cost effective. In fact, due to the low energy content of ethanol and biodiesel, the net cost to drive a specific distance in the flexible-fuel vehicles is higher than the current gasoline prices. Almost 80% of the cost for biofuel production is incurred for farming, transport, and conversion of feedstocks (Demirbas 2007). For the operation of a successful commercial biofuel plant, it is essential that the overall energy balance of the process is positive. Energy balance is determined by measuring the difference between the amount of energy input to the manufacture of fuel and the energy released when it is used for combustion in the vehicle. Depending upon the type of biomass feedstock, the energy balances for biofuel production tend to show large differences. Moreover, since most of the fossil fuel energy is still invested in irrigation, transportation, electricity generation, and so on during biofuel production, the sustainability of these technologies seems questionable. The energy and economic analysis of different biofuel production processes such as bioethanol, biodiesel, biomethane, and so on are provided in details in Chapter 15.

## 1.6 CHALLENGES AND OPPORTUNITIES

The growing attention to renewable energy production has brought significant developments in biomass conversion technologies. Biofuels, such as bioethanol and biodiesel, produced using first-generation energy crops, such as sugarcane, corn, soybean, and oil palms, help in reducing global warming. Similarly, second-generation feedstock refers to the biofuels derived from microbial generation feedstocks, such as farm wastes, municipal solid wastes, and so on, seem to be promising low cost substitutes for biofuel production. Despite several benefits, the commercial exploitation of biofuels still suffers from several disadvantages. The low-energy content of bioethanol and its high corrosive properties can easily damage a combustion engine (Devarapalli and Atiyeh 2015). Alternative liquid biofuel like biobutanol is an attractive alternative to bioethanol because it has high energy content comparable to gasoline (Bharathiraja et al. 2017). With the continuing advancements in biofuel technologies, various government policies have made possible the development of improved facilities for biofuel production and usage. Apart from being a renewable energy source, biofuels have several beneficial implications. Since biofuels are biodegradable, they reduce the risk of groundwater and soil pollution (Naik et al. 2010). In addition, biofuels provide job opportunities, generate independent income, provide financial independence, and ensure energy security in the developing countries. Biofuel production in a biorefinery approach, involving integration of different bioconversion processes to produce fuels and chemicals, can be a possible solution to improve the process economy (Schlosser and Blahusiak 2011). The biorefinery concept is similar to the current day petro-refineries that produce multiple products from petroleum oil. Biorefinery is especially attractive for rural communities if all the waste generated is recycled. Therefore, the growing global biofuel markets and the industrial biomass-based biorefineries can be valuable tools for socio-economic and sustainable development of a country.

## 1.7  BOOK OVERVIEW

Biofuels exhibit enormous potential as a renewable and sustainable source of energy. To keep abreast of this rapidly evolving technology, this book is intended to provide a detailed description of fundamental concepts of different biofuels including their biochemistry, microbiology, bioreactor engineering, and so on. This textbook consists of fifteen chapters that cover each type of biofuel in details. Chapter 1 includes the basic introduction and history of biofuels with an insight into the recent global trends of biofuel production. Chapter 2 focuses on different renewable feedstocks used for biofuel production along with traditional and modern biomass conversion technologies. In Chapter 3, different genera of microbial strains that are involved in biofuel production are discussed and the methodology of their selection, optimization, and improvement for high rate biofuel production are highlighted. Chapters 4 and 5 describe the various biochemical pathways that are currently being exploited for biofuels production, with a special focus on the possible metabolic engineering and synthetic biology approaches to enhance biofuels yield. Chapters 6 and 7 are exclusively dedicated to biohydrogen production by dark and photofermentative technologies. Chapter 8 provides a brief overview of the basic concepts of the biomethanation process and presents the readers with perspective ideas to address the challenges of large-scale biogas production. In Chapter 9, the major focus is on the commercial bioethanol production with a key emphasis on the economics and the environmental impact of the different ethanol conversion processes. Chapter 10 provides a comprehensive review on the biobutanol production and its potential use as a biofuel. In addition, the various ongoing studies for the improvement of the acetone-butanol-ethanol (ABE) fermentation process are highlighted. Chapter 11 reviews the different processes and feedstocks utilized for biodiesel production and describes the fuel properties and specifications required for global biodiesel standards. In Chapter 12, the principles, mechanisms, and technical aspects of microbial electrochemical cells are explored along with their applications. In Chapter 13, a comprehensive overview of the various reactor configurations employed for the gaseous biofuels production is presented along with their potential benefits and disadvantages. Chapter 14 discusses the different technical obstacles that challenge the expansion of biofuel production and, in addition, certain case studies relating to industrial biofuels production are emphasized. Finally, in Chapter 15, the various approaches to assess the energy and economics of the biofuel production technologies are described. It is hoped that the topics discussed in these chapters will provide some useful tips for creating possible new avenues for biofuel researchers.

## REFERENCES

Bharathiraja B, Jayamuthunagai J, Sudharsanaa T, Bharghavi A, Praveenkumar R, Chakravarthy M, Yuvaraj D (2017) Biobutanol—An impending biofuel for future: A review on upstream and downstream processing tecniques. *Renew Sustain Energy Rev* 68:788–807.

British Petroleum (2018) BP Statistical Review of World Energy (https://www.bp.com/content/dam/bp/en/corporate/pdf/energy-economics/statistical-review/bp-stats-review-2018-full-report.pdf).

Chandel AK, Silva SS, Singh OV (2011) Detoxification of lignocellulosic hydrolysates for improved bioconversion of bioethanol. In: Bernardes MAS (Ed.), *Biofuel Production-Recent Developments and Prospects*. InTech, Rijeka, Croatia.

Cherubini F (2010) The biorefinery concept: Using biomass instead of oil for producing energy and chemicals. *Energy Convers Manag* 51:1412–1421.

Ciais P, Sabine C, Bala G, Bopp L, Brovkin V, Canadell J, Chhabra A et al. (2013) The physical science basis. Contribution of working group I to the fifth assessment report of the Intergovernmental Panel on Climate Change. *Chang IPCC Clim* 465–570.

Coelho ST (2012) Traditional biomass energy: Improving its use and moving to modern energy use. In: Edited by Secretariat of the International Conference for Renewable Energies, Bonn 2004, *Renewable Energy: A Global Review of Technologies, Policies and Markets*. Hoboken, NJ: Taylor & Francis Group, pp. 230–261.

Cremonez PA, Feroldi M, Nadaleti WC, De Rossi E, Feiden A, De Camargo MP, Cremonez FE, Klajn FF (2015) Biodiesel production in Brazil: Current scenario and perspectives. *Renew Sustain Energy Rev* 42:415–428.

Demirbas A (2007) Importance of biodiesel as transportation fuel. *Energy Policy* 35:4661–4670.

Demirbas A (2010) Use of algae as biofuel sources. *Energy Convers Manag* 51:2738–2749.

Devarapalli M, Atiyeh HK (2015) A review of conversion processes for bioethanol production with a focus on syngas fermentation. *Biofuel Res J* 2:268–280.

Fischer CR, Klein-Marcuschamer D, Stephanopoulos G (2008) Selection and optimization of microbial hosts for biofuels production. *Metab Eng* 10:295–304.

Guo M, Song W, Buhain J (2015) Bioenergy and biofuels: History, status, and perspective. *Renew Sustain Energy Rev* 42:712–725.

Lamonica M (2008) GM invests in "trash to ethanol" start-up—CNET. https://www.cnet.com/news/gm-invests-in-trash-to-ethanol-start-up/. Accessed November 12, 2018.

Larson ED (2008) Biofuel production technologies: Status, prospects and implications for trade and development. *United Nations Conf Trade Dev* 1–41.

McLamb E (2010) The Secret World of Energy|Ecology Global Network. http://www.ecology.com/2010/09/15/secret-world-energy/. Accessed November 6, 2018.

Naik SN, Goud V V., Rout PK, Dalai AK (2010) Production of first and second generation biofuels: A comprehensive review. *Renew Sustain Energy Rev* 14:578–597.

Nigam PS, Singh A (2011) Production of liquid biofuels from renewable resources. *Prog Energy Combust Sci* 37:52–68.

Peralta-Yahya PP, Zhang F, Del Cardayre SB, Keasling JD (2012) Microbial engineering for the production of advanced biofuels. *Nature* 488:320–328.

Sandhu HS, Gilbert RA, Kingston G, Subiros JF, Morgan K, Rice RW, Baucum L, Shine JM, Davis L (2013) Effects of sugarcane harvest method on microclimate in Florida and Costa Rica. *Agric For Meteorol* 177:101–109.

Schlosser S, Blahusiak M (2011) Biorefinery for production of chemicals, energy and fuels. *Elektroenergetika* 4:8–15.

Singh S, Adak A, Saritha M, Sharma S, Tiwari R, Rana S, Arora A, Nain L (2017) Bioethanol production scenario in India: Potential and policy perspective. In: Chandel AK and Sukumaran RK (Eds.), *Sustainable Biofuels Development in India*. Springer International Publishing, Cham, Switzerland, pp. 21–37.

Webb A (2013) A brief history of biofuels: From ancient history to today. In: *BioFuelNet*. http://www.biofuelnet.ca/nce/2013/07/31/a-brief-history-of-biofuels-from-ancient-history-to-today/. Accessed November 5, 2018.

World Bioenergy Association (2017) WBA Global Bioenergy Statistics 2017. *World Bioenergy Association* 1–80.

World Energy Council (2016) World Energy Resources Bioenergy (https://www.worldenergy.org/wp-content/uploads/2017/03/WEResources_Bioenergy_2016.pdf).

Zhang J, Zhang W (2012) Controversies, development and trends of biofuel industry in the world. *Environ Skept Critics* 1:48–55.

# 2 Biofuels Production from Renewable Energy Sources

## 2.1 INTRODUCTION

The accelerating rate of world population causes the global energy demands. This demand usually is fulfilled by using fossil fuels, which are a limited reserve. So the demand is diminishing the existing fossil fuels (coal, petroleum, and natural gas) to a great extent. With the increasing concerns over climate change and the world $CO_2$ emissions, most countries now have shifted their focus to carbon neutral renewables as an alternative source of energy (Herzog et al. 2001). In this respect, biomass energy appears to be the most abundant renewable form of energy and has tremendous potential to substitute for conventional fossil fuels (Christian 2000). It is estimated that biomass contributes to 14% of the total 18% renewables in the energy mix, thus accounting for 10% of the total global energy supply (World Energy Council 2016). Traditionally biomass has been used for local energy demands such as direct combustion, cooking, drying, and charcoal production in developing countries (Coelho 2012). However, with the energy independency and the environmental concerns, the modern use of biomass has made it a global trade commodity. Biomass feedstocks can be readily used as energy carriers for electricity generation sectors as well as for transportation sectors. Biofuels represent a generic term for the transportation fuels obtained from the biomass feedstock (Woiciechowski et al. 2016). Different varieties of fuels can be obtained from biomass feedstock, such as ethanol, methanol, butanol, biodiesel, hydrogen, and methane. The procedure for conversion of feedstock varies with the potential fuel that is produced. Based on feedstock, state, type of conversion, and source, biofuels have been classified into various forms as shown in Figure 2.1.

Biofuels are the most viable and sustainable option for replacing the dependency on fossil fuels due to their inherent advantages (i.e., low greenhouse gas [GHG] emissions, carbon neutral, easily available, replenishable, biodegradable, cost effective, etc.). Furthermore, biofuels help to resolve the energy security, foreign exchange savings, and socio-economic issues for developing and industrialized countries. It is expected that, in the coming decades, the share of the biofuel market will be widespread in future energy systems. This chapter deals with the various renewable sources that have been used to produce biofuels with a major focus on the current research and the expected future outcomes of these technologies.

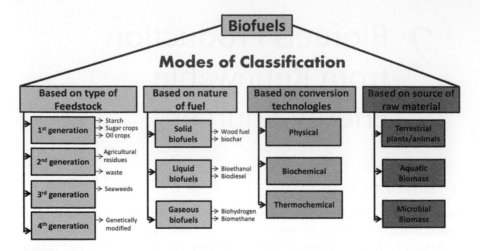

**FIGURE 2.1**  Different modes of classification for biofuels.

## 2.2  FEEDSTOCK FOR BIOFUELS

As shown in Figure 2.1, biofuels can be categorized into different generations based on the type of feedstock used for production. The increase in demand for fuel crops, the debate over the food vs. fuel crops, lower GHG mitigation capabilities, and sustainability issues have resulted in the search for alternate forms of biomass for biofuel production. Figure 2.2 shows the current scenario of the biomass-based biofuel production. It can be observed that the conversion technology and the type of fuel generated are dependent on the initial feedstock. Several factors are essential for determining an ideal biofuel crop such as its availability, sustainability, perennial nature, degradability, composition, and GHG mitigation capability. Knowing these factors along with the ability to develop improved varieties and management practices, the suitable feedstock can be selected. This section presents an overview of different feedstocks that have been used for biofuel production so far and provides an insight on the advent of next generation biofuels from the conventional methodologies.

### 2.2.1  First-Generation Biofuels

The first-generation biofuels are the most conventional form of biofuels. They are primarily produced using edible food crops such as grains, sugar crops, and oil seed crops. The annual yield of these crops varies from region to region and thus the global energy supply depends upon their availability (Table 2.1). The major examples of first-generation biofuels include bioethanol and biodiesel, which are the only biofuels that have been successfully commercialized so far.

Bioethanol is usually produced by the yeast *Saccharomyces cerevisiae* using an anaerobic fermentation process (Demirbas 2008). Apart from the yeast, certain Gram negative bacteria species such as *Escherichia coli* and *Zymomonas mobilis* also have

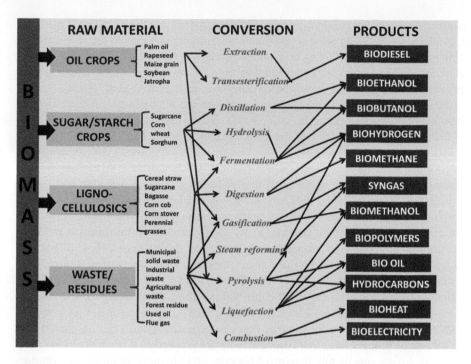

**FIGURE 2.2**   Production of biofuels from various energy crops.

**TABLE 2.1**
**Yields of Major Crops and Their Theoretical Energy Potential in the World**

| Crop | Yield (tons/ha) | | | | | World | |
|---|---|---|---|---|---|---|---|
| | Africa | America | Asia | Europe | Oceania | Yield (tons/ha) | Energy Potential (EJ) |
| Maize | 2.10 | 7.72 | 5.0 | 6.89 | 5.62 | 5.62 | 5.05 |
| Rice | 2.31 | 4.15 | 4.65 | 7.33 | 3.82 | 3.89 | 2.05 |
| Wheat | 1.75 | 2.92 | 3.13 | 4.25 | 3.31 | 3.31 | 2.46 |
| Barley | 1.37 | 3.28 | 1.78 | 3.69 | 2.91 | 2.92 | 0.42 |
| Millet | 0.63 | 1.75 | 1.36 | 1.40 | 0.90 | 0.90 | 0.16 |
| Sorghum | 0.99 | 3.79 | 1.30 | 3.53 | 1.53 | 1.53 | 0.33 |
| Rapeseed | 1.42 | 1.92 | 1.57 | 3.16 | 2.04 | 2.04 | 0.91 |
| Soybean | 1.26 | 2.95 | 1.31 | 1.75 | 2.61 | 2.62 | 1.61 |
| Oil palm | 4.19 | 14.9 | 18.3 | 0.00 | 14.7 | 14.7 | 0.13 |
| Cassava | 11.2 | 8.42 | 13.3 | 21.9 | 10.2 | 10.2 | 0.12 |
| Sugarcane | 64.4 | 67.8 | 67.8 | 81.0 | 69.5 | 69.5 | 0.29 |
| Sugar beet | 53.8 | 53.1 | 53.1 | 62.0 | 60.3 | 60.3 | 0.21 |

*Source:* World Bioenergy Association, WBA Global Bioenergy Statistics 2014, In *World Bioenergy Association*, http://worldbioenergy.org/uploads/WBA GBS 2017_hq.pdf, Accessed May 3, 2018.

been reported as potent microorganisms for bioethanol production (Sims et al. 2008; Chandel et al. 2011). At present, the major bioethanol producers in the world are the United States and Brazil using corn and sugarcane as feedstocks, respectively (Naik et al. 2010). The United Kingdom and Australia on the other hand use wheat as their primary feedstock for the starch-based ethanol industry (Long et al. 2015).

Research has shown that barley and sorghum are alternate potent feedstock for bioethanol production (Drapcho et al. 2008). Barley can be grown as a winter crop to enhance the ethanol yields of the United States by up to 10% (Hattori and Morita 2010). However, barley has several disadvantages. It has low starch content compared to corn and it produces β-glucan as a by-product which makes the fermentation broth viscous, thereby hampering the mixing process. Sorghum on the other hand has a starch content like corn and thus can be used as an alternative to corn. Furthermore, sorghum has an advantage of being drought and heat resistant which makes it suitable for growing in arid regions (Drapcho et al. 2008). India is the third largest producer of sorghum and, due to its attractive fuel properties, the National Policy of Biofuels, Government of India, has identified sweet sorghum as an alternative feedstock for ethanol production in India (Singh et al. 2017). Besides India, sweet sorghum-based ethanol production is successfully established in China, the Philippines, and Brazil.

Another starch-based feedstock for ethanol production is cassava. It is a tropical root crop which contains 22% starch that can be easily hydrolyzed for the ethanol fermentation process. Due to its drought resistant properties, it has been used extensively in Africa, specifically in parts of Nigeria, and Thailand for bioethanol production (Drapcho et al. 2008).

The potential of these starch- and sugar-based crops for biobutanol production also has been investigated (Tashiro and Sonomoto 2010; Chandel et al. 2011; Nigam and Singh 2011). Butanol has more advantage over ethanol as it has a higher octane number, higher energy density, and it is a direct substitute for gasoline. Butanol is produced by fermentative bacteria such as *Clostridium* sp. using the acetone-butanol-ethanol (ABE) pathway. The production pathways for butanol will be discussed in detail in Chapter 10.

In contrast to the bioalcohols, biodiesel is produced by the transesterification of animal fats or plant oil in the presence of a homogeneous or a heterogeneous catalyst (Demirbas 2008). Biodiesel can serve as a direct substitute for petroleum diesel and can be used in diesel engines either directly or with minute modifications. Germany and France are the two largest producers of biodiesel using rapeseed oil as feedstock, whereas the United States and Canada use soybeans for biodiesel production (Long et al. 2015). However, it is observed that production of both crops might not be enough to meet the energy demands of the future (Demirbas 2007).

Palm oil is another potential feedstock for biodiesel used by Malaysia and Indonesia that has brought economic benefits to both countries (Mekhilef et al. 2011). Selection of feedstock for biodiesel production is much more difficult compared to bioalcohols production because several criteria must be met. For example, the water content must be less that 1% to avoid soap production during transesterification process. The free fatty acids must be less than 0.5%. The phosphorus and sulfur content must be less than 10 and 15 ppm, respectively. High contents of saturated and mono-unsaturated fatty acids must be present with low contents of polyunsaturated fatty acids (PFA) (Drapcho et al. 2008). Most biodiesel that is available commercially today meet these criteria.

Despite the successful application of first-generation biofuels, they are associated with a number of constraints and concerns that restrict their use in the commercial market. For example, the growing population and the demand for energy bring out the continuing food versus fuel debate, which concerns negligent use of food crops for fuel production. Furthermore, this use of food crops for fuel production contributes to higher world prices for food and animal feeds. Another major issue is the judicial use of land and water resources for biofuel production. The current production processes use fossil-based power to grow, collect, and process the feedstock, thereby reducing the impact of GHG reduction. The production of first-generation biofuels also has a potential negative impact on the biodiversity and accelerates deforestation. Current production costs make biofuels a more expensive option compared to gasoline even when used in the blended form and thus biofuels are not economically favorable (Chandel et al. 2011). Since the productions costs and food demands are expected to increase over the coming years, there is limited scope for improvement in the production of first-generation biofuels.

## 2.2.2 SECOND-GENERATION BIOFUELS

To overcome the limitations of first-generation biofuels, second-generation biofuels were developed using forest residues and non-food crop based feedstock. These feedstocks either do not require additional land for producing them or they can be grown using marginal or degraded lands. Thus, they do not compete for arable lands with food or fiber production, thus limiting their impact on edible crops (Sims et al. 2008). Furthermore, their impact on $CO_2$ concentrations is expected to be carbon neutral or carbon negative (Naik et al. 2010). Apart from the above characteristic features, the second-generation energy crop must meet several other criteria such as (1) being substantially produced, (2) not affecting the biodiversity ecosystem, (3) using water resources efficiently, (4) cultivation being free of exploitation of land owners, and (5) ensuring benefit to the national economy of a country (Sims et al. 2008). Satisfying all of the above criteria is a challenge for the selection of the second-generation energy crop. Broadly, the feedstocks used for second-generation biofuel production can be categorized as lignocellulosic feedstocks, dedicated energy crops or short rotation crops, and other waste/residues.

### 2.2.2.1 Lignocellulosic Feedstock

Lignocellulosic feedstock is the most abundant renewable feedstock available worldwide with an annual worldwide production of 10–20 billon dry tons (Aro 2016). They include biomass-obtained agricultural wastes, wood and forest residues, perennial grasses or trees, and landfill wastes (such as municipal, commercial, and industrial solid wastes). Biochemical conversion of these feedstocks requires pretreatment prior to fermentation so that the complex carbohydrates are accessible to the hydrolytic enzymes or microorganisms. Different pretreatment techniques have been used so far as shown in Figure 2.3. However, most of these technologies drastically increase the cost of biofuel production.

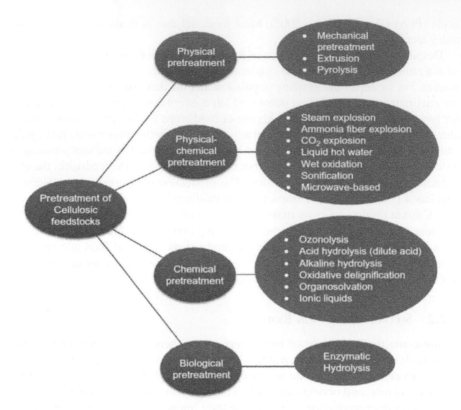

**FIGURE 2.3**  Available pre-treatment techniques for biodegradation of cellulosic wastes.

Lignocellulosic feedstocks are mainly composed of cellulose and hemicellulose polymers interlinked with lignin in a heterogeneous matrix. The combined mass of cellulose and hemicellulose varies from species to species and typically accounts to about 50%–70% of the total dry mass with a significant amount of lignin. The chemical composition of certain agricultural and forest-based lignocellulosic biomass is provided in Table 2.2. Both cellulose and hemicellulose are polymers of sugar moieties which can be hydrolyzed by hydrolytic enzymes. However, lignin is composed of phenolic compounds that inhibit the hydrolysis process (Sims et al. 2008). Thus, a biomass with high lignin content is not suitable for biochemical conversion.

Agricultural feedstocks offer significant quantities of low-lignin containing residues with a high potential for biofuel production. These agricultural feedstocks include cereal straw, wheat chaff, rice husk, corn cobs, corn stover, sugarcane bagasse, nut shells, and so on. Biofuels from these residues can be obtained at relatively reasonable costs with limited infrastructure compared to the dedicated cultivation of energy crops with their associated costs such as labor and land. Some of these residues can be concentrated at processing plants (e.g., bagasse, sawmill residues, etc.), while others must be collected and transported (e.g., cereal straw, rice husk, etc.). To further ensure the economic viability of the process, whole

**TABLE 2.2**

**Chemical Composition of Selected Lignocellulosic Feedstock**

| Biomass | Cellulose (%) | Hemicellulose (%) | Lignin (%) | References |
|---|---|---|---|---|
| Corn stover | 36.9 | 21.3 | 12.5 | Drapcho et al. (2008) |
| Corn cob | 45 | 35 | 15 | Saratale et al. (2008) |
| Barley straw | 37 | 44 | 11 | Chandel et al. (2011) |
| Wheat straw | 34 | 27.6 | 18 | Chandel et al. (2011) |
| Rice husk | 50 | – | 25–30 | Ummah et al. (2015) |
| Rice straw | 37 | 22.7 | 13.6 | Chandel et al. (2011) |
| Bagasse | 39.7 | 24.6 | 25.2 | Chandel et al. (2011) |
| Oat straw | 37.1 | 24.9 | 15.4 | Chandel et al. (2011) |
| Sorghum straw | 35.87 | 26.04 | 7.52 | Cardoso et al. (2013) |
| Switch grass | 31 | 20.4 | 17.6 | Chandel et al. (2011) |
| Bamboo dust | 41–49 | – | 25–28 | Tan et al. (2015) |
| Sawdust | 31–64 | 71–89 | 14–34 | Tan et al. (2015) |
| Paddy straw | 28–48 | – | 12–16 | Tan et al. (2015) |
| Maize stalk straw | 38 | 26 | 11 | Chandel et al. (2011) |
| Nut shells | 25–30 | 25–30 | 30–40 | Saratale et al. (2008) |
| Coconut fiber | 36–43 | 1.5–2.5 | 41–45 | Saratale et al. (2008) |
| Hardwood stems | 40–45 | 18–40 | 18–28 | Biswas et al. (2015) |
| Softwood stems | 34–50 | 21–35 | 28–35 | Biswas et al. (2015) |
| Municipal solid waste | 21–64 | 5–22 | 3–28 | Biswas et al. (2015) |
| Poplar | 49 | 17 | 18 | Chandel et al. (2011) |
| Eucalyptus | 43.3 | 31.8 | 24.7 | Chandel et al. (2011) |
| Miscanthus straw | 44.7 | 29.6 | 21 | Chandel et al. (2011) |

crop biorefineries for harvesting value-added products is suggested. For example, the whole crop harvesting of oil seed rape can provide oil for cooking, high protein meal for poultry, and straw for second-generation biofuel production (Sims et al. 2008).

Wood and forest residues mainly include primary wastes from forestry areas such as bark, off cuts, sawdust, and shavings. Unlike agricultural residues, wood and forest residues contain a high content of lignin and a low moisture content. These characteristics attribute to thermochemical processing of these feedstock for biofuel production. At present, most forest-based residues recovered are burned to generate heat and electricity (Drapcho et al. 2008). For example, black liquor, a by-product of paper and pulp industries (consists of lignin in the pulp form), is burned to generate thermal energy. Similarly, wood pellets, such as the wood chips, sawdust, and round wood, are co-fired to generate green electricity. Although these wastes can be obtained at affordable prices, their sustainability is debatable since the extraction of forest residues can lead to a reduction in soil fertility, lower the soil carbon, enhance soil erosion, and indirectly promote deforestation. Furthermore, their availability is concentrated to certain areas, which would lead to additional transportation costs to deliver the feedstock to the processing site.

Another major source of lignocellulosic wastes includes the landfill wastes generated from municipal, commercial, or industrial solid wastes. These wastes are also known as tertiary cellulosic wastes (Woiciechowski et al. 2016). They can be used to produce conventional biofuels such as bioethanol (Matsakas et al. 2017). Another strategy is use of these wastes during the anaerobic digestion process for biogas production (Mata-Alvarez et al. 2000). Many biogas plants already have been commercialized and the global market for biogas plants is estimated to reach US$10 billion by 2022 (Global Industry Analysts Inc.). Besides biogas production, these feedstock have shown tremendous potential for biohydrogen production via the fermentation processes such as photo and dark fermentation (Meherkotay and Das 2008; Saratale et al. 2008; Perera et al. 2012). These conversion technologies provide the advantage of simultaneous waste treatment along with energy generation. In recent years, the two-stage anaerobic digestion (i.e., biohythane) process for biohydrogen and biomethane production using organic wastes has gained interest as an alternate renewable source of energy (Roy and Das 2016). Although the waste to energy concept is highly attractive, a major hurdle is the separation and the collection of useful residues from the landfill, which makes the process tedious.

### 2.2.2.2 Dedicated Energy Crops or Short Rotation Crops

Energy crops or short rotation crops (SRC) are a subcategory of lignocellulosic feedstocks that are exclusively for accumulating biomass for biofuel production. Most of these crops are densely cultivated to produce ethanol and/or biogas or are combusted to generate heat and electricity. They include the SRCs or short rotation forestry (SRFs) crops such as eucalyptus, poplar, willow, and robinia (Long et al. 2015). They are typically grown in marginal or degraded lands using high yielding varieties. The SRCs are grown in 2- and 4-year rotations with an annual yield of 10 ton hac$^{-1}$ y$^{-1}$. These yields can be improved further using proper crop management practices. SRCs have several advantages over conventional crops for biofuel production; for example, they can be quickly regenerated after harvesting, they can be steadily supplied to the processing plants avoiding the need for storage, they can provide natural filters to the soil for managing floods or bioremediation of water, and, since the wood processing and harvesting is well established, they can be harvested and transported using the existing technologies. In India, extensive research is being conducted on *Jatropha curcas* plant seeds (non-edible oil seed) for biodiesel production using SRF-based agroforestry (Achten et al. 2010). These plant seeds have several advantages such high oil content (40%), carbon neutral nature, and ability to grow in dry marginal, non-agricultural lands unlike soybean, rapeseed, or palm oil. Thus, *Jatropha curcas* has the potential to provide economic benefits to the country. However, even with these advantages, none of the crops mentioned in this chapter have been used fully for biofuel production so far because the yields required to reach the desired production scales require substantial time and production costs. A major limitation is the available water content at the cultivation site, which lowers the yielding capacity of these plants. Furthermore, continuous planting and harvesting leads to depletion of soil nutrients.

Like the SRCs, perennial grasses such as miscanthus, switch grass, and prairie grass have been considered as dedicated energy crops. Like SRCs, they can be grown on marginal grazing lands and in arid climatic zones. Additionally, they have beneficial soil effects on degraded lands (Sims et al. 2008). They require limited inputs for growing and thus their cultivation is affordable. However, like SRCs, the establishment of dedicated grasslands will require significant time and set up costs. Another major hurdle is the control of pests and diseases, which lower the biomass yields. Also, the selection of a specific species is subjected to biodiversity issues.

### 2.2.2.3 Other Waste/Residues/By-products

Besides the feedstock sources already mentioned, certain wastes and residues such as animal wastes, domestic and industrial wastewater, and by-products of industrial processing plants such as glycerol and molasses are used as feedstocks for second-generation biofuel production. Animal wastes are composed of undigested grains and straw which are normally used as fertilizers in farms (Drapcho et al. 2008). The residual animal wastes can be subjected to an anaerobic digestion process for methane production. This form of energy can be readily used on site to satisfy the on-site farm energy requirements thereby reducing the transportation costs. Similarly, the wastewater from sewage or domestic or industrial plants can be used for biogas or biohydrogen production (Woiciechowski et al. 2016). Molasses, a low-value co-product of raw sugar obtained from sugarcane and sugar beets is extensively used for ethanol production in India, Indonesia, the Caribbean, and other countries (Gopal and Kammen 2009; CIBE 2017). Likewise, vinasse, a by-product of the ethanol distillation process, has been considered a potential feedstock for biogas or biohydrogen production (Woiciechowski et al. 2016). Glycerol, a by-product of the transesterification process for biodiesel production, is another important feedstock for hydrogen production or for the production of other commercial products (Adhikari et al. 2009). Using these industrial wastes and by-products as feedstocks can provide significant cost reductions to the biofuel production process. However, the main challenge is the composition of these wastes which would vary depending upon the initial source. Thus, reproducibility of the yields obtained is a major concern when these feedstocks are used for biofuel production.

From the previous discussion, it can be speculated that considerable attention has been given over the past few decades to develop second-generation biofuels due to their intrinsic advantages over the first-generation feedstocks. However, although several investments have been made on pilot and demonstration plants, the commercialization of second-generation biofuels is still questionable. It is observed that the feedstock production costs for the harvesting, treating, transporting, storing, and so on and the necessary conversion technologies are the main hurdles that must be overcome for successful implementation of this technology (Naik et al. 2010). In addition, the quality and the properties of the biofuel obtained from the second-generation feedstock does not meet the requirement of the consumer and the industry. Since the development and breeding of high yielding second-generation energy crops is currently under extensive research, there is still scope to minimize the production costs and improve the process efficiencies with forthcoming scientific advances.

## 2.2.3 THIRD-GENERATION BIOFUELS

The second-generation biofuels address most of the limitations associated with the first-generation biofuels as described in the previous section; however, the concerns over their efficient use of land and water resources continue to prevail. Also, first-generation biofuels cannot compensate for the biodiesel production using vegetable oils or other oil crops. This lack has prompted the research focus towards the use of oleaginous microbial biomass such as algae, yeast, and bacteria as feedstocks for biodiesel production (Nigam and Singh 2011). Oleaginous microbial biomass is characterized by a lipid content of more than 20% (Meng et al. 2009). These organisms can be grown at higher yields using limited resources and can be used to produce both the feedstock as well as fuel. They have a capability to accumulate large amounts of fatty acids in their biomass which can be extracted and used to produce biodiesel. Besides oils, certain species are rich in carbohydrates and proteins which can be used as feedstocks for the anaerobic digestion process for biogas or biohydrogen production. Furthermore, by using biomass biorefinery approaches, certain species can be used to generate high value products that could compensate for the cost of biofuel production.

Algal-based biofuels are the most widely studied group of third-generation biofuels. Algae are primitive photosynthetic organisms that lack roots, stems, or leaves. They assimilate atmospheric $CO_2$ and use light energy for their growth. In fact, they absorb $CO_2$ much more efficiently compared to terrestrial plants (Aro 2016). Depending upon the cell size, they can be micro (ranging from a few micrometers to a few hundreds of micrometers) or macroscopic (extending up to 100 feet) in nature. Due to their biodiversity, a range of fuels, such as ethanol, butanol, biodiesel, biohydrogen, and biogas, can be obtained using algal feedstocks, thus projecting their versatility for biofuel production (Nigam and Singh 2011). They can be grown in autotrophic (inorganic), heterotrophic (organic), and mixotrophic (combination of autotrophic and heterotrophic) modes and cultivated in outdoor raceway ponds or indoor photobioreactors (Nigam and Singh 2011). These different culture techniques allow the use of different combinations for obtaining high yields of biomass and bioproducts. Algae have several inherent advantages for biofuel production; for example, they can produce lipids, carbohydrates, and proteins in large quantities, they can synthesize 100 times more oil per acre of land than any other plant, they can absorb $CO_2$ from discharge gases, thus aiding in reduced GHG emissions, and they can be grown in wastewaters and saline or brackish waters rather than being dependent on the water used for irrigation or human consumption. Certain species, when grown in nutrient-limiting conditions, can convert the free fatty acids (FFA) into triacylglycerol (TAG) accounting for 20%–50% of their dry weight, which in turn can be converted to biodiesel using the transesterification process. However, the nutrient-limiting condition could affect the growth and biomass production. Thus, the selection of the most efficient algal strain with the desired composition is essential to achieve high yields of biofuels. Table 2.3 provides the carbohydrate, protein and lipid content of selected microalgal species. It can be observed that certain species such as *Nanochloropsis oceanica* and *Chlamydomonas reinhardtii* are rich in lipid contents and thus are ideal for biodiesel production. Similarly, *Scenedesmus* sp. and *Arthrospira* sp. are

**TABLE 2.3**
**Compositional Analysis of Selected Algal Biomass**

| Algae | Carbohydrate (g/g) | Protein (g/g) | Lipid (g/g) | References |
|---|---|---|---|---|
| **Macroalgae** | | | | |
| *Laminaria japonica* | 0.56 | 0.08 | 0.01 | Shi et al. (2011) |
| *Ecklonia stolonifera* | 0.48 | 0.13 | 0.02 | Jung et al. |
| *Undaria pinnatifida* | 0.40 | 0.15 | 0.02 | (2011) |
| *Hijikia fusiforme* | 0.28 | 0.05 | 0.01 | |
| *Gelidium amansi* | 0.61 | 0.16 | 0.03 | |
| *Porphyra tenera* | 0.35 | 0.38 | 0.04 | |
| *Gracilaria verrucosa* | 0.33 | 0.15 | 0.03 | |
| *Codium fragile* | 0.32 | 0.10 | 0.01 | |
| **Microalgae** | | | | |
| *Chlorella vulgaris* | 0.38 | 0.49 | 0.007 | Yun et al. (2012) |
| *Chlorella* sp. | 0.25 | 0.41 | 0.11 | Sun et al. (2011) |
| *Chlorella sorokiniana* | 0.14 | 0.14 | | Roy et al. (2014) |
| *Nanochloropsis oceanica* | 0.33 | 0.10 | 0.34 | Xia et al. (2013) |
| *Chlamydomonas reinhardtii* | 0.17 | 0.48 | 0.21 | Becker (2007) |
| *Porphyridium cruentum* | 0.40–0.57 | 0.28–0.39 | 0.09–0.14 | |
| *Dunaliella bioculata* | 0.04 | 0.49 | 0.08 | Demirbas |
| *Tetraselmis maculata* | 0.15 | 0.52 | 0.03 | (2010) |
| *Prymnesium parvum* | 0.25–0.33 | 0.28–0.45 | 0.22–0.38 | |
| *Scendesmus quadricauda* | 0.23 | 0.47 | 0.019 | |
| *Scendesmus dimorphous* | 0.21–0.52 | 0.08–0.18 | 0.16–0.40 | |
| *Spirogyra* | 0.33–0.64 | 0.06–0.20 | 0.11–0.21 | |
| *Scendesmus obliquus* | 0.10–0.17 | 0.50–0.56 | 0.12–0.14 | Becker (2007) |
| *Chlorella vulgaris* | 0.12–0.17 | 0.51–0.58 | 0.14–0.22 | |
| *Spirulina maxima* | 0.13–0.16 | 0.60–0.71 | 0.06–0.07 | |
| *Spirulina platenensis* | 0.08–0.14 | 0.46–0.63 | 0.04–0.09 | |
| *Anabaena cylindrica* | 0.25–0.30 | 0.43–0.56 | 0.04–0.07 | |
| *Synechococcus* sp. | 0.15 | 0.63 | 0.11 | |
| Dunaliella saliana | 0.32 | 0.57 | 0.06 | |
| *Chlorella pyrenoidosa* | 0.26 | 0.57 | 0.02 | |
| Euglena gracilis | 0.14–0.18 | 0.39–0.61 | 0.14–0.20 | |
| Arthrospira platensis | 0.44 | 0.45 | | Cheng et al. (2012) |

rich in carbohydrates and are ideal feedstocks for ethanol, biogas, or biohydrogen production via fermentative or anaerobic digestion processes.

At present, only four algal strains are grown in large scale for commercial purposes. These strains include *Spirulina* sp. and *Chlorella vulgaris*, which are cultivated in large scale for their biomass which is used in food and health industries, and *Dunaliella salina* and *Haemotococcus pluvialis*, which are cultivated to extract carotenoids (like beta-carotene and astaxanthin) for use as health supplements (Long et al. 2015). However, the use of algal strains for biofuel production is still far from reality. Despite many technical and molecular advances, several discrepancies emerge between the projected outcomes based on extrapolations and actual experimental data (Long et al. 2015). The algal oils produced are mostly in unsaturated forms that tend to be volatile at high temperatures and thus are more prone to degradation (Long et al. 2015). Cultivation in open ponds is another major hurdle because it leads to cross contamination and seasonal variation issues that must be carefully addressed. The concept of an algal biorefinery to generate multifaceted products from a single species is presumptuous because it is impossible for a single strain to have all the necessary properties. Although genetic modification to induce desirable properties is possible, the economic viability and the environmental impacts of such species are questionable. Major research and development (R&D) in this area is focused on redirecting metabolic pathways to generate the desired products without compromising growth, development of cost-effective bioreactors, coproduction of high-value chemicals, and the optimization of the harvesting or extraction process to minimize the recovery costs. These developments could make the algal biofuel production profitable in the near future. The life cycle analysis of algal-based biofuels show that it would take more than 25 years for this technology to be commercially available (Wahlen et al. 2013).

In addition to algae, however, yeast, filamentous fungi, and bacteria also have the tendency to accumulate oils at high concentration (Wahlen et al. 2013). These organisms are capable of metabolizing a diverse range of carbohydrates and can grow at faster rates with higher cell densities compared to algal cells. Bacteria are the most fast-growing and easily cultivable organisms. Only a few species belonging to the group actinomycetes have the capability to accumulate lipid in the form of polyhydroxyalkanoate (PHA) (Meng et al. 2009). Although a few studies have been reported for extraction of PHA for biopolymer production, no study has been conducted for using these organisms for biodiesel production. Yeasts and filamentous fungi (molds) on the other hand have been used to produce oleochemicals such as fuels, chemicals, food, and feed ingredients for several decades (Sitepu et al. 2014). The commercial production of these oils demonstrates the feasibility of using yeasts and filamentous fungi for biodiesel production. Yeasts can store up to 70% weight per weight (% weight/weight) lipid in various forms such as triacylglycerides, diacylglycerides, monoacylglycerides, fatty acids, sterol esters, glycolipids, and polyprenols (Sitepu et al. 2014). At present, a few species such as *Rhodosporidium toruloids*, and *Trochosporon pullulans* have been reported for biodiesel production and several new species continue to be discovered. The fatty acid profile and the storage lipid are bound to be varied depending on the type of species. So, like microalgae, strain selection and strain improvement must be the primary concern for biodiesel production. At present, the excessive production costs seem to be the hurdle for this technology to reach the commercial market. At present, using this biomass to produce chemicals, food, and health care supplements seems more economically viable than for biodiesel production. However, these barriers can be overcome by developing high-yielding strains through genetic modifications with simultaneous extraction of multiple products such as vitamins, pigments, and proteins.

## 2.2.4 FOURTH-GENERATION BIOFUELS

The availability of raw materials used for first-, second-, and third-generation biofuels is geographically limited and thus is not sustainable on a global scale. The use of synthetic biology and metabolic engineering tools to derive biofuels from sunlight and $CO_2$ as the inexhaustible and inexpensive resources has been categorized as fourth-generation biofuels. The intent of fourth-generation biofuels is to avoid biomass destruction with complete capturing and storing of $CO_2$ to make the process carbon neutral or carbon negative. By applying this technology, new artificial biological systems can be developed or the metabolic pathways in the existing biological systems can be reconstructed to generate high quality biofuels (Aro 2016). These fourth-generation biofuels can be categorized as solar biofuels, electrobiofuels, and synthetic biofuels depending upon the nature of the raw materials.

The solar biofuels are obtained by the genetic modification of the existing photosynthetic organisms such as unicellular microalgae or cyanobacteria to produce the desired fuels. Not only hydrogen can be generated by the photosynthetic water splitting but also reduced carbon compounds such as methane or ethanol can be extracted using $CO_2$ as raw material. The production of solar biofuels requires careful understanding of the natural light harvesting mechanisms, the associated enzymes, and the means of carbon metabolism of the photosynthetic organisms to alter the pathways for generating valuable chemicals (that are currently obtained via fossil fuels) and biofuels. Successful demonstrations have already been performed using cyanobacteria by introducing various fermentative metabolic pathways into its genome to obtain valuable products such as hydrogen, ethanol, butanol, and lactic acid (Aro 2016). Since this technology is fast emerging, new scientific breakthroughs are expected which may be a promising solution to the world's energy crisis.

Electrofuels are obtained by the combination of photovoltaic cells or solar cells and the bioelectrochemical systems. In case of microbial electrosynthesis, reduced carbon-based chemicals and fuels can be generated using $CO_2$ and electrons from electrodes as carbon and energy sources, respectively (Patil et al. 2015; Bajracharya et al. 2017). These systems are a reverse modification of microbial fuel cells wherein the microorganisms known as electrotrophs that can uptake electron from solid electrodes are used. Currently, this technology is in its nascent stage of research and only a few proof of principle studies have been carried out (Rabaey et al. 2010). Although promising results have been observed, several technical barriers must be overcome for this technology to reach the commercial market.

Synthetic biofuels are developed by microfabricating the biological system at the laboratory scale and developing new metabolic routes to obtain the product of interest (Aro 2016). These systems require modeling and designing of the targeted biological organelles with proper simulation and standardization for the targeted products. It provides a platform to develop new synthetic biology tools (based on the principles of biology and biochemistry) to engineer. The technology, however, is still in the developmental stages and there is lot of scope to construct small synthetic factories or organelles to generate cost-efficient biofuels.

## 2.3 BIOMASS CONVERSION TECHNOLOGIES

Depending upon the simplicity or complexity of the available feedstock and the desired end products, different biomass conversion technologies have been developed over the past few decades. Technologies such as yeast fermentation, transesterification, and distillation for the production of first-generation biofuels have already been successfully implemented on a commercial scale. However, new technologies that are emerging with the different generation of feedstocks are still at experimental or demonstration stages. To make biofuel prices competitive with the traditional fossil fuels, the production technology must be as cost effective as possible. Moreover, it must not have any negative impact on the environment. The current existing bioconversion technologies can be broadly classified into three main categories (i.e., physical, biochemical, and thermochemical methods) as shown in Figure 2.4.

### 2.3.1 PHYSICAL METHODS

#### 2.3.1.1 Biomass Briquetting

Biomass briquettes are alternatives to coal briquettes that are made using densification of agricultural or forestry residues and other waste biomass (Naik et al. 2010). They are also known as bio-coals. They can be produced in compact well-defined shapes such as cubic, prismatic, or cylindrical as per the requirement of the consumer. The densification is performed to enable the use of less volume of the biomass for the same amount of energy output. Besides improving the energy density per unit volume, briquetting helps in handling otherwise bulky or uneven biomass,

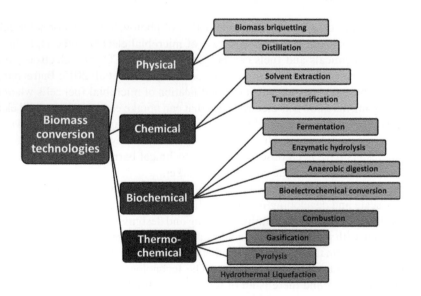

**FIGURE 2.4**   Types of biomass conversion technologies.

reducing its water content, increasing the calorific fraction, lowering the ash content, and providing high homogeneity to the feedstock (Solano et al. 2016). The major goals for producing biomass briquettes are to provide high value to an existing product and lower the transportation costs. The process of densification is two steps—compaction to reduce the raw material volume and sealing to ensure the stability of the final product. The biomass briquettes can be used in several domestic and industrial applications such as cooking fuel as a substitute to coal, firing in industrial furnaces, steam and heat generation in boilers, and for residential heating.

#### 2.3.1.2 Steam Distillation

Distillation is a method of separating the components of a mixture based on boiling points by means of evaporating and then condensing the vapor into liquid. In steam distillation, steam is supplied as a means of heat to evaporate heat sensitive compounds at lower temperatures. The end product is a two-phase system with a fraction of water and the organic distillate. The desired component from the two-phase system can be separated using partitioning, decantation, or any other suitable method. This process is extensively used in industries to extract oils from plants (e.g., eucalyptus oil and orange oil are produced commercially using the steam distillation method). Distillation also is used for obtaining distilled beverages from fermented products or for producing bio-oil after pyrolysis as explained in the following sections (Naik et al. 2010).

### 2.3.2 CHEMICAL METHODS

#### 2.3.2.1 Solvent Extraction

Solvent extraction is a method of separation based on the solubility of a component in two immiscible liquids. It mainly is a two-phase system (i.e., an aqueous phase [polar like water] and an organic phase [non-polar like organic solvent]). The solvent rich in the solute (i.e., the desired component) is called the extract whereas the solvent depleted with solute is called the raffinate (Topare et al. 2011). Hexane is the most commonly used solvent for industrial purposes. Usually, for obtaining the desired product, the solvent extraction is accompanied by evaporation and distillation processes. It is used for the processing of perfumes, vegetable oils, and biodiesel. Furthermore, it can be used for the extraction of value-added compounds from biomass such as trepenoids, sterols, and waxes. The extracted biomass can later be used for biofuel production (Naik et al. 2010).

#### 2.3.2.2 Transesterification

Animal or plant fats oils are usually composed of triglycerides which are the esters formed by the free fatty acids and glycerol (Suppes 2004). These triglycerides cannot be used as a fuel due to their high viscous properties that cause incomplete combustion, carbon deposition, and so on. The viscosity of oils can be reduced using different methods such as blending with petroleum diesel, microemulsion, thermal cracking, and transesterification reaction. Among these, transesterification is the most common method of producing biodiesel from oil seed crops. It involves a reaction of primary alcohols with triglycerides of fatty acids (oil from energy crops) in

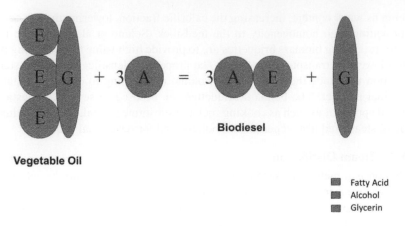

**FIGURE 2.5**   The transesterification reaction for the production of biodiesel.

the presence of a homogeneous or a heterogeneous catalyst to form fatty acid ethyl or methyl esters (biodiesel) and glycerol (Keera et al. 2011) (Figure 2.5).

It is estimated that transesterification can covert 98% of triglycerides into biodiesel (Naik et al. 2010). The advantage of using ethyl or methyl esters is that they can be directly used as a substitute for diesel. Also, since the transesterification process is well established in existing petroleum refineries, the technology can be easily implemented for biomass-based biodiesel production.

### 2.3.3   BIOCHEMICAL METHODS

#### 2.3.3.1   Fermentation

Fermentation involves the use of fermentative microorganisms to convert the biomass into liquid (ethanol, butanol) or gaseous (hydrogen) biofuels along with value-added chemicals such as lactic acid, acetic acid, and butyric acid. The complex biomass is initially pulverized or pretreated to convert it into simple monomeric units. These monomers are then finally converted to the target biofuel by the action of specific microbes. Ethanol fermentation using the yeast *Saccharomyces cerevisiae* is the only distinguished process that has been commercialized so far and is discussed in detail in Chapter 9. Besides ethanol, different microbial fermentative pathways for butanol (the ABE pathway) and hydrogen (mixed acid pathway) production have also been discovered in the metabolic pathways of different bacteria and extensive research is being conducted for commercialization of these products (Datta and Zeikus 1985; Levin 2004; Enzymes 2008). Figure 2.6 shows the major pathways to produce fermentation products. It can be observed that most of the glycolysis pathway is common for all the fermentative organisms and the pyruvate acts as a precursor for the generation of different products based on the enzymatic machinery available. These biochemical pathways can be redirected to obtain the desired biofuel at high concentrations using metabolic engineering techniques. The various biochemical pathways and the redirection of these pathways for the targeted products are discussed in Chapter 4.

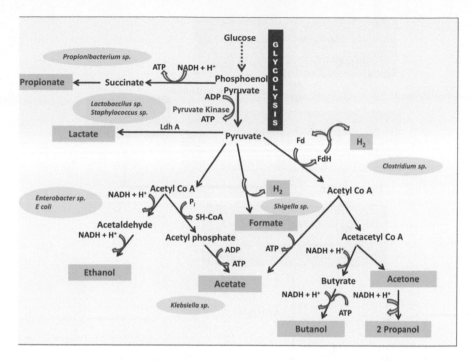

**FIGURE 2.6** Fermentative pathways of different microorganisms for biofuel production.

## 2.3.3.2 Anaerobic Digestion

Anaerobic digestion (AD) is the breakdown of organic matter in the absence of oxygen to produce biogas or biomethane. The advantage of this technology is that it can use a wide range of feedstock. The major steps involved in AD are hydrolysis, acidogenesis, acetogenesis, and methanogenesis (Figure 2.7). The technology is cheaper and much simpler. It is proved to be a promising alternative to the conventional activated sludge process because it has the added advantage of energy generation along with wastewater treatment (Mata-Alvarez et al. 2000). Biogas can be a potential substitute for liquified petroleum gas (LPG) and has several advantages such as its recovery is simpler because the product (gas) automatically separates from the substrates. In addition, biogas produces enriched organic manure which can supplement or even replace chemical fertilizers. It provides a source for decentralized power generation and it can provide employment opportunities in rural areas. Furthermore, household wastes and bio-wastes can be disposed of usefully and in a healthy manner using this process. However, there are certain environmental and economic barriers that must to be addressed to make this process an efficient alternative to the existing fossil-based fuels.

## 2.3.3.3 Bioelectrochemical Systems

A bioelectrochemical system (BES) is a combination of biological systems with the electrochemical technologies to convert the energy stored in the biomass directly into electricity (Sleutels et al. 2012). The system uses a specific type of bacteria

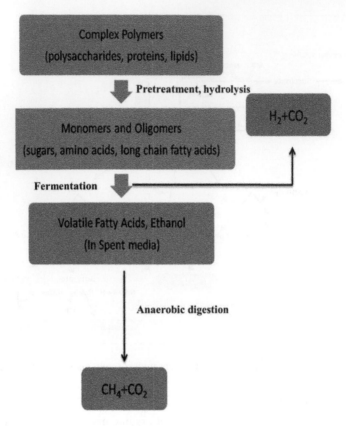

**FIGURE 2.7**   Steps involved in the anaerobic digestion process.

known as exoelectrogens that can interact and donate electrons to the solid electrode surfaces. Conventional fuel cells are converted to microbial fuel cells such that the organic substrate or biomass acts as the fuel and the microorganisms act as catalysts. It is mainly two half cells (anode and cathode) that are connected by an ion exchange membrane. By the action of the microbes, the energy stored in the biomass is oxidized to $CO_2$, protons, and electrons that when connected to an external circuit recombine at the cathode to generate electricity (microbial fuel cells) or hydrogen (microbial electrolysis cell) depending upon the terminal electron acceptor present at the cathode. The principles and the working of BES are discussed in detail in Chapter 12.

### 2.3.4   THERMOCHEMICAL METHODS

#### 2.3.4.1   Combustion

Solid biomass or low moisture containing biomass has been subjected to combustion for ages to supply heat. The complete combustion takes place in the presence of oxygen leaving $CO_2$ and water as the final products. The incomplete combustion produces charcoal, which can further be burned in a forced air supply to produce more heat. The combustion of biomass has lower GHG emissions compared to conventional fossil fuels, it has a high combustion efficiency, and the process is economically feasible. It is readily used in domestic stoves or boilers for heat generation. The combustion gases produced from the biomass also can be passed through a heat exchanger to produce hot air, hot water, or steam. The co-firing of biomass with the existing fossil fuels to reduce the effect of coal GHG emissions in thermal power plants is another strategy that has been implemented successfully with promising results (Nussbaumer 2003). A further advantage of co-firing is that it requires a lower investment for adapting to existing coal-fired power plants compared to standalone biomass combustion systems. However, certain biomass tends to produce smoke, tar, or ash which can have detrimental impact on the environment. Thus, it is essential to accommodate the possible side effects of the by-products in the biomass combustion systems.

### 2.3.4.2 Gasification

Gasification traditionally has been used to convert carbonaceous fossil fuels to gaseous compounds by supplying limited amounts (compared to combustion) of oxygen, air, or steam at high temperatures (Bhowmik and Roy 1983). The same process also is implemented for biomass conversion to syngas or producer gas. The composition of the produced gas is dependent upon the raw material, the process type, and operating conditions. Syngas is composed of carbon monoxide (CO) and hydrogen ($H_2$), whereas the producer gas is several gases such as CO, $CO_2$, $H_2$, $CH_4$, and $N_2$ (Molino et al. 2016). Both gases can be used for the production of bioenergy and bioproducts. The most direct application involves use in dual-mode engines to produce heat, steam, and electricity. It also can be used in (combined heat and power) CHP generating systems such as combustion engines, steam turbines, or fuel cells (Drapcho et al. 2008). The biomass-derived syngas also can be used to produce biodiesel and bioethanol (Munasinghe and Khanal 2010). The major limitation for this process, however, is the moisture content of the biomass, which can increase the energy requirement of the process (i.e., for producing dry biomass). The biomass particle emissions also can cause impurities in the syngas produced that might require further processing steps. Thus, it is essential to select the appropriate biomass to minimize the energy losses.

### 2.3.4.3 Pyrolysis

Pyrolysis is the thermal decomposition of biomass that occurs in the absence of oxygen (Chandel et al. 2011). It has been used widely to produce charcoal (from wood) and coke (from coal). The process can be used to produce biochar, bio-oil, and flue gas from biomass. During pyrolysis, the long chains of carbon, hydrogen, and oxygen compounds present in the biomass break down into smaller molecules. The decomposition starts at 300°C and proceeds to 900°C and above. The efficiency of the process and the final nature of the product or products generated are influenced by various operational parameters such as the reaction temperature, feedstock, biomass heating rate, and pressure. The particle size of the feed is the most influencing factor. The smaller the particle size, the higher the heat transfer rate from particle to particle. The large size biomass can cause incomplete decomposition or high ash formation, which is undesirable. Thus, it is necessary to resize the biomass particle to the appropriate range to obtain a high yield of the desired products.

Depending on the operational parameters, the pyrolysis process can be divided into three categories—slow pyrolysis, fast pyrolysis, and flash pyrolysis. Slow pyrolysis occurs at low temperatures ranging from 300°C to 700°C, with a very low heating rate. It is used generally to produce biochar. It usually occurs at very high solid retention times (SRT) ranging from few hours to days. Fast pyrolysis, on the other hand, occurs at high temperatures (>800°C) with high heating rates (>10–200°C/second (s)) in shorter SRTs (2–10 s). This process effectively yields 50%–60% bio-oil, 30%–40% biochar, and 10%–20% gaseous side products such as hydrogen, methane, carbon monoxide, and carbon dioxide. These gases can be upgraded to various chemicals and other transportation fuels by applying various techniques such as steam reforming ($H_2$), the Fischer-Tropsch reaction (methanol, hydrocarbon), and syngas fermentation (ethanol). Flash pyrolysis is like fast pyrolysis; however, extremely high heating rates (103–104°C/s) with extremely low retention times (<0.5 s) is applied to obtain high yields of bio-oil (70%–80%) (Kan et al. 2016). The various steps involved in pyrolysis are illustrated in Figure 2.8. The advantage of this technology is that it can be performed at small scale and in remote locations and thus greatly reduces the transportation and handling costs. Also, the bio-oil produced has a higher energy density compared to the solid biomass and thus can be upgraded for use in special engine fuels. It also can be used to produce a wide range of organic compounds and speciality chemicals.

### 2.3.4.4 Hydrothermal Liquefaction

In hydrothermal liquefication (HTL), the biomass is depolymerized thermally to produce bio-oil under moderate temperature and pressure conditions. The advantage of this process compared to other thermochemical biomass conversion techniques is that wet biomass can be processed with the improvement of energy per unit mass of biomass and reducing the drying costs. The bio-oil generated from the HTL process can be used as it is in heavy engines or can be upgraded to biodiesel, gasoline, or other biofuels. The chemical reaction involves the breakdown of long chain biomolecules into smaller constituents with the removal of oxygen in the form of water (dehydration) or carbon dioxide (decarboxylation) (Goudriaan and Peferoen 1990; Akhtar and Amin 2011). A homogeneous or a heterogeneous catalyst can be used to improve the yields and quality of the produced bio-oil. Figure 2.9 provides a schematic representation of the HTL process.

Several factors influence the yields of the HTL process such as the feedstock composition, temperature and heating rate, pressure, solvents, residence time, and catalyst. The most critical factor is the temperature that influences the polymerization and repolymerization reactions. Beyond the optimal temperature, repolymerization occurs, which leads to the formation of char (Akhtar and Amin 2011).

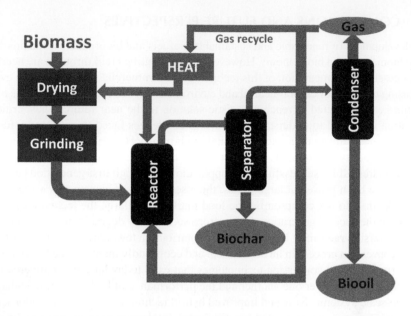

**FIGURE 2.8** Overview of the pyrolysis process.

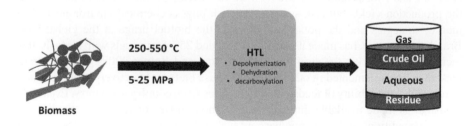

**FIGURE 2.9** Schematic representation of the hydrothermal liquefication process.

Similarly, non-optimum conditions of heat can lead to secondary reactions forming undesirable products. The type of catalyst used also has a significant contribution to the yields of biodiesel. Different catalysts such as KOH, $Na_2CO_3$, and transition metals such as nickel, palladium, and platinum have been used as catalysts for the HTL process (Goudriaan and Peferoen 1990). These catalysts enhance the yields by converting the proteins and carbohydrate components of biomass into bio-oil in addition to the fats and oils. Besides the use of wet biomass, HTL has distinct advantages such as it does not produce harmful products such as ammonia, NOx, and SOx, it can produce twice the amount of energy density of bio-oil compared to pyrolysis process, and it can be directly implemented to the existing petroleum infrastructure (Toor et al. 2011).

## 2.4    CONCLUSIONS AND FUTURE PERSPECTIVES

The development of renewable and sustainable biofuels and bioproducts is the mission of the biomass-based bioeconomy. However, many scientific breakthroughs are needed for successful implementation of this technology at commercial scales. With the ever-increasing demand for carbon-neutral and environmentally friendly resources, the biofuel market is expected to reach a realistic solution in the near future. Thus, critical evaluation is needed to understand the various challenges faced to produce biofuels. The inadequate information, lack of capital, high transportation costs, and the inconsistency in quality and quantity of the produced biofuels are the major concerns that must be addressed for successful global application. Although first-generation biofuels already have been proven at large scales, the research must focus on new varieties of feedstocks that do not compromise the food market. Similarly, the techno-economic hurdles for the second-generation energy crops such as low degradation efficiency and lower conversion rates must be overcome for contribution towards an effective waste to energy conversion process in an economical and ecofriendly manner. The biochemical and thermochemical conversion technologies that exist today have no commercial or technical advantage over one another and the performance of both remain speculative for large-scale systems. New and improved hybrid techniques with several pilot scale demonstrations are pre-requisites for successful implementation of these technologies. Process integration must be adapted to minimize the energy demands and reduce the overall steps in the biofuel production process. At the molecular level, cheaper enzymes and robust industrial strains must be developed that can not only improve the production yields but also sustain the harsh physicochemical environments. The market strategies and the polices also impact the biofuel usage at the global level. Brazil, for example, has made it mandatory to blend 25% fossil fuels with biofuels, thus increasing the market scenario for biofuels (Rutz and Rainer 2007). Similar strategies must be implemented in other countries to use biofuels effectively. Another major hurdle is the availability of feedstock which varies from country to country. Even if the feedstock is made available through trading among different countries, the different climatic conditions and economic barriers such as handling, storing, and transporting still prevail to impede the economy of biofuel production. Thus, new technologies must be developed that focus on sustainable worldwide distribution of feedstocks to expand the global production and use of biofuels. Biomass refinery is another attractive approach for concomitant production of biofuel and value-added co-products that can aid in improving the economic prospects of these technologies. The life-cycle assessment of these biorefinery systems can help in understanding the possible positive or negative impacts to the existing petro-refinery systems. Another major barrier is the limited knowledge on the casualties of the biofuels. Therefore, current research on biofuel production must be linked with available and future engine technologies to ensure their accessibility. At present, biofuels can be used only with certain modifications in the conventional combustion engines. Although biofuels have shown tremendous potential for use in electric engines or fuel cells, the technology remains unproven so far. With significant advancements in the techno-economic and environmental challenges discussed in this chapter, the replacement of a fossil fuel-based economy with biofuel economy seems inevitable.

## REFERENCES

Achten WMJ, Almeida J, Fobelets V, Bolle E, Mathijs E, Singh VP, Tewari DN, Verchot L V., Muys B (2010) Life cycle assessment of Jatropha biodiesel as transportation fuel in rural India. *Appl Energy* 87:3652–3660.

Adhikari S, Fernando SD, Haryanto A (2009) Hydrogen production from glycerol: An update. *Energy Convers Manag* 50:2600–2604.

Akhtar J, Amin NAS (2011) A review on process conditions for optimum bio-oil yield in hydrothermal liquefaction of biomass. *Renew Sustain Energy Rev* 15:1615–1624.

Aro EM (2016) From first generation biofuels to advanced solar biofuels. *Ambio* 45:24–31.

Bajracharya S, Yuliasni R, Vanbroekhoven K, Buisman CJN, Strik DPBTB, Pant D (2017) Long-term operation of microbial electrosynthesis cell reducing $CO_2$ to multi-carbon chemicals with a mixed culture avoiding methanogenesis. *Bioelectrochemistry* 113:26–34.

Becker EW (2007) Micro-algae as a source of protein. *Biotechnol Adv* 25:207–210.

Bhowmik T, Roy GK (1983) Biothermal gasification—A critical review. *J Inst Eng* 64:1–4.

Biswas R, Uellendahl H, Ahring BK (2015) Wet explosion: A universal and efficient pretreatment process for lignocellulosic biorefineries. *Bioenergy Res* 8:1101–1116.

Cardoso WS, Tardin FD, Tavares GP, Queiroz PV, Mota SS, Kasuya MCM, de Queiroz JH (2013) Use of sorghum straw (Sorghum bicolor) for second generation ethanol production: Pretreatment and enzymatic hydrolysis. *Quim Nova* 36:623–627.

Chandel AK, Silva SS, Singh OV (2011) Detoxification of lignocellulosic hydrolysates for improved bioconversion of bioethanol. In: Bernardes MAS (Ed.), *Biofuel Production-Recent Developments and Prospects*, In Tech, Rijeka, Croatia, pp. 606.

Cheng J, Xia A, Liu Y, Lin R, Zhou J, Cen K (2012) Combination of dark- and photo-fermentation to improve hydrogen production from Arthrospira platensis wet biomass with ammonium removal by zeolite. *Int J Hydrog Energy* 37:13330–13337.

Christian DG (2000) Biomass for ren ewable energy, fuels, and chemicals. *J Environ Qual* 29:662.

Coelho ST (2012) Traditional biomass energy: Improving its use and moving to modern energy use. In: *Renewable Energy: A Global Review of Technologies, Policies and Markets*. Taylor & Francis Group, Hoboken, NJ, pp. 230–261.

Datta R, Zeikus JG (1985) Modulation of acetone-butanol-ethanol fermentation by carbon monoxide and organic acids. *Appl Environ Microbiol* 49:522–529.

Demirbas A (2007) Importance of biodiesel as transportation fuel. *Energy Policy* 35:4661–4670.

Demirbas A (2008) Biofuels sources, biofuel policy, biofuel economy and global biofuel projections. *Energy Convers Manag* 49:2106–2116.

Demirbas A (2010) Use of algae as biofuel sources. *Energy Convers Manag* 51:2738–2749.

Drapcho CM, Nhuan NP, Walker TH (2008) *Biofuels Feedstocks in Biofuels Engineering Process Technology*, McGraw-Hill, New York, pp. 69–104.

Gopal AR, Kammen DM (2009) Molasses for ethanol: The economic and environmental impacts of a new pathway for the lifecycle greenhouse gas analysis of sugarcane ethanol. *Environ Res Lett* 4:44005.

Goudriaan F, Peferoen DGR (1990) Liquid fuels from biomass via a hydrothermal process. *Chem Eng Sci* 45:2729–2734.

Hattori T, Morita S (2010) Energy crops for sustainable bioethanol production which, where and how? *Plant Prod Sci* 13:221–234.

Herzog AV, Lipman TE, Kammen DM (2001) Renewable energy sources. *Encycl Life* 1–63.

International Confederation of European Beet Growers (CIBE) (2017) Molasses for bioenergy and bio-based products. Fact sheet, Brussels, Belgium. https://www.cibe-europe.eu/img/user/174-17%20CIBE%20CEFS%20-%20Fact%20sheet%20on%20Molasses%2027%20September%202017(1).pdf, September 27, 2017.

Jung K-W, Kim D-H, Shin H-S (2011) Fermentative hydrogen production from Laminaria japonica and optimization of thermal pretreatment conditions. *Bioresour Technol* 102:2745–2750.

Kan T, Strezov V, Evans TJ (2016) Lignocellulosic biomass pyrolysis: A review of product properties and effects of pyrolysis parameters. *Renew Sustain Energy Rev* 57:1126–1140.

Keera ST, El Sabagh SM, Taman AR (2011) Transesterification of vegetable oil to biodiesel fuel using alkaline catalyst. *Fuel* 90:42–47.

Levin D (2004) Biohydrogen production: Prospects and limitations to practical application. *Int J Hydrogen Energy* 29:173–185.

Long SP, Karp A, Buckeridge MS, Davis SC, Jaiswal D, Moore PH, Moose SP, Murphy DJ, Onwona-Agyeman S, Vonshak A (2015) Chapter 10: Feedstocks for biofuels and bio-energy. In: Somerville CR and Van Sluys M-A (Eds.), *Bioenergy Sustain Bridg Gaps*, pp. 302–346.

Mata-Alvarez J, Macé S, Llabrés P (2000) Anaerobic digestion of organic solid wastes. An overview of research achievements and perspectives. *Bioresour Technol* 74:3–16.

Matsakas L, Gao Q, Jansson S, Rova U, Christakopoulos P (2017) Green conversion of municipal solid wastes into fuels and chemicals. *Electron J Biotechnol* 26:69–83.

Meherkotay S, Das D (2008) Biohydrogen as a renewable energy resource—Prospects and potentials. *Int J Hydrogen Energy* 33:258–263.

Mekhilef S, Siga S, Saidur R (2011) A review on palm oil biodiesel as a source of renewable fuel. *Renew Sustain Energy Rev* 15:1937–1949.

Meng X, Yang J, Xu X, Zhang L, Nie Q, Xian M (2009) Biodiesel production from oleaginous microorganisms. *Renew Energy* 34:1–5.

Molino A, Chianese S, Musmarra D (2016) Biomass gasification technology: The state of the art overview. *J Energy Chem* 25:10–25.

Munasinghe PC, Khanal SK (2010) Biomass-derived syngas fermentation into biofuels: Opportunities and challenges. *Bioresour Technol* 101:5013–5022.

Naik SN, Goud VV., Rout PK, Dalai AK (2010) Production of first and second generation biofuels: A comprehensive review. *Renew Sustain Energy Rev* 14:578–597.

Nigam PS, Singh A (2011) Production of liquid biofuels from renewable resources. *Prog Energy Combust Sci* 37:52–68.

Nussbaumer T (2003) Combustion and co-combustion of biomass: Fundamentals, technologies, and primary measures for emission reduction. *Energy Fuels* 17:1510–1521.

Patil SA, Gildemyn S, Pant D, Zengler K, Logan BE, Rabaey K (2015) A logical data representation framework for electricity-driven bioproduction processes. *Biotechnol Adv* 33(6):736–744.

Perera KRJ, Ketheesan B, Arudchelvam Y, Nirmalakhandan N (2012) Fermentative biohydrogen production II: Net energy gain from organic wastes. *Int J Hydrogen Energy* 37:167–178.

Rabaey K, Johnstone A, Wise A, Read S, Rozendal RA (2010) Microbial electrosynthesis: From electricity to biofuels and biochemicals. *Bio Tech Int* 22:6–8.

Roy S, Das D (2016) Biohythane production from organic wastes: Present state of art. *Environ Sci Pollut Res* 23:9391–9410.

Roy S, Kumar K, Ghosh S, Das D (2014) Thermophilic biohydrogen production using pretreated algal biomass as substrate. *Biomass Bioenergy* 61:157–166.

Rutz D, Rainer J (2007) *Technology Handbook*. Elsevier, Amsterdam, the Netherlands, pp. 1–149.

Saratale GD, Chen S Der, Lo YC, Saratale RG, Chang JS (2008) Outlook of biohydrogen production from lignocellulosic feedstock using dark fermentation—A review. *J Sci Ind Res (India)* 67:962–979.

Shi S, Valle-Rodríguez JO, Siewers V, Nielsen J (2011) Prospects for microbial biodiesel production. *Biotechnol J* 6(3):277–285.

Sims R, Taylor M, Jack S, Mabee W (2008) From 1st to 2nd generation bio fuel technologies: An overview of current industry and R&D activities. *IEA Bioenergy* 1–124.

Singh S, Adak A, Saritha M, Sharma S, Tiwari R, Rana S, Arora A, Nain L (2017) Bioethanol production scenario in India: Potential and policy perspective. In: *Sustainable Biofuels Development in India*. Springer International Publishing, Cham, Swizerland, pp. 21–37.

Sitepu IR, Garay LA, Sestric R, Levin D, Block DE, German JB, Boundy-Mills KL (2014) Oleaginous yeasts for biodiesel: Current and future trends in biology and production. *Biotechnol Adv* 32:1336–1360.

Sleutels THJA, Ter Heijne A, Buisman CJN, Hamelers HVM (2012) Bioelectrochemical systems: An outlook for practical applications. *ChemSusChem* 5:1012–1019.

Solano D, Vinyes P, Arranz P (2016) *Biomass Briquetting Process, A Guideline Report*, United Nations Development Programme, http://www.lb.undp.org/

Sun J, Yuan X, Shi X, Chu C, Guo R, Kong H (2011) Fermentation of Chlorella sp. for anaerobic bio-hydrogen production: Influences of inoculum–substrate ratio, volatile fatty acids and NADH. *Bioresour Technol* 102:10480–10485.

Suppes G (2004) Transesterification of soybean oil with zeolite and metal catalysts. *Appl Catal A Gen* 257:213–223.

Tan WC, Kuppusamy UR, Phan CW, Tan YS, Raman J, Anuar AM, Sabaratnam V (2015) Ganoderma neo-japonicum Imazeki revisited: Domestication study and antioxidant properties of its basidiocarps and mycelia. *Sci Rep* 5:12515.

Tashiro Y, Sonomoto K (2010) Advances in butanol production by clostridia. *Technol Educ Top Appl* 1383–1394.

Toor SS, Rosendahl L, Rudolf A (2011) Hydrothermal liquefaction of biomass: A review of subcritical water technologies. *Energy* 36:2328–2342.

Topare NS, Raut SJ, Renge VC, Khedkar S V, Chavan YP, Bhagat SL (2011) Extraction of oil from algae by solvent extraction and oil expeller method. *Int J Chem Sci* 9:1746–1750.

Ummah H, Suriamihardja DA, Selintung M, Wahab AW (2015) Analysis of chemical composition of rice husk used as absorber plates sea water into clean water. *ARPN J Eng Appl Sci* 10:6046–6050.

Wahlen BD, Morgan MR, McCurdy AT, Willis RM, Morgan MD, Dye DJ, Bugbee B, Wood BD, Seefeldt LC (2013) Biodiesel from microalgae, yeast, and bacteria: Engine performance and exhaust emissions. *Energy Fuels* 27:220–228.

WBA (2014) WBA Global Bioenergy Statistics 2014. In: *World Bioenergy Association*. http://worldbioenergy.org/uploads/WBA GBS 2017_hq.pdf. Accessed May 3, 2018.

Woiciechowski AL, Bianchi A, Medeiros P, Rodrigues C, Porto L, Vandenberghe DS (2016) *Green Fuels Technology*. Springer International Publishing, Cham, Switzerland.

World Energy Council (2016) World Energy Resources: Bioenergy 2016, p. 60. https://www.worldenergy.org/wp-content/uploads/2017/03/WEResources_Bioenergy_2016.pdf

Xia A, Cheng J, Lin R, Lu H, Zhou J, Cen K (2013) Comparison in dark hydrogen fermentation followed by photo hydrogen fermentation and methanogenesis between protein and carbohydrate compositions in Nannochloropsis oceanica biomass. *Bioresour Technol* 138:204–213.

Yun Y-M, Jung K-W, Kim D-H, Oh Y-K, Shin H-S (2012) Microalgal biomass as a feedstock for bio-hydrogen production. *Int J Hydrogen Energy* 37:15533–15539.

# 3 Microorganisms Involved in Biofuel Production Processes

## 3.1 INTRODUCTION

Rapid industrialization and increasing population has raised various concerns over the sustainability of existing water and energy resources. Energy, specifically, is still dependent upon exhaustible, non-renewable resources such as coal and petroleum. Research on renewable energy is fast expanding and giving rise to solar, wind, geothermal, and biomass energy. Among these, biomass is the most abundant natural form of energy. Biomass can be used as it is or can be converted to biofuels for bio-energy generation. Conventionally, dry biomass was used for energy production in the form of heat and electricity.

The current research, however, is targeted towards developing clean and inexpensive means of biomass conversion processes. Biomass can be converted to biofuels through biochemical and thermochemical processes (Naik et al. 2010). The biochemical processes have an advantage over thermochemical processes because they are less energy intensive and are environmentally friendly. Microorganisms play a crucial role in the biochemical conversion of biomass. Different group of microbes such as bacteria, yeasts, and fungi have innate capabilities to generate different value-added products through their metabolic mechanisms (Figure 3.1).

Currently, the most predominant biofuel is the sugar- or starch-based ethanol produced using yeasts (Munasinghe and Khanal 2010). However, interest is gaining towards biofuel production from non-edible crops and has shifted the research focus towards new advanced biofuels, feedstocks, and microbial hosts. One of the major challenges for the commercialization of biofuel industries is the lack of high-yielding robust strains that can thrive in harsh environmental conditions. Several advancements have been made in the screening, isolation, and genetic modification of the microbes for obtaining potent biofuel producing strains. However, most of the biofuel research remains practically implausible. This chapter describes the different genera of microbial strains involved in the high-yield biofuel production and discusses the means of their selection, optimization, and improvement for enhanced biofuel production.

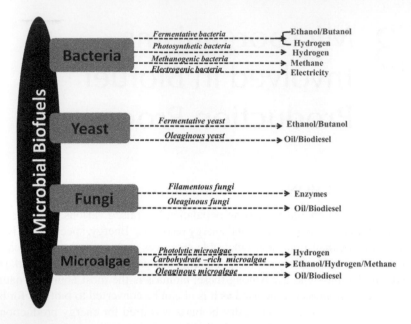

**FIGURE 3.1**   Various means of microbial biofuel production.

## 3.2   SELECTION OF SUITABLE MICROBES FOR BIOFUEL PRODUCTION

As, described in Chapter 2, a plethora of biomass feedstocks are available to produce biofuels ranging from simple sugars to complex organic sources. Microbial fermentation is the most convenient method for the conversion of these feedstocks to biofuel with less energy input (compared to thermochemical processes). However, the realization of high quantities of targeted biofuel is greatly dependent upon the type of microorganisms used. The desirable characteristics of an ideal microbial host for biofuel production process include (1) high biodegradability (i.e., ability to degrade different type of carbohydrates—hexoses and pentoses) to achieve high conversion efficiencies, (2) high yielding (i.e., ability to achieve high concentrations of targeted product per gram of biomass), (3) not susceptible to end-product toxicity (i.e., ability to tolerate high concentrations of end products), (4) not susceptible to environmental stress (i.e., can thrive in harsh environmental conditions), (5) substrate and product flexibility (i.e., ability to utilize a wide variety of substrates and generate multiple value-added products to improve the economy of the process), and (6) genetically competent (i.e., ability to incorporate foreign DNA into its genome to obtain mutants with desired characteristics) (Fischer et al. 2008). It is impossible for one strain to have all the desired characteristics and thus a screening is required for the selection of the most promising microbial species for biofuel production. The screening process is dependent upon the type of feedstock and the targeted biofuel (e.g., if the product of interest is biodiesel, then the microbial species (yeast or microalgae) would be screened based on their lipid content (Liu et al. 2010). Similarly, if the target biofuel is ethanol or butanol, along with the high-yielding varieties, then the microbial species tolerant to higher concentration of these alcohols will be chosen (Elshahed 2010). Likewise, for producing biofuels from complex cellulosic materials, microbes with active cellulolytic enzymes in their metabolic framework would be most effective (Maki et al. 2009).

After the initial screening process, the selected microbial species is subjected further to strain improvement techniques to maximize the overall yields of the biofuels. These strain improvements are performed through process design strategies or by using genetic and/or metabolic engineering strategies (Lancini and Lorenzetti 1993). By optimizing suitable physicochemical parameters such as media composition, pH, temperature, and reactor configurations, the performance of the strain can be enhanced significantly (Fischer et al. 2008). El-Gendy and others (2013) showed improved performance of *Saccharomyces cerevisiae* for ethanol production from sugarcane molasses by optimizing different process parameters such as incubation period, initial pH, incubation temperature, and substrate. Similarly, Ranjan and others (2013) used process optimization strategy to enhance butanol production by *Clostridium acetobutylicum* MTCC 481 from rice straw hydrolysate, using optimal levels of temperature, pH, inoculum size, inoculum age, and agitation. The hydrogen production capability of the strain *Thermoanaerobacterium thermosaccharolyticum* IIT BT- ST1 was studied by Roy and others (2014) and was further improved by optimizing the most influential parameters such as temperature, pH, glucose, $FeSO_4$, and yeast extract in a batch process. These studies suggest that maintaining ideal operational conditions specific to the microbial species is essential for a successful biofuel production process.

Besides process engineering techniques, however, mutation or genetic recombination strategies have been successfully used for obtaining competent industrial strains for biofuel production (Elshahed 2010). Maki and others (2009) documented the use of different mutagenic techniques that helped in improving lignocellulosic bioconversion capabilities of cellulase-producing bacteria. Various metabolic schemes were selected to overcome the poor solvent resistance in acetone and butanol-producing microorganisms (Ezeji et al. 2010). Such mutated or genetically modified strains have been considered preferable to the wild species (unmodified microbial species) for biofuel production due to their advanced characteristic features. A more detailed aspect on strain improvement technologies and the various molecular approaches for improvement of biofuel yields is presented in Chapter 5.

## 3.3 BACTERIAL BIOFUELS

Bacteria are the most widely studied group of microorganisms that have been used extensively for biofuel production. They are the most primitive organisms that are found abundantly in the Earth's atmosphere. Their faster growth rates, flexibility to use different types of substrates, and ability to generate multiple metabolic products make them attractive hosts for biofuel production. Their metabolic pathways can be exploited easily, enhanced, or modified to obtain the desired end products. Based on mechanisms involved in biofuel production, the bacterial genera can be broadly classified as photosynthetic bacteria, fermentative bacteria, methanogenic bacteria, and electrogenic bacteria (Table 3.1). This section deals briefly with the various groups of bacteria involved in biofuel production. The mechanisms and the metabolic pathways of these organisms will be discussed in detail in the subsequent chapters.

## TABLE 3.1
## Different Genera of Bacteria Used for Biofuel Production

| Organism | Type | Mechanism | Biofuel | References |
|---|---|---|---|---|
| *Synechocystis* PCC 6803 | Cyanobacteria | Direct biophotolysis | Hydrogen | Khanna and Lindblad (2015) |
| *Plectonema boryanum* | Cyanobacteria | Indirect biophotolysis | Hydrogen | Hallenbeck (2012) |
| *Anabaena variabilis* | Cyanobacteria | Indirect biophotolysis | Hydrogen | Tamagnini et al. (2002) |
| *Chloroflexus aurantiacus* | Green-gliding bacteria | Photofermentation | Hydrogen | Gogotov et al. (1991) |
| *Rhodobacter sphaeroides* | Purple non-sulfur bacteria | Photofermentation | Hydrogen | Trchounian (2015) |
| *Thiocapsa roseopersicina* | Purple non-sulfur bacteria | Photofermentation | Hydrogen | Harai et al. (2010) |
| *Zymomonas mobilis* | Fermentative bacteria | Ethanol fermentation | Ethanol | Müller (2001) |
| *Thermoanaerobacter ethanolicus* | Fermentative bacteria | Ethanol fermentation | Ethanol | Lacis and Lawford (1991) |
| *Clostridium thermohydrosulfuricum* | Fermentative bacteria | Ethanol fermentation | Ethanol | Lovitt et al. (1988) |
| *Butyribacterium methylotrophicum* | Fermentative bacteria | Mixed acid fermentation | Ethanol | Köpke et al. (2011) |
| *Clostridium acetobutylicum* | Fermentative bacteria | Acetone-Butanol-Ethanol fermentation | Butanol | Yen et al. (2011) |
| *Clostridium beijerinckii* BA101 | Fermentative bacteria | Acetone-Butanol-Ethanol fermentation | Butanol | Ezeji et al. (2003) |
| *Clostridium toanum* | Fermentative bacteria | Mixed acid fermentation | Butanol | Jones and Woods (1986) |
| *Clostridium butyricum* strain 4P1 | Fermentative bacteria | Mixed acid fermentation | Methanol | Schink and Zeikus (1980) |
| *Clostridium propionicum* | Fermentative bacteria | Mixed acid fermentation | n-Propanol | Hwang et al. (2015) |
| *Clostridium toanum* | Fermentative bacteria | Acetone-Butanol-Isopropanol fermentation | Isopropanol | Jones and Woods (1986) |

*(Continued)*

**TABLE 3.1 (*Continued*)**
**Different Genera of Bacteria Used for Biofuel Production**

| Organism | Type | Mechanism | Biofuel | References |
|---|---|---|---|---|
| *Clostridium* sp. strain NJP7 | Fermentative bacteria | Acetone-Butanol-Isopropanol fermentation | Isopropanol | Nigam and Singh (2011) |
| *Clostridium thermocellum* 27405 | Fermentative bacteria | Mixed acid fermentation | Hydrogen | Levin et al. (2006) |
| *Enterobacter cloacae* IIT-BT 08 | Fermentative bacteria | Mixed acid fermentation | Hydrogen | Kumar and Das (2000) |
| *Thermoanaerobacterium thermosaccharolyticum* IIT BT-ST1 | Fermentative bacteria | Mixed acid fermentation | Hydrogen | Roy et al. (2014) |
| *Methanosarcina barkeri* | Acetoclastic Methanogen | Methanogenesis | Methane | Ali Shah et al. (2014) |
| *Methanobacterium byrantii* | Hydrogenotrophic Methanogen | Methanogenesis | Methane | Balch et al. (1979) |
| *Methanobrevibacter smithii* | Hydrogenotrophic Methanogen | Methanogenesis | Methane | Balch et al. (1979) |
| *Methanomicrobium mobile* | Hydrogenotrophic Methanogen | Methanogenesis | Methane | Balch et al. (1979) |
| *G. sulfurreducens* PCA (ATCC 51573) | Exoelectrogen | Electrogenesis (MEC) | Hydrogen | Call et al. (2009) |
| *Methanobacterium palustre* | Exoelectrogen | Electrogenesis (MEC) | Methane | Booth (2009) |

### 3.3.1 PHOTOSYNTHETIC BACTERIA

Photosynthetic bacteria can produce hydrogen by using sunlight, water, and inorganic or organic substrates. Like plants, these bacteria have light-absorbing pigments and reaction centers that make them capable of converting light energy to chemical energy (McKinlay and Harwood 2010). Two types of mechanisms have been reported for photobiological hydrogen production—biophotolysis and photofermentation (Manish and Banerjee 2008b).

### 3.3.1.1 Biophotolysis

Biophotolysis is the process in which water is split into molecular hydrogen and oxygen by the action of sunlight (Equation 1.1). It is carried out by cyanobacteria, which can performing oxygenic photosynthesis (McKinlay and Harwood 2010). Based on whether light is irradiated during hydrogen evolution, biophotolysis is divided into two categories—direct and indirect biophotolysis (Yu and Takahashi 2007).

$$2H_2O + hv \text{ (light energy)} \rightarrow 2H_2 + O_2 \qquad (3.1)$$

### 3.3.1.1.1 Direct Biophotolysis

Direct photolysis uses photosynthesis-derived electrons to reduce directly the hydrogen-producing enzyme (hydrogenase) without intermediate $CO_2$ fixation (Figure 3.2a). It usually occurs in $N_2$-fixing and non-$N_2$-fixing cyanobacteria (Tamagnini et al. 2002). The mechanism involves the use of bi-directional hydrogenases (*Hox* hydrogenase) that take up electrons from the ferredoxin/NADPH reductants and catalyse the hydrogen evolution reaction. The mechanisms and functioning of these enzymes are not clearly understood and only a few cyanobacterial species have been reported that exhibit direct photolysis by the action of these enzymes. Khanna and Lindblad (2015) elucidated the role of ferredoxin as an electron donor to the bidirectional *Hox* hydrogenase in the *Synechocystis* PCC 6803 species. These organisms were able to generate hydrogen through their photosynthetic cellular mechanisms. Similarly, light-dependent hydrogen production was observed in the filamentous cyanobacterium *Oscillatoria limnetica*, which could produce hydrogen in aerobic and anaerobic environments (Markov et al. 1995). Although the direct conversion of sunlight to hydrogen seems attractive, the extreme sensitivity of the hydrogenase enzyme to oxygen limits the maximum conversion efficiencies achieved. Several strategies have been implemented over the past few decades to improve the oxygen sensitivity of the hydrogenase enzymes; however, a technological breakthrough is awaited (Tamagnini et al. 2002).

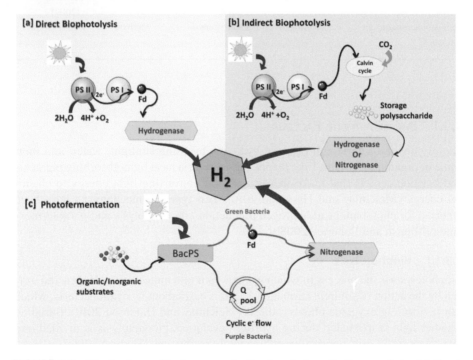

**FIGURE 3.2** Mechanisms of hydrogen production by photosynthetic organisms: (a) direct biophotolysis, (b) indirect biophotolysis, and (c) photofermentation.

### 3.3.1.1.2    Indirect Biophotolysis

In indirect biophotolysis, the water splitting photosynthesis and the consequent reductants (Fd/NADPH) are used up for $CO_2$ fixation which forms reduced organic carbon compounds ($[CH_2O]_n$). These organic compounds are further used to drive hydrogen production using different enzymes (i.e., hydrogenases and/or nitrogenases) (Figure 3.2b). The nitrogenases-dependent hydrogen production is the most extensively studied mechanism in the $N_2$-fixing cyanobacteria. Hydrogenases and nitrogenases are also inhibited in the presence of oxygen. However, most cyanobacterial species have inherent capabilities to separate the oxygenic photosynthesis with the anoxygenic nitrogenase (and simultaneous hydrogen production) either in a single stage heterocystous system (spatial variation) or in a two-stage (dark/night) non-heterocystous system (temporal variation) (e.g., *Nostoc* sp. and *Anabaena* sp. comprising of heterocysts are able to produce hydrogen in the range of 0.17 mol $H_2$ mg chla$^{-1}$ h$^{-1}$ to 4.2 mol $H_2$ mg chla$^{-1}$ h$^{-1}$) (Levin 2004). Whereas, *Arthospira* sp., a non-heterocystous marine filamentous cyanobacterium, was able to produce hydrogen at a rate of 5.91 $\mu$mol $H_2$ mg chla$^{-1}$ h$^{-1}$ (Tamagnini et al. 2002). Besides $N_2$-fixing cyanobacteria, non-$N_2$-fixing cyanobacteria such as *Gloeocapsa alpicola*, *Synechococcus* sp., *Microcystis* sp., and *Gloebacter* sp. can undergo indirect photolysis with the help of bidirectional or uptake hydrogenases (Dutta et al. 2005). In these organisms, the electrons obtained from the photosystems are diverted to a Calvin-Benson-Bassham cycle with the help of Fd/NADPH redox shuttles. During the night, the reduced carbon compound is used to generate energy by fermentation and the extra electrons are accepted by hydrogenases (Fe-Fe/Ni-Fe) that in turn catalyse the hydrogen evolution reaction. The fermentation pathways in certain cyanobacteria can give rise to other compounds besides hydrogen which include acetate (*Anabaena azollae*, *Nostoc* sp.), ethanol (*Cyanothece*, *Microcoleus chthonoplastes*), and lactate (*Oscillatoria limnetica*, *Spirulina mimosa*) (Stal and Moezelaar 2006). These extracellular products are dependent on the predominant fermentation pathway existing in the cyanobacterial species. Several technological advancements have been made to improve the conversion efficiencies of indirect photolysis (Markov et al. 1995; Dasgupta et al. 2010; Raksajit et al. 2012); however, the extremely low rates (compared to dark fermentation in heterotrophic bacteria) restrict the practical feasibility of the process.

### 3.3.1.2   Photofermentation

Unlike cyanobacteria, other photosynthetic bacteria such as purple and green bacteria are anoxygenic in nature. These organisms have innate capabilities to manifest $H_2$ production through photofermentation by consuming a wide variety of substrates ranging from inorganic to organic substrates in the presence of light (Figure 3.2c). The photosynthetic bacteria that undergo photofermentation belong to the group of green bacteria or purple bacteria. These bacterial groups contain bacteriochlorophyll which has plant-like light-harvesting properties. However, no oxygen is produced during the process. The green bacteria such as *Chlorobium limicola* and *Chloroflexus aurantiacus* contain a photosystem I (PSI)-like photosystem which takes up light energy from sunlight and drives the photosynthesis reaction (McKinlay and Harwood 2010). These bacteria derive electrons for hydrogen production through oxidation of inorganic or organic substrates, which are further used to reduce ferredoxin. The reduced ferredoxin in turn acts as an electron donor for the nitrogenase to perform the dark reaction as well as hydrogen production. On the other hand, the purple bacteria such as *Rhodobacter sphaeroides* and *Rhodopseudomonas palustris* contain a photosystem II (PSII)-like photosystem and thus do not have the capability to reduce the ferredoxin. In these organisms, the electrons needed for nitrogenase-mediated hydrogen production are derived from the cyclic electron flow through the quinone pool (Figure 3.2c). Often, the photofermentation process is used for the treatment of wastewater and many studies suggest the integration of this technology with the dark fermentative hydrogen production process (discussion following) to enhance the overall hydrogen yields (Dasgupta et al. 2010; Harai et al. 2010; Xia et al. 2013). Despite the various efforts to improve the photofermentative hydrogen yields, the low hydrogen production rates and the economic barriers associated with the design and scale up of photobioreactors impose challenge for the commercialization of this technology.

### 3.3.2 Heterotrophic Fermentative Bacteria

The versatility of the end products obtained from the bacterial fermentation pathways has enabled the use of fermentative microorganisms for biofuel production (Figure 2.6). Their simple metabolic machinery enables targeting of a specific product of interest and improves the yields by process engineering or genetic manipulations. Furthermore, by using the same bacteria under different growth conditions, different metabolic products of high economic value can be generated. Figure 3.3 shows the distinct fermentation pathways that have been discovered among the

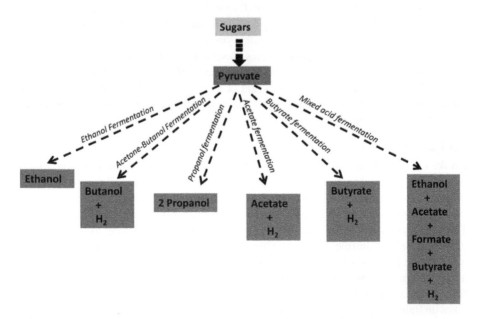

**FIGURE 3.3**  Different fermentative pathways for biofuel production.

various bacterial groups, most of which are named after their main end products such as ethanol fermentation, acetone-butanol fermentation, homoacetic acid fermentation, and mixed acid fermentation (Müller 2001).

### 3.3.2.1 Ethanol Fermentation

Among the various biofuels, ethanol has been the most successful biofuel produced commercially so far. Although most of the bioethanol is obtained using the yeast *Saccharomyces cerevisiae*, several bacteria also have been identified as potential candidates for ethanol production. The most notable bioethanol-producing bacteria is *Zymomonas mobilis*, which follows the Entner–Doudoroff pathway (Rogers et al. 1982). It has advantage over yeasts because it does not require additional oxygen for its growth, it does not produce high biomass, and it has an ability to tolerate high titers of ethanol. However, despite the favorable attributes of *Zymomonas* sp. towards high-yield ethanol production (reaching the theoretical maximum), its usage at a commercial level is restricted because *Zymomonas* sp. can convert ethanol to acetic acid in the presence of the slightest amount of oxygen. This conversion in turn reduces the overall ethanol yields and hinders ethanol separation from the fermentation broth. Furthermore, it can use only a limited range of substrates, which can result in high economic losses. Besides *Zymomonas* sp., other organisms such as *Salmonella* sp. and *Escherichia coli* and certain thermophilic bacteria such as *Thermoanaerobacter ethanolicus* also have been found to produce ethanol; however, the yields obtained are too low to be scaled up. To improve the bioethanol yields and the substrate flexibility, different genetic engineering techniques have been used in a manner attractive for industrial ethanol production (He et al. 2014). Also, the advent of advanced biofuel production by introducing new high-flux pathways in model organisms such as *Escherichia coli* has shown a promising future for a bioethanol-based bioeconomy (Elshahed 2010; Chandel et al. 2011).

### 3.3.2.2 Acetone-Butanol Fermentation

Though ethanol is the most widely produced biofuel, it has several drawbacks; for example, it has a lower energy content than gasoline, it cannot be directly used in the internal combustion engines, and it is hygroscopic in nature (Kang and Lee 2015). Butanol is an alternate biofuel with better fuel properties and is highly compatible with the existing infrastructure. The butanol production by bacterial fermentation was reported during World War I by the chemist Chaim Weizmann who mainly targeted acetone production with butanol as a by-product (Sauer 2016). However, with the rise of the petrochemical industries, the research focus on bacterial butanol production was significantly reduced. In recent years, due to the increasing demands for renewable energy generation and low carbon emission fuels, biobutanol production again has gained considerable attention. Butanol is naturally produced by the *Clostridia* species through the acetone-butanol-ethanol (ABE) fermentation pathway. These species are mostly obligate anaerobes (i.e., highly sensitive to the presence of oxygen). They can use a wide variety of substrates ranging from simple starch and sugars to inexpensive abundant renewable raw materials (Ndaba et al. 2015). The most notable organisms include *C. acetobutylicum*, *C. beijerinckii*, and *C. saccharobutylicum*. These organisms, however, are associated with several disadvantages.

They are slow-growing organisms that slow down the upstream processes and the strict anaerobic growth conditions they require limit any kind of genetic manipulations. Furthermore, the organisms are highly sensitive to the high concentrations of butanol. Thus, to make the acetone-butanol fermentation economically feasible, most of the research is targeted on in situ recovery of butanol during the production process. Also, genetic and metabolic engineering approaches for producing high-yielding butanol-resistant organisms have gained considerable interest in recent years.

Besides acetone and ethanol, it has been observed that during butanol production, concomitant production of hydrogen is obtained, which is also an attractive source of renewable energy. Since butanol is produced in the stationary phase of the growth cycle and hydrogen in the log phase, it is possible to separate and extract both the products by carefully monitoring the growth patterns of the *Clostridial* species. So, simultaneous production of hydrogen as well as butanol is considered to improve the energy recovery from the biomass (Valdez-Vazquez et al. 2015; Mitra et al. 2017). However, this integrated technology is still at its nascent phase of research and many technological hurdles affecting the yields and productivity of both the products need to be overcome.

### 3.3.2.3   Mixed Acid Fermentation

Certain fermentative bacteria tend to follow a mixed acid pathway generating multifaceted end products. It usually occurs when two or more different pathways are used to attain the required reducing equivalents for energy generation during the terminal steps of fermentation. It can be seen in Enterobacteriaceae such as *E. coli* and *Enterobacter aerogenes*. By scrutinizing the metabolic pathway and providing the favorable physicochemical conditions, the final concentration of the desired end product can be significantly enhanced. Mostly, among the alcohols produced during mixed acid fermentation, ethanol and butanol are more prevalent due to their characteristic fuel properties as discussed previously. However, recent advances have directed the focus on alternate alcohols (such as methanol and propanol), which are plausible alternatives to ethanol or butanol and are less explored as biofuels. Methanol is the major end product of different pectinolytic strains such as *Clostridium*, *Erwinia*, and *Pseudomona*s species (Schink and Zeikus 1980). Certain methanotrophs such as *Methylococcus capsulatus* can also produce methanol using methane as substrate (Hwang et al. 2015). Like ethanol, it can be used by blending with gasoline or directly in its pure form. It has less carbon emission compared to ethanol; however, it is highly volatile and toxic in nature. Besides methanol, propanol also is gaining attention as a promising biofuel. Propanol is widely used as a chemical solvent and is currently manufactured through petrochemical routes. It occurs in two isomeric forms—n-propanol and isopropanol. It is observed that n-propanol is produced naturally during mixed acid fermentation processes, which can be exploited for renewable biopropanol production. Bacteria such as *Clostridium propionicum* and *Clostridium neopropionicum* produce propionic acid as the major end product and n-propanol as the by-product using the Wood-Werkman pathway (Hwang et al. 2015). These organisms have been considered for producing n-propanol. However, the n-propanol yields are too low for commercial exploitation. On the other hand, isopropanol is produced by the organisms that follow the acetone-butanol-isopropanol pathway (Jones and Woods 1986). Moreover, different synthetic biology approaches have led to isopropanol production in engineered strains of *E. coli* (Hwang et al. 2015). Although several developments are being made for biosynthesis of propanol, the use of propanol as a fuel still remains debatable. Due to lower energy density and octane numbers of propanol, butanol is preferred. Only with the discovery of high-yielding strains and the identification of low-cost feedstocks can the feasibility of biopropanol production be achieved.

Besides bioalcohols, the other important by-product of mixed acid fermentations is biohydrogen, which can be used as a clean and renewable biofuel. Dark fermentative biohydrogen production has been successfully demonstrated by various scientists as an efficient means of energy generation using organic wastes. It can be produced by both obligate and facultative anaerobic bacteria. Facultative bacteria can produce 2 moles hydrogen per mole of glucose, while the strictly anaerobic bacteria can produce 4 moles of hydrogen (Das and Veziroglu 2008). The majority of the hydrogen production is driven by the anaerobic metabolism of pyruvate catalyzed by either *pyruvate formate lyase* (PFL) or *pyruvate ferredoxin oxidoreductase* (PFOR) (Manish and Banerjee 2008a). Different bacteria such as *Enterobacter aerogenes*, *Clostridium butyricum*, *Bacillus coagulans*, and *Citrobacter* sp. have been reported to yield a high rate of hydrogen production (Das and Veziroglu 2008). Currently, feasibility of dark fermentative hydrogen production has already been successfully demonstrated at pilot scales. However, various techno-economic issues are yet to be addressed for the commercialization of this technology.

### 3.3.3 METHANOGENIC BACTERIA

Biomethane is conventionally produced through the anaerobic digestion process and has been used as a domestic fuel for ages. In this process, methanogens play a pivotal role in converting the organic acids (produced after the hydrolysis and fermentation of organic wastes) to methane. Most methanogens are found in extreme environments and at higher pH (ranging from 6 to 9). Methogens are obligate anaerobes and, based on the substrates used, they can be broadly classified into two different groups—acetoclastic and hydrogenotrophic. Acetoclastic methanogens use acetate as a substrate for methane production. They are highly diverse in nature and are usually found in syntrophic associations (Balch et al. 1979). *Methanosarcina barkeri* and *Methanosarcina* sp. are a few examples of acetoclastic methanogens that are found in anaerobic digesters (Ali Shah et al. 2014). On the other hand, hydrogeneotrophic methanogens use $CO_2$ and hydrogen to produce methane. It is speculated that only 30% of the methanogenic consortia of anaerobic digesters are hydrogenotrophs suggesting dominance of acetoclastic methanogens (Ali Shah et al. 2014). The hydrogenotrophic methanogens include *Methanobacterium byrantii*, *Methanobrevibacter arboriphilus*, and *Methanobacterium thermoautotrophicum* (Balch et al. 1979). Certain species of *Methanosarcina* can use the acetoclastic methanogenesis pathway and the hydrogenotrophic methanogenesis pathway (Ali Shah et al. 2014). The major disadvantage of these organisms is their slow growth rates that prolong the methanogenesis process. Moreover, the low biogas yields, tedious methane recovery, unwanted odor nuisance, and the large capital investments are the major challenges of anaerobic digestion process that need to be addressed.

### 3.3.4 ELECTROGENIC BACTERIA

With the discovery of certain bacteria capable of generating electricity (Potter 1911), several attempts have been made to use these bacteria for direct power generation. These organisms often are used in different applications such as microbial fuel cells and microbial electrolysis to harness electricity and fuel (hydrogen), respectively (Varanasi and Das 2017a). The electron-generating bacteria, known as exoelectrogens or simply electrogens, can interact and shuttle electrons outside their cell to the solid metal surfaces. Depending upon the organism, this electron transfer can either be mediated (with the aid of exogenous or endogenous shuttles) or a direct electron transfer. Early reports focused on exogenous mediated electron transfer using organisms such as *Enterobacter aerogenes* and *E.coli*. Later on, organisms such as *Pseudomonas aeruginosa* (shuttle: pyocyanin) and certain *Shewanella* sp. (shuttle: flavins) were discovered which had innate capabilities to produce endogeneous shuttles (Varanasi and Das 2017b). Direct electron transfer is reported in organisms like *Geobacter sulfurreducens*, which rely on c-Cyts and/or pili for the biofilm formation and electron transfer to solid electrodes (Kumar et al. 2016). Direct electron transfer is expected to be much efficient than the mediated electron transfer because of minimal mass transfer losses in the former. So far, maximum current is produced using *Geobacter* sp. However, the ongoing research has brought about several new species whose efficiency still remains to be examined. Furthermore, recent studies have shown the potential of these organisms to convert $CO_2$ to fuels, thereby, providing nexus of applications in the coming years (Nevin et al. 2010).

## 3.4   YEAST BIOFUELS

Yeast, mostly *Saccharomyces cerevisiae*, has been traditionally used for commercial bioethanol production for several decades. It is used because of its characteristic features such as low cost, high ethanol productivity, easy availability, high tolerance, and flexibility of substrates (Jansen et al. 2017). Besides *S. cerevisiae*, other various types of yeast strains such as *Pichia stipites* (NRRL-Y-7124) and *Kluyveromyces fagilis* (Kf1) also have been reported for high-yield ethanol production (Mohd Azhar et al. 2017). Like bacteria, yeasts produce ethanol through the sugar fermentation process by following the Embden–Meyerhof pathway. The major challenge of yeast fermentation is that it cannot convert pentoses to ethanol and thus lowers the substrate conversion efficiency when using organic wastes as substrates. Over the years, various new recombinant strains have been developed. However, no breakthrough has been observed yet. Besides bioethanol, yeasts also have been explored for use in alternate advanced biofuel production. The yeast cell factory, *Saccharomyces cerevisiae*, has been metabolically engineered to produce other higher alcohols like butanol and isobutanol (Buijs et al. 2013).

Besides bioalcohols, several oleaginous yeasts have been considered as suitable feedstocks for biodiesel production. Moreover, certain yeast species tend to produce oleochemicals from plant biomasses (Sitepu et al. 2014). The major oleaginous yeasts include *Y lipolytica*, *Rhodotorula graminis*, and *Cryptococcus albidicus*, and several new strains have been discovered over the years, whose capabilities are yet to be examined (Sitepu et al. 2014). Usually yeasts that accumulate 20% of oils per unit dry cell weight are considered to be oleogenic. Different types of lipids are present inside the yeast cells that can be extracted and converted to fatty acid methyl/ethyl esters (i.e., biodiesel). The ideal characteristics of lipid-producing yeast strains include high cell density and fast growth, high substrate utilization efficiency, higher lipid content, and high tolerance for inhibitors. Although the yeast biodiesel production is still in its nascent phase of research, the improvements in cultivation and processing of yeast oils as well as a biorefinery approach targeting extraction of other valuable products such as polyunsaturated fatty acids (PUFA), solvents, and adhesives can make the yeast-based biodiesel production economically viable in the near future.

## 3.5 BIOFUELS FROM FILAMENTOUS FUNGI

Although yeasts are also a subclass of fungi, the characteristic difference between them and the filamentous fungi has brought a massive distinction among these organisms. This section mainly deals with the use of various filamentous fungi for biofuel production. Like yeasts certain filamentous fungi exhibit the capacity to accumulate intracellular lipids during metabolic stress periods (Abu-elreesh and Abd-el-haleem 2014). Most of the yeasts contain lipids in the form of triacylglycerols (TAG). By altering the growth conditions, the accumulation of lipids in the yeasts can be enhanced significantly. The different filamentous fungal species that have been reported for biodiesel production include *Cunninghamella echinulata*, *Mortierella isabellina*, *Mucor circinelloides*, *Mortierella alpine*, *Aspergillus niger*, and *Rhizopus oryzae* (Zheng et al. 2012). Filamentous fungi can accumulate up to 80% weight per weight (%w/w) lipids. However, not all these lipids are useful for biodiesel production. It was suggested that only saponifiable lipids and free fatty acids can be converted to fatty acid methyl esters (Zheng et al. 2012). Using filamentous fungi for biodiesel production has several advantages; for example, it has high growth rates and thus can rapidly accumulate storage lipids, it shows good lipid profiles and thus can be converted to high quality biodiesel, it can use a variety of carbon sources, and due to its tendency to form cell pellets, the harvesting process can be simplified. The major hurdle is the cost of the production process which needs to be addressed for the commercial viability of the process. Furthermore, the strains available have only 20%–25% usable lipids for biodiesel production (Huang et al. 2016). Thus, new strains need to be screened and tested for their potential for biodiesel production. Another major challenge is the extraction and standardization of lipid oils, which needs the utmost concern.

## 3.6 MICROALGAL BIOFUELS

Microalgae are microscopic algae that are found in freshwater and marine water ecosystems. Because of similar properties, cyanobacteria also are characterized often as microalgae. Like green plants, microalgae are capable of photosynthesis but are not differentiated with roots, stems, or leaves. Moreover, they have higher growth rates and can be grown in non-arable lands (Demirbas 2010). Certain microalgae have the capability to alter their chemical composition according to the change in the environmental conditions. This property has marked potential for use of these organisms in biofuel production. Microalgae can be converted to bioethanol, biohydrogen, biomethane, and biodiesel using different biochemical and thermochemical processes (Demirbas 2010). Microalgae has starch as its storage carbohydrate, which can be used as a feedstock for the bacterial or the yeast fermentation process for bioalcohols or biogas production. Unlike other renewable feedstocks such as lignocellulosic biomass, algal feedstocks require mild pre-treatment to extract the carbohydrates. Moreover, certain green microalgae such as *Chlamydomonas reinhardtii*, *Chlorella fusca*, and *Scenedesmus obliquus* have innate capabilities to produce hydrogen through direct photolysis (Singh and Das 2018). However, the production observed is too low compared to fermentative hydrogen production.

In recent years, significant attention has been given to microalgae-based biodiesel production. Like fungi, microalgae also possess storage lipids in the form of TAG that can be enhanced by using suitable growth conditions. However, microalgae have an advantage over fungi because they can be grown photosynthetically. Furthermore, the ability of microalgae to fix atmospheric $CO_2$ can aid in lowering the greenhouse gas emissions, which is environmentally sustainable (Mata et al. 2010). *Chlorella sorokiniana*, *Chlamydomonas reinhardtii*, *Dunaliella bioculata*, *Scendesmus quadricauda*, and *Nanochloropsis oceanica* are a few oleaginous microalgal species that have been reported to contain high-lipid contents over other species (Faried et al. 2017). Microalgae is a versatile feedstock that can be used for obtaining multiple valuable products in a biorefinery approach (Chew et al. 2017). After biodiesel is extracted, the defatted biomass can be used for the extraction of high-value pigments (such as astraxanthine carotene) followed by biofuel production as explained previously. By using such an approach, the economic viability of the process can be significantly improved.

Although microalgal technologies show immense potential, cultivation and harvesting pose several challenges. Microalgae are cultivated using closed photobioreactors or using open pond systems. Both these systems have inherent advantages and disadvantages. Open pond systems can be economically attractive but are prone to contamination, while closed photobioreactor systems can ensure sterile environments but require stringent operating conditions. Since the production costs of algae are too high, its use as a feedstock for liquid or gaseous biofuels is still debatable.

## 3.7 CONCLUSIONS AND FUTURE PERSPECTIVES

Microbial biofuels are emerging as potential substitutes for conventional petroleum-derived fuels because they are renewable, sustainable, and environmentally friendly. Most of these microbes produce biofuels using first-generation feedstocks such as corn, maize, and sugarcane. However, these organisms cannot completely use second-generation feedstocks such as lignocellulosic biomasses. This inability is because of the complex lignin content of second-generation feedstocks, which are not biodegradable. In addition, several microbial species such as bacteria and yeasts cannot use pentose sugars efficiently. Different genera of microbes have been discovered over the years that can use and break these sugars, and various metabolic strategies have been developed to enhance the substrate degradability of existing species. Although several advancements have been made recently, the various techno-economic barriers still limit the practical scalability of microbial biofuel technologies. With novel synthetic biology approaches to directly convert solar energy to biofuels using genetically engineered cyanobacterial species, the dependence of rich sugar-based feedstock can be eliminated. Furthermore, a recent breakthrough in bio-electrofuel technology can bring about new unknown pathways for the conversion of $CO_2$ to advance biofuels. These exciting new microbial species and synthetic pathways can create multiple avenues for the establishment of biofuel technologies in the near future.

## REFERENCES

Abu-elreesh G, Abd-el-haleem D (2014) Promising oleaginous filamentous fungi as biodiesel feed stocks: Screening and identification. *Eur J Exp Biol* 4:576–582.

Ali Shah F, Mahmood Q, Maroof Shah M, Pervez A, Ahmad Asad S (2014) Microbial ecology of anaerobic digesters: The key players of anaerobiosis. *Sci World J* 2014:183752.

Balch WE, Fox GE, Magrum LJ, Woese CR, Wolfe RS (1979) Methanogens: Reevaluation of a unique biological group. *Microbiol Rev* 43:260–296. doi: 10.1016/j.watres.2010.10.010.

Booth B (2009) Electromethanogenesis: The direct bioconversion of current to methane. *Environ Sci Technol* 43:4619–4619.

Buijs NA, Siewers V, Nielsen J (2013) Advanced biofuel production by the yeast *Saccharomyces cerevisiae.Curr Opin Chem Biol* 17:480–488.

Call DF, Wagner RC, Logan BE (2009) Hydrogen production by geobacter species and a mixed consortium in a microbial electrolysis cell. *Appl Environ Microbiol* 75:7579–7587.

Chandel AK, Silva SS, Singh OV (2011) Detoxification of lignocellulosic hydrolysates for improved bioconversion of bioethanol. In: Bernardes MAS (Ed.), *Biofuel Production-Recent Developments and Prospects*. InTech, Rijeka, Croatia.

Chew KW, Yap JY, Show PL, Suan NH, Juan JC, Ling TC, Lee DJ, Chang JS (2017) Microalgae biorefinery: High value products perspectives. *Bioresour Technol* 229:53–62.

Das D, Veziroglu T (2008) Advances in biological hydrogen production processes. *Int J Hydrogen Energy* 33:6046–6057.

Dasgupta CN, Jose Gilbert J, Lindblad P, Heidorn T, Borgvang SA, Skjanes K, Das D (2010) Recent trends on the development of photobiological processes and photo-bioreactors for the improvement of hydrogen production. *Int J Hydrogen Energy* 35:10218–10238.

Demirbas A (2010) Use of algae as biofuel sources. *Energy Convers Manag* 51:2738–2749.

Dutta D, De D, Chaudhuri S, Bhattacharya S (2005) Hydrogen production by Cyanobacteria. *Microb Cell Fact* 4:36.

El-Gendy NS, Madian HR, Amr SSA (2013) Design and optimization of a process for sugar-cane molasses fermentation by saccharomyces cerevisiae using response surface methodology. *Int J Microbiol* 2013:815631.

Elshahed MS (2010) Microbiological aspects of biofuel production: Current status and future directions. *J Adv Res* 1:103–111.

Ezeji T, Milne C, Price ND, Blaschek HP (2010) Achievements and perspectives to overcome the poor solvent resistance in acetone and butanol-producing microorganisms. *Appl Microbiol Biotechnol* 85:1697–1712.

Ezeji TC, Qureshi N, Blaschek HP (2003) Production of acetone, butanol and ethanol by Clostridium beijerinckii BA101 and in situ recovery by gas stripping. *World J Microbiol Biotechnol* 19:595–603.

Faried M, Samer M, Abdelsalam E, Yousef RS, Attia YA, Ali AS (2017) Biodiesel production from microalgae: Processes, technologies and recent advancements. *Renew Sustain Energy Rev* 79:893–913.

Fischer CR, Klein-Marcuschamer D, Stephanopoulos G (2008) Selection and optimization of microbial hosts for biofuels production. *Metab Eng* 10:295–304.

Gogotov IN, Zorin NA, Serebriakova LT (1991) Hydrogen production by model systems including hydrogenases from phototrophic bacteria. *Int J Hydrogen Energy* 16:393–396.

Hallenbeck PC (2012) Microbial technologies in advanced biofuels production. *Microb Technol Adv Biofuels Prod* 9781461412:1–272.

Harai É, Kapás Á, Lányi S, Ábrahám B, Nagy I, Muntean O (2010) Biohydrogen production by photofermentation of lactic acid using *Thiocapsa roseopersicina*. *UPB Sci Bull*, Ser B 72.

He M, Wu B, Qin H, Ruan Z, Tan F, Wang J, Shui Z, Dai L, Zhu Q, Pan K, Tang X, Wang W, Hu Q (2014) Zymomonas mobilis: A novel platform for future biorefineries. *Biotechnol Biofuels* 7:101.

Huang G, Zhou H, Tang Z, Liu H, Cao Y, Qiao D, Cao Y (2016) Novel fungal lipids for the production of biodiesel resources by Mucor fragilis AFT7-4. *Environ Prog Sustain Energy* 35:1784–1792.

Hwang IY, Hoon Hur D, Hoon Lee J, Park CH, Chang IS, Lee JW, Yeol Lee E (2015) Batch conversion of methane to methanol using methylosinus trichosporium OB3B as biocatalyst. *J Microbiol Biotechnol* 25:375–380.

Jansen MLA, Bracher JM, Papapetridis I, Verhoeven MD, De Bruijn H, De Waal PP, Van Maris AJA, Klaassen P, Pronk JT (2017) *Saccharomyces cerevisiae* strains for second-generation ethanol production: From academic exploration to industrial implementation. *FEMS Yeast Res* 17:5.

Jones DT, Woods DR (1986) Acetone-butanol fermentation revisited. *Microbiol Rev* 50:484–524.

Kang A, Lee TS (2015) Converting sugars to biofuels: Ethanol and beyond. *Bioengineering* 2:184–203.

Khanna N, Lindblad P (2015) Cyanobacterial hydrogenases and hydrogen metabolism revisited: Recent progress and future prospects. *Int J Mol Sci* 16:10537–10561.

Köpke M, Mihalcea C, Bromley JC, Simpson SD (2011) Fermentative production of ethanol from carbon monoxide. *Curr Opin Biotechnol* 22:320–325.

Kumar N, Das D (2000) Enhancement of hydrogen production by Enterobacter cloacae IIT-BT 08. *Process Biochem* 35:589–593.

Kumar R, Singh L, Zularisam AW (2016) Exoelectrogens: Recent advances in molecular drivers involved in extracellular electron transfer and strategies used to improve it for microbial fuel cell applications. *Renew Sustain Energy Rev* 56:1322–1336.

Lacis LS, Lawford HG (1991) Thermoanaerobacter ethanolicus growth and product yield from elevated levels of xylose or glucose in continuous cultures. *Appl Environ Microbiol* 57:579–85.

Lancini G, Lorenzetti R (1993) Strain improvement and process development. In: *Biotechnology of Antibiotics and Other Bioactive Microbial Metabolites*. Springer US, Boston, MA, pp. 175–190.

Levin D (2004) Biohydrogen production: Prospects and limitations to practical application. *Int J Hydrogen Energy* 29:173–185.

Levin D, Islam R, Cicek N, Sparling R (2006) Hydrogen production by Clostridium thermocellum 27405 from cellulosic biomass substrates. *Int J Hydrogen Energy* 31:1496–1503.

Liu G-Q, Du Li, Chao-Yang Zhu, Kuan Peng, Huai-Yun Zhang (2010) Screening of ole-aginous microorganisms for microbial lipids production and optimization. In: *2010 International Conference on Bioinformatics and Biomedical Technology*. IEEE, pp. 149–152.

Lovitt RW, Shen GJ, Zeikus JG (1988) Ethanol-production by thermophilic bacteria—biochemical basis for ethanol and hydrogen tolerance in Clostridium-Thermohydrosulfuricum. *J Bacteriol* 170:2809–2815.

Maki M, Leung KT, Qin W (2009) The prospects of cellulase-producing bacteria for the bioconversion of lignocellulosic biomass. *Int J Biol Sci* 5:500–516.

Manish S, Banerjee R (2008a) Comparison of biohydrogen production processes. *Int J Hydrogen Energy* 33:279–286.

Manish S, Banerjee R (2008b) Comparison of biohydrogen production processes. *Int J Hydrogen Energy* 33:279–286.

Markov SA, Bazin MJ, Hall DO (1995) The potential of using cyanobacteria in photobioreac-tors for hydrogen production. *Adv Biochem Eng/Biotechnol* 52:59–86.

Mata TM, Martins AA, Caetano NS (2010) Microalgae for biodiesel production and other applications: A review. *Renew Sustain Energy Rev* 14:217–232.

McKinlay JB, Harwood CS (2010) Photobiological production of hydrogen gas as a biofuel. *Curr Opin Biotechnol* 21:244–251.

Mitra R, Balachandar G, Singh V, Sinha P, Das D (2017) Improvement in energy recovery by dark fermentative biohydrogen followed by biobutanol production process using obli-gate anaerobes. *Int J Hydrogen Energy* 42:4880–4892.

Mohd Azhar SH, Abdulla R, Jambo SA, Marbawi H, Gansau JA, Mohd Faik AA, Rodrigues KF (2017) Yeasts in sustainable bioethanol production: A review. *Biochem Biophys Reports* 10:52–61.

Müller V (2001) Bacterial fermentation. In: *Encyclopedia of Life Sciences*. Wiley Basingstoke, Nature Publishing Group.

Munasinghe PC, Khanal SK (2010) Biomass-derived syngas fermentation into biofuels: Opportunities and challenges. *Bioresour Technol* 101:5013–5022.

Naik SN, Goud V V., Rout PK, Dalai AK (2010) Production of first and second generation biofuels: A comprehensive review. *Renew Sustain Energy Rev* 14:578–597.

Ndaba B, Chiyanzu I, Marx S (2015) n-Butanol derived from biochemical and chemical routes: A review. *Biotechnol Rep* 8:1–9.

Nevin KP, Woodard TL, Franks AE (2010) Microbial electrosynthesis: Feeding microbial electrosynthesis: Feeding microbes electricity to convert carbon dioxide and water to multicarbon extracellular organic. *MBio* 1:e00103–e00110.

Nigam PS, Singh A (2011) Production of liquid biofuels from renewable resources. *Prog Energy Combust Sci* 37:52–68.

Potter MC (1911) Electrical effects accompanying the decomposition of organic compounds. *Proc R Soc B Biol Sci* 84:260–276.

Raksajit W, Satchasataporn K, Lehto K, Mäenpää P, Incharoensakdi A (2012) Enhancement of hydrogen production by the filamentous non-heterocystous cyanobacterium Arthrospira sp. PCC 8005. *Int J Hydrogen Energy* 37:18791–18797.

Ranjan A, Mayank R, Moholkar VS (2013) Process optimization for butanol production from developed rice straw hydrolysate using *Clostridium acetobutylicum* MTCC 481 strain. *Biomass Convers Biorefinery* 3:143–155.

Rogers PL, Lee KJ, Skotnicki ML, Tribe DE (1982) Ethanol production by *Zymomonas mobi-lis*. In: *Microbial Reactions*. Springer, Berlin, Germany, pp. 37–84.

Roy S, Vishnuvardhan M, Das D (2014) Improvement of hydrogen production by newly iso-lated *Thermoanaerobacterium thermosaccharolyticum* IIT BT-ST1. *Int J Hydrogen Energy* 39:7541–7552.

Sauer M (2016) Industrial production of acetone and butanol by fermentation-100 years later. *FEMS Microbiol Lett* 363:1–4.

Schink B, Zeikus JG (1980) Current microbiology microbial Methanol formation: A major end product of pectin metabolism. *Curr Microbio* 4:387–389.

Singh H, Das D (2018) *Biofuels from Microalgae: Biohydrogen.* Springer, Cham, Switzerland, pp. 201–228.

Sitepu IR, Garay LA, Sestric R, Levin D, Block DE, German JB, Boundy-Mills KL (2014) Oleaginous yeasts for biodiesel: Current and future trends in biology and production. *Biotechnol Adv* 32:1336–1360.

Stal LJ, Moezelaar R (2006) Fermentation in cyanobacteria. *FEMS Microbiol Rev* 21:179–211.

Tamagnini P, Axelsson R, Lindberg P, Oxelfelt F, Wünschiers R, Lindblad P (2002) Hydrogenases and hydrogen metabolism of cyanobacteria. *Microbiol Mol Biol Rev* 66:1–20.

Trchounian A (2015) Mechanisms for hydrogen production by different bacteria during mixed-acid and photo-fermentation and perspectives of hydrogen production biotech-nology. *Crit Rev Biotechnol* 35:103–113.

Valdez-Vazquez I, Pérez-Rangel M, Tapia A, Buitrón G, Molina C, Hernández G, Amaya-Delgado L (2015) Hydrogen and butanol production from native wheat straw by syn-thetic microbial consortia integrated by species of *Enterococcus* and *Clostridium*. *Fuel* 159:214–222.

Varanasi JL, Das D (2017a) Bioremediation and power generation from organic wastes using microbial fuel cell. In: *Microbial Fuel Cell: A Bioelectrochemical System That Converts Waste to Watts.* Springer International Publishing, Cham, Switzerland, pp. 285–306.

Varanasi JL, Das D (2017b) Characteristics of microbes involved in microbial fuel cell. In: *Microbial Fuel Cell: A Bioelectrochemical System That Converts Waste to Watts.* Springer International Publishing, Cham, Switzerland, pp. 43–62.

Xia A, Cheng J, Lin R, Lu H, Zhou J, Cen K (2013) Comparison in dark hydrogen fermenta-tion followed by photo hydrogen fermentation and methanogenesis between protein and carbohydrate compositions in *Nannochloropsis oceanica* biomass. *Bioresour Technol* 138:204–213.

Yen H-W, Li R-J, Ma T-W (2011) The development process for a continuous acetone–butanol–ethanol (ABE) fermentation by immobilized *Clostridium acetobutylicum*. *J Taiwan Inst Chem Eng* 42:902–907.

Yu J, Takahashi P (2007) Biophotolysis-based hydrogen production by Cyanobacteria and Green Microalgae. *Commun Curr Res Educ Top Trends Appl Microbiol* 1:79–89.

Zheng Y, Yu X, Zeng J, Chen S (2012) Feasibility of filamentous fungi for biofuel production using hydrolysate from dilute sulfuric acid pretreatment of wheat straw. *Biotechnol Biofuels* 5(1):50.

# 4 Biochemical Pathways for the Biofuel Production

## 4.1 INTRODUCTION

Biofuels are renewable fuels that are produced using energy stored in biomass in the form of carbohydrates, lipids, and proteins. These biomolecules can be directly extracted and used for the production of fuels (e.g., biodiesel production from lipids extracted from oil crops and algae) or can be converted through microbial metabolism to liquid and gaseous biofuels (e.g., ethanol production from sugarcane using yeasts) (Demirbas 2008; Nigam and Singh 2011). Microbial metabolism occurs in two phases of biochemical processes—anabolism (synthesis of new compounds) and catabolism (breakdown of complex biomolecules for energy generation). Both these phases involve a series of interconnected biochemical reactions that are known as biochemical pathways.

The understanding of the fundamental biochemical pathways is essential for designing an efficient biofuel production process. The biochemical pathways differ with respect to the type of microorganism under consideration. In heterotrophic microorganisms, there are three major biochemical pathways for the conversion of organic substrates to energy—aerobic respiration, anaerobic respiration, and fermentation (Müller 2001). In contrast, autotrophic microbes derive energy directly from sunlight using the photosynthesis process. Microorganisms tend to modify their metabolic pathways to acclimatize to their surrounding environment. This property of microbes can be exploited for obtaining the desired product of interest (such as biomass or biofuel) by providing controlled growth conditions. Knowledge of the metabolic pathways is thus essential to enhance the yield of the desired product and to improve process efficiency. Furthermore, once known, these pathways can be redefined and reconstructed into suitable hosts using various metabolic engineering approaches. This chapter provides an insight into the various biochemical pathways that are currently exploited for biofuel production.

## 4.2 BIOCHEMICAL PATHWAYS FOR BIODIESEL PRODUCTION

Biodiesel is produced by the transesterification of vegetable oils, animal fats, or microbial lipids. The main component of these oils and fats is triacylglycerols or triglycerides (TAGs), which are the fatty acyl esters of glycerol (i.e., they comprise of three fatty acids bound to a glycerol backbone). Most plants have the ability to synthesize TAGs in the form of storage lipids, while few oleaginous fungi, yeast, and microalgae have the capabilities to store TAGs (Abu-elreesh and Abd-el-haleem 2014; Patel et al. 2016; Faried et al. 2017). Although most research on TAGs suggests their occurrence in eukaryotic organisms, certain prokaryotic bacteria, mainly belonging to the actinomycetes group, also possess the ability to accumulate TAGs (Alvarez and Steinbüchel 2003). This section mainly deals with the biosynthesis of TAGs in different organisms. The conversion technologies for TAG to biodiesel are discussed in Chapter 11.

## 4.2.1 PLANT TAG SYNTHESIS

TAGs are the predominant component of the oils extracted from the seeds or fruits of oleaginous plants like sunflower, oilseed rape, maize, and soybean. TAGs of higher plants contain acyl groups of palmitic acid (16:0), stearic acid (18:0), oleic acid (18:1), linolenic acid (18:2), and α-linolenic acid (18:3). In these organisms, the fatty acid biosynthesis occurs in the stroma of plastids. The subsequent incorporation of fatty acids into the glycerol backbone leads to the biosynthesis of TAGs in the endoplasmic reticulum (ER) (Figure 4.1a). During fatty acid synthesis, the acetyl-CoA carboxylases (ACC) catalyze the irreversible carboxylation of acetyl-CoA to form malonyl-CoA. The malonyl group is then transferred to the acyl carrier protein (ACP), giving rise to malonyl-ACP, which is the primary substrate for fatty acid synthase complex. Repeated condensation of malonyl-ACP by the action of fatty acid synthase complex with the subsequent addition of two carbon units after each elongation leads to the formation of fatty acid-ACP complexes (like 16:0 ACP, 18:0 ACP, and 18:1 ACP). The type of acyl chain bound to ACP depends upon the type of enzyme present (mostly ketoacyl synthases I, II, and III) and varies in different plant species. The termination of fatty acid elongation is catalyzed by ACP thioesterases, which catalysis the hydrolysis of acyl-ACP to produce free fatty acids that are able to cross the plastidial envelope. These free fatty acids are then reactivated as acyl-CoAs in the cytosol and form the acyl-CoA pool, which provides the acyl donor for the acyl transferase reactions of the TAG assembly in the ER.

In the ER, the glycerol backbone for TAG synthesis is in the form of glycerol-3-phosphate, which is generated by the catalytic action of glycerol-3-phosphate dehydrogenase on dihydroxyacetone phosphate derived from glycolysis. The most generalized pathway for TAG synthesis in the ER is the acyl-CoA dependent pathway or the Kennedy pathway. In this pathway, three sequential acyl-CoA dependent acylations are performed by different acyl transferases, which exhibit preference to a specific acyl-CoA. The removal of phosphate from the glycerol backbone occurs prior to the final acylation step (Figure 4.1a). The first acylation is catalyzed by acyl-CoA and sn-glycerol-3-phosphate acyl transferase (GPAT) which converts glycerol-3-phosphate to lyso-phosphatidic acid. The second acylation is performed by acyl-CoA and lyso-phosphatidic acid acyl transferase (LPAAT), which forms phosphatidic acid. The phosphate group from the phosphatidic acid is removed by the enzyme phosphatidic acid phosphatase (PAP), which forms the sn-1, 2 diacylglycerol (DAG). The third and final acylation leads to the conversion of DAG to TAG by the catalytic action of acyl-CoA and diacylglycerol acyl transferase (DGAT). This step is the rate limiting step of lipid biosynthesis and is the potential target for genetic modification in higher plants to increase the TAG accumulation. These TAGs form droplets around the outer leaflet of the ER and eventually develop into oil bodies ranging from 0.5 to 2.5 μm.

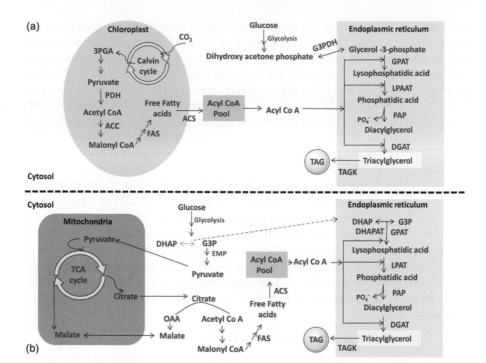

**FIGURE 4.1** (a) TAG biosynthesis in green plants and algae and (b) TAG biosynthesis in yeasts and fungi. (PDH: pyruvate dehydrogenase; ACC: acetyl CoA carboxylase; ACS: acyl CoA synthase; FAS: Fatty acid synthase; G3PDH: glyceraldehyde 3-phosphate dehydrogenase; GPAT: glyceraldehyde 3-phosphate acyltransferase;, LPAAT: lyso-phosphatidic acid acyl transferase; PAP: phosphatidic acid phosphatase; DGAT; diacylglycerol acyltransferase; TAGK: triacylglycerol kinase; DHAP: dihydroxyacetone phosphate; G3P: glyceraldehyde 3-phosphate; DHAPAT: dihydroxyacetone phosphate acyl transferase)

Certain fatty acids incorporate into membrane lipids of plastids and the ER and can later accumulate as TAGs. The conversion of the membrane lipids to TAG is carried out by several acyl-CoA independent pathways, which utilize different enzymes that can synthesize TAGs using alternate reactions (Cagliari et al. 2011). The enzyme diacylglycerol transacylase (DGTA) transfers acyl molecules between two successive DAGs to form TAG and monoacylglycerol (MAG) as a co-product. Alternately, the enzyme phospholipid diacylglycerol acyltransferase (PDAT) can convert the *sn*-phospholipids to DAG which are subsequently converted to TAG through DGAT or DGTA. Another enzyme, choline phosphotransferase (CPT) can convert phospholipids into DAG in a reversible reaction and thus in turn aid in TAG biosynthesis through DGAT or PDAT pathways.

## 4.2.2 TAG BIOSYNTHESIS IN ALGAE

Algae comprise of diverse composition of acyl lipids and unusual fatty acids that are not found in other phyla (Cagliari et al. 2011). Due to this extreme diversity, these organisms have been exploited excessively for obtaining different forms of saturated and unsaturated fatty acids for several applications such as nutrient supplements, paints, and lubricants. Due to the capability of algae to adapt to different environmental conditions and higher growth rates (compared to higher plants), algae are considered as potential feedstocks for biodiesel production. The basic biochemical pathways for TAG synthesis in algae are analogous to the higher plants (i.e., predominantly by the Kennedy pathway). However, there are few differences. The TAG accumulation in algae occurs in a single cell in contrast to higher plants where lipids are synthesized and localized in a specific cell, tissue, or organ (Cagliari et al. 2011). Another special feature in algae is that it can accumulate a large number of carbon chain fatty acids (>20), while higher plants can synthesize only up to C18. However, the TAG accumulation in algae is species or strain specific. It is observed that less TAG is synthesized during the normal favorable growth conditions of algae. During this period, most of the fatty acids are stored in the form of membrane lipids. On the other hand, in unfavorable and adverse environmental conditions, certain algal strains shift their metabolism from membrane lipids to TAGs. It is estimated that in such species, TAG contribute about 80% of the total lipid content of the algal cell. So, the lipid productivity and in turn biodiesel production can be enhanced significantly by using such algal strains and providing suitable environmental stress.

### 4.2.3  TAG Biosynthesis in Yeast and Fungi

Like algae, recent interest in single cell oil has provided greater insights for the use of yeast and fungal-based lipids for human exploitation. Biodiesel production using yeast and fungal cells can be a promising sustainable substitute for vegetable oils and algae because they have higher growth rates, they do not require arable land for growth, and they can utilize a plethora of hydrophobic and hydrophilic substrates (Athenaki et al. 2018). Unlike higher plants and algae, the fatty acid synthesis and the acyl-CoA pool formation in yeast and fungi are confined to mitochondria and cytosol, whereas the TAG synthesis occurs in various organelles. The ER is the major site for synthesis of TAG, which contains all the four major enzymes needed for TAG synthesis (Figure 4.1b). Other organelles include lipid particles (TAG storage systems), cytosol, mitochondria, and vacuoles, which contain few of the enzymes required for TAG synthesis (Sorger and Daum 2003). Similar to the higher plants and algae, the glycerol backbone for TAG synthesis in yeast and fungi is either glycerol-3-phosphate or dihydroxyacetone phosphate. By the action of the respective enzymes (i.e., GPAT and dihydroxyacetone phosphate acyl transferase (DHAPAT)), these substrates are converted to phosphatidic acid. Phosphatidic acid can be formed also from phospholipids by the action of Phospholipase D. Subsequently, phosphatidic acid is converted to DAG by PAP. Two proteins are responsible mainly for acylation of DAG to TAG namely DRA1 and LRO1. These enzymes use acyl-CoA and phosphatidylcholine as acyl donors, respectively (Sorger and Daum 2003). Besides these, several other minor alternative routes of TAG synthesis are reported which include action of acyl-CoA: cholesterol acyltransferase (ACAT) which uses sterols as substrate, phospholipid diacylglycerol acyltransferase (PDAT) which uses sn-2 acyl groups of phosphatidylcholines as an acyl donor. Another major route is by the phosphorylation of DAG through diacylglycerol kinase which forms phosphatidic acid that can be subsequently converted to DAG and TAG as explained previously. The existence of these alternative pathways, however, is still debatable and current research is focused on understanding the origin of these enzymes. Once TAGs are synthesized, they are exported out to the cytosol using TAG kinases and are stored in lipid particles. These lipid particles not only served as TAG storage molecule but also comprise alternative enzyme systems which can produce de novo neutral lipid synthesis (Sorger and Daum 2003). As in algae, TAG synthesis is not essential in yeast during growth but under environmental stress conditions, such as nitrogen stress, these organisms enhance lipid production to survive in the stationary phase of growth. Thus, these organisms are attractive sources for obtaining TAGs for biodiesel production.

### 4.2.4 BACTERIAL TAG SYNTHESIS

Recently, TAGs production and accumulation also have been observed in prokaryotic organisms such as actinomycetes (Alvarez and Steinbüchel 2003). Most bacteria that accumulate storage lipids can produce specialized storage polyesters such as poly (3-hydroxybutyric acid) and poly-hydroxyalkanoic acids (Alvarez and Steinbüchel 2003). These organisms are known to produce TAGs using enzymes (GPAT, LPAT, and PAP) comparable to those found in other organisms. However, the occurrence of DGAT is restricted and, if present, has dual function (i.e., esterification of diacylglycerols and long chain fatty alcohols for the synthesis of TAGs and wax esters, respectively). Depending upon which intermediates are present in the organisms, the function of DGAT can vary and the bacteria that lack this enzyme are unable to produce neutral lipids.

## 4.3 BIOCHEMICAL PATHWAYS FOR BIOHYDROGEN PRODUCTION

Biohydrogen is considered as the fuel for the future because it has the highest energy density (per unit mass basis). It does not contribute to greenhouse emissions. It can be produced using different renewable feedstocks and it is environment friendly (Meherkotay and Das 2008). Biohydrogen can be produced using different phototrophic and heterotrophic organisms by means of various biochemical processes such as biophotolysis, photofermentation, and dark fermentation. A brief description of these processes follows.

### 4.3.1 BIOPHOTOLYSIS

Biophotolysis is the mechanism by which the photosynthetic microorganisms such as algae and cyanobacteria use the energy derived from sunlight to split water into molecular hydrogen and oxygen (Yu and Takahashi 2007). The photosynthetic machinery of the eukaryotic algae is embedded in the thylakoid membranes of chloroplast, whereas, in the prokaryotic cyanobacteria, it is present in the cytosol. Unlike in photosynthesis, where the reductants obtained during the water splitting reaction are channeled towards the Calvin cycle to fix carbon for cell growth and development, in biophotolysis, these reductants are used up for producing hydrogen by the action of different enzymes. Depending upon the source of reducing equivalents, biophotolysis can be broadly classified into two categories: direct and indirect biophotolysis (Dasgupta et al. 2010). Three major enzymes are involved in biophotolysis that drive the hydrogen production reaction which include (1) Fe-Fe hydrogenases, (2) Ni-Fe hydrogenases, and (3) nitrogenases. The hydrogenases (both Fe-Fe and Ni-Fe) are involved mostly in direct biophotolysis. The Fe-Fe hydrogenases are predominant in eukaryotes such as green plants and algae, while the Ni-Fe hydrogenases are found only in the prokaryotic organisms like cyanobacteria, bacteria, and archaea (Singh and Das 2018). In contrast, nitrogenases are more actively involved during indirect biophotolysis. Mostly these enzymes are inactivated in the presence of $O_2$. Besides these enzymes, another important class of hydrogenases (i.e., uptake hydrogenases) also plays a crucial role in some nitrogen-fixing cyanobacteria.

In direct biophotolysis, reduced electrons produced by photosynthesis during water splitting are used directly by $H_2$-producing hydrogenase without intermediate $CO_2$ fixation. In this process, the light energy is captured by the photochemical reaction centers PSI and PSII which consist of distinct photochemical reaction centers P700 and P680, respectively (Figure 4.2). These light-harvesting photosystems use the solar energy to split water into hydrogen and oxygen. The electrons generated through this process are channeled towards Ferredoxin (Fd) and further are directly taken up by hydrogenases under suitable conditions (i.e., anaerobic conditions). The major disadvantage of this process is the extreme sensitivity of hydrogenase to oxygen. Hydrogenase gets deactivated at the $O_2$ partial pressure of less than 2% (Singh and Das 2018). Thus, the hydrogen productivity is too low in this process due to the oxygen inhibition. To improve the hydrogen yields, a two-stage direct photolysis is performed under sulfur-deprivation conditions (Das and Veziroglu 2008).

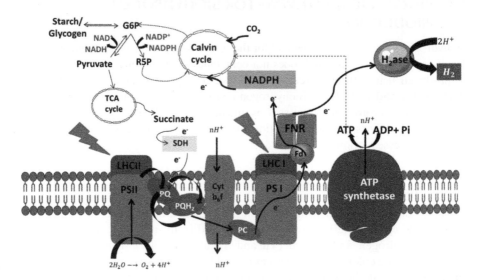

**FIGURE 4.2**  Mechanism of hydrogen production by direct biophotolysis.

In such conditions, the PSII activity (which requires sulphur for its biosynthesis) is partially suppressed and, in turn, lowers the oxygen evolution. This process helps in the development of an anaerobic condition inside the cell and helps the hydrogenases to be active for longer duration of time. The organisms such as *Synechocystis* sp. (cyanobacteria) and *Chlamydomonas reinhardtii* (green algae) produce hydrogen through direct photolysis mechanism.

Indirect biophotolysis involves the use of reducing equivalents derived from the endogenously stored carbohydrates. These carbohydrates are formed during the $CO_2$ fixation of the photosynthesis process thus involving the indirect use of sunlight. This process is mostly predominant in cyanobacteria. Like direct biophotolysis, indirect biophotolysis also can be performed in single-stage and two-stage processes (Figure 4.3). In the single-stage indirect biophotolysis, the oxygen evolution reaction and the hydrogen evolution reaction are separated by spatial arrangements (Manish and Banerjee 2008). It is carried out usually by nitrogen-fixing cyanobacteria comprised of specialized cells known as heterocysts (Figure 4.3a). The $CO_2$ fixation occurs in the vegetative cells, while the nitrogen fixation and hydrogen evolution occurs in the heterocysts. Due to the absence of $O_2$-evolving PSII in heterocysts, an anaerobic condition prevails which helps in nitrogen fixation and hydrogen production by oxygen sensitive nitrogenases. Moreover, the cell walls of heterocysts are impermeable to oxygen and thus do not allow the oxygen coming

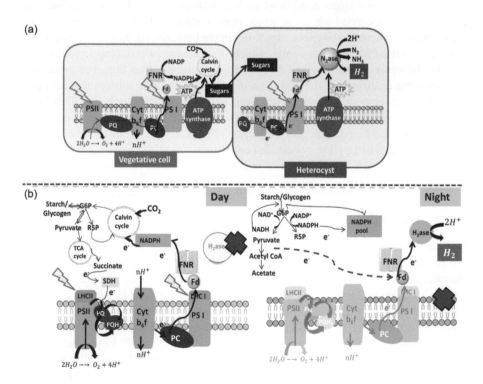

**FIGURE 4.3** (a) Spatial and (b) temporal separation during indirect biophotolysis in cyanobacteria.

from the vegetative cells to enter inside. Therefore, the environment of heterocyst is most suitable for hydrogen production. *Anabaena* sp. and *Nostoc* sp. are a few examples of heterocystous cyanobacteria. Two-stage indirect biophotolysis mostly occurs in non-$N_2$-fixing non-heterocystous cyanobacteria such as *Synechocystis* sp., *Synechococcus* sp., and *Gloebacter* sp. In these organisms, hydrogen production is mediated by hydrogenases by maintaining temporal separation (separation depending on time of expression) of hydrogen production from the oxygen-evolving photosynthesis (Figure 4.3b). This process takes place by using the diurnal (day and night) cycle (i.e., during the day $CO_2$ fixation occurs through photosynthesis and during the night hydrogen is produced through fermentation of this accumulated carbohydrate.

### 4.3.2 PHOTOFERMENTATION

Photofermentation is carried out by phototrophic bacteria that are capable of anoxygenic photosynthesis. These organisms mainly belong to two different classes of bacteria (i.e., purple-bacteria and green bacteria) (Harai et al. 2010). The purple bacteria are subdivided into purple-sulphur bacteria or purple non-sulphur bacteria, while the green bacteria are grouped into green sulphur and green-gliding bacteria (Meherkotay and Das 2008). Usually these bacteria produce biohydrogen under environmental stress conditions such as nitrogen or sulphur deprivation. The photosynthesis is carried out using single photosystems, which are similar to those found in cyanobacteria and green algae. In purple bacteria, the photosystem present is similar to the PSII reaction center, which is incapable of reducing ferredoxin on its own. However, it can generate ATP through cyclic electron flow. For producing hydrogen, the electrons are derived from inorganic or organic sources that accumulate at the quinone pool through the reaction center bacteriochlorophyll (P870). From the quinone pools, the electrons are directed to reduce ferredoxin and ultimately lead to nitrogenase-mediated hydrogen production (Figure 4.4a). In contrast, green bacteria consist of a PSI-like photosystem that can directly reduce ferredoxin through FeS protein (Figure 4.4b). This reduced ferredoxin is used as an electron donor for $CO_2$ fixation as well as hydrogen production.

### 4.3.3 DARK FERMENTATION

The dark fermentation process, carried out by heterotrophic fermentative bacteria, produces biohydrogen at much higher rates compared to the biophotolysis or photofermentation processes (Manish and Banerjee 2008). Any organic substrates (from simple sugars to complex wastes) can be used for dark-fermentative hydrogen production. It is observed that facultative and obligate anaerobic bacteria can produce hydrogen using different metabolic pathways (Elsharnouby et al. 2013). The facultative anaerobes follow the pyruvate formate lyase (PFL) pathway, while the obligate anaerobes follow the pyruvate ferredoxin oxidoreductase (PFOR) pathway (Figure 4.5). In these pathways, the bacteria use the proton ($H^+$) as the electron acceptor and disposes of the excess electrons in the form of molecular hydrogen (Das and Veziroglu 2008).

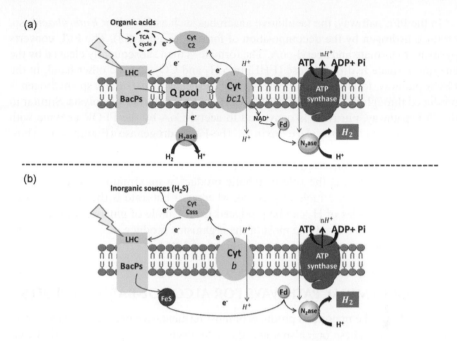

**FIGURE 4.4** Photofermentative hydrogen production in (a) purple bacteria and (b) green bacteria.

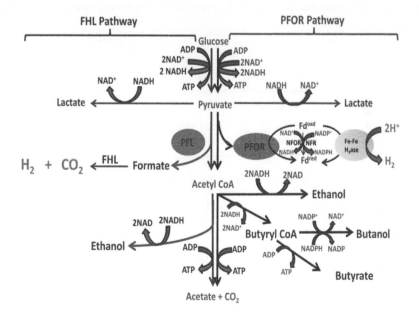

**FIGURE 4.5** Dark-fermentative hydrogen production in facultative and obligate anaerobes (FHL: formate hydrogen lyase; PFL: pyruvate formate lyase; ATP: adenosine triphosphate; ADP: adenosine diphosphate; NADH: nicotinamide adenine dinucleotide; Fd: ferredoxin; PFOR: pyruvate-Fd oxidoreductase; NFOR: NADH-Fd oxidoreductase; LDH: lactate dehydrogenase).

In the PFL pathway, the facultative anaerobes such as *E. coli* or *Enterobacter* sp. produce hydrogen by the decomposition of formate (Figure 4.5). The PFL converts pyruvate to formate and acetyl-coA. The formate is then subsequently cleaved by the enzyme formate hydrogen lyase (FHL) into $H_2$ and $CO_2$. On the other hand, in the PFOR pathway, followed by obligate anaerobes such as *Clostridium* sp., hydrogen is produced through the re-oxidation of NADH generated during glycolysis. Similar to the PFL pathway, puruvate is converted to acetyl-CoA by the PFOR enzyme with simultaneous oxidation of ferredoxin by [Fe-Fe] hydrogenase (Figure 4.5). Thus, oxidation of ferredoxin leads to hydrogen production in these organisms. The acetyl-CoA can be further converted to acetate or butyrate depending upon the surrounding conditions. If acetate is the sole metabolic product, a maximum of 4 moles of $H_2$ can be produced from 1 mole of glucose, whereas if butyrate is the sole metabolic product, only 2 moles of $H_2$ can be produced from 1 mole of glucose. In the practical scenario, it is observed that most of the organisms produce mixed acids yielding 2–4 moles of $H_2$ per mole of glucose (Trchounian 2015).

## 4.4 BIOCHEMICAL PATHWAYS FOR ALCOHOL-BASED BIOFUELS

Bioalcohols are the major end products of most fermentative organisms such as bacteria and yeasts. These organisms use alternate electron acceptors (instead of oxygen) to obtain energy in anaerobic conditions. The most common reduced alcoholic end products of anaerobic fermentation include ethanol, butanol, and isopropanol (Figure 4.6). In anaerobic fermentation, pyruvate is the major intermediate of the

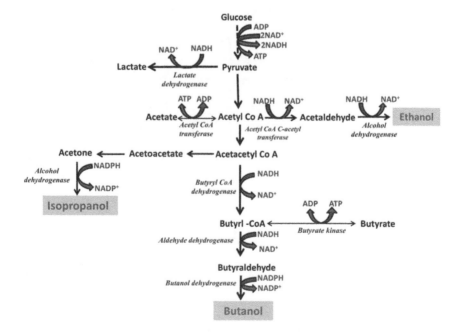

**FIGURE 4.6** Fermentative pathways for bioalcohols production.

glycolysis process, which is further converted to acetyl CoA and $CO_2$ by the action of pyruvate dehydrogenase. The fate of acetyl CoA is dependent upon the type of enzymes available. It can be converted reversibly to acetate by the action of acetyl CoA transferase or can lead to formation of acetaldehyde, which is ultimately converted to ethanol by the alcohol dehydrogenase. Certain species comprised of the acetyl CoA C-acetyl transferase converts acetyl CoA to acetoacetyl CoA, which can be channeled towards isopropanol or butanol production (Figure 4.6). The favorable end product is dependent upon the type of organism and can differ among different strains of the same species. The net energy yield of anaerobic fermentation is only 2 ATP molecules as opposed to aerobic respiration which yields 38 ATP.

## 4.5 BIOCHEMICAL PATHWAYS FOR BIOMETHANE PRODUCTION

Methane is the main component of biogas obtained from the conventional anaerobic digestion process (Ariunbaatar et al. 2014). It involves the metabolic interaction of various groups of microorganisms which are categorized into four different groups—hydrolyzers (which hydrolyze complex organics to simple sugars), acidogens (which convert simple sugars to organic acids), acetogens (which convert organic acids to acetate), and methanogens (which form methane) (Ali Shah et al. 2014). So, the key organisms for methane production are methanogens that can derive energy by converting restricted number of substrates to methane. Based on the substrate used, methanogens can be classified as (1) acetoclastic methanogens, which convert acetate to methane and (2) hydrogenotrophic methanogens, which convert hydrogen and carbon dioxide to methane (Figure 4.7). Besides these common substrates, certain methanogen also can convert methanol and methylamines to methane (Balch et al. 1979).

Most of the methane produced in nature originates from acetate. The pathway for methane generation from acetate includes three major following steps: (1) activation of acetate to acetyl CoA, (2) carbon-carbon and carbon-sulfur bond cleavage of acetyl-CoA, and (3) methyl transfer to methyl-S-CoM and reductive demethylation of methyl-S-CoM to methane (Figure 4.7). In this pathway, acetate is first activated to acetyl-CoA and then followed by the decarbonylation (cleavage of the carbon-carbon and carbon sulfur bonds). This reaction is catalyzed by the CO dehydrogenase (CODH) enzyme complex. CODH oxidizes the carbonyl group to carbon dioxide and reduces a ferredoxin. The methyl group is then transferred to the corrinoid/iron-sulfur component within the complex and finally to coenzyme M. This reaction is catalyzed by more than one methyl transferase. In contrast, hydrogenotrophic methanogens include seven major steps as illustrated in Figure 4.7. In this pathway, mainly two types of [Ni- Fe] hydrogenases are involved: coenzyme $F_{420}$-reducing hydrogenase and coenzyme $F_{420}$-nonreducing hydrogenase. The latter is also known as methyl viologen-reducing hydrogenase. The terminal step for acetoclastic and hydrogenotrophic methanogenesis involves methane formation whereby the methyl group carried by coenzyme M is reduced to methane by the enzyme methyl-coenzyme M reductase (MCR). This enzyme catalyses the reaction of methyl CoM with coenzyme B to produce methane and heterodisulfide (CoM-S-S-CoB) (Karthikeyan et al. 2017).

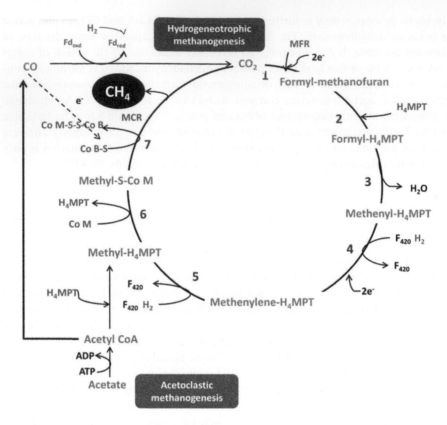

**FIGURE 4.7** Hydrogenotrophic and acetoclastic methanogenesis (MFR: methanofuran; H$_4$MPT: tetrahydromethanopterin; CoM: Coenzyme M; MCR: methyl CoM reductase; CoB: Coenzyme B; CoM-S-S-CoB: heterodisulfide).

## 4.6  SUMMARY AND CONCLUSION

Most of the fermentative organisms such as bacteria and yeasts have natural tendencies to produce bioalcohols and biogas as major end products under suitable environmental conditions. These end products can be targeted as potential biofuels that can replace the conventional petroleum-based fossil fuels. Similarly, certain photosynthetic microorganisms can store lipids or carbohydrates, which can be either extracted or used as a feedstock for biofuel generation. The knowledge of the metabolic pathways is essential to enhance the yield of the desired product and to improve the process efficiency. To improve lipid productivities, it is essential to enhance the TAGs by providing suitable environmental stress conditions. Similarly, to improve hydrogen production, the activity of hydrogenases or nitrogenases needs to be improved. It is often desirable to target specific end products from a series of intermediates of a metabolic pathway. Often, the productivities are lowered due to the channeling of reductants towards unwanted compounds. For example, if butanol is the major end product, the reducing equivalents used up for ethanol production are undesirable and hence this pathway can be blocked to channelize all the reducing equivalents towards butanol formation. Such strategies often are employed in various bioprocesses to enhance the overall yield of the product of interest. Thus, understanding the biochemical make up of an organism is essential to estimate its potential for biofuel production. This chapter thus provides the fundamental basis for the subsequent chapters that deal with the production strategies and the genetic manipulations for improving biofuel production on a large scale.

## REFERENCES

Abu-elreesh G, Abd-el-haleem D (2014) Promising oleaginous filamentous fungi as biodiesel feed stocks: Screening and identification. *Eur J Exp Biol* 4:576–582.

Ali Shah F, Mahmood Q, Maroof Shah M, Pervez A, Ahmad Asad S (2014) Microbial ecology of anaerobic digesters: The key players of anaerobiosis. *Sci World J* 2014:183752.

Alvarez HM, Steinbüchel A (2003) Triacylglycerols in prokaryotic microorganisms. *Appl Microbiol Biotechnol* 60:367–376.

Ariunbaatar J, Panico A, Esposito G, Pirozzi F, Lens PNL (2014) Pretreatment methods to enhance anaerobic digestion of organic solid waste. *Appl Energy* 123:143–156.

Athenaki M, Gardeli C, Diamantopoulou P, Tchakouteu SS, Sarris D, Philippoussis A, Papanikolaou S (2018) Lipids from yeasts and fungi: Physiology, production and analytical considerations. *J Appl Microbiol* 124:336–367.

Balch WE, Fox GE, Magrum LJ, Woese CR, Wolfe RS (1979) Methanogens: Reevaluation of a unique biological group. *Microbiol Rev* 43:260–296.

Cagliari A, Margis R, Dos Santos Maraschin F, Turchetto-Zolet AC, Loss G, Margis-Pinheiro M (2011) Biosynthesis of triacylglycerols (TAGs) in plants and algae. *Int J Plant Biol* 2:40–52.

Das D, Veziroglu T (2008) Advances in biological hydrogen production processes. *Int J Hydrogen Energy* 33:6046–6057.

Dasgupta CN, Jose Gilbert J, Lindblad P, Heidorn T, Borgvang SA, Skjanes K, Das D (2010) Recent trends on the development of photobiological processes and photobioreactors for the improvement of hydrogen production. *Int J Hydrogen Energy* 35:10218–10238.

Demirbas A (2008) Biofuels sources, biofuel policy, biofuel economy and global biofuel projections. *Energy Convers Manag* 49:2106–2116.

Elsharnouby O, Hafez H, Nakhla G, El Naggar MH (2013) A critical literature review on biohydrogen production by pure cultures. *Int J Hydrogen Energy* 38:4945–4966.

Faried M, Samer M, Abdelsalam E, Yousef RS, Attia YA, Ali AS (2017) Biodiesel production from microalgae: Processes, technologies and recent advancements. *Renew Sustain Energy Rev* 79:893–913.

Harai É, Kapás Á, Lányi S, Ábrahám B, Nagy I, Muntean O (2010) Biohydrogen production by photofermentation of lactic acid using thiocapsa roseopersicina. *UPB Sci Bull*, Ser B 72.

Karthikeyan R, Cheng KY, Selvam A, Bose A, Wong JWC (2017) Bioelectrohydrogenesis and inhibition of methanogenic activity in microbial electrolysis cells—A review. *Biotechnol Adv* 35:758–771.

Manish S, Banerjee R (2008) Comparison of biohydrogen production processes. *Int J Hydrogen Energy* 33:279–286.

Meherkotay S, Das D (2008) Biohydrogen as a renewable energy resource—Prospects and potentials. *Int J Hydrogen Energy* 33:258–263.

Müller V (2001) Bacterial fermentation. In: *Encyclopedia of Life Sciences*, Nature Publishing Group.

Nigam PS, Singh A (2011) Production of liquid biofuels from renewable resources. *Prog Energy Combust Sci* 37:52–68.

Patel A, Arora N, Sartaj K, Pruthi V, Pruthi PA (2016) Sustainable biodiesel production from oleaginous yeasts utilizing hydrolysates of various non-edible lignocellulosic biomasses. *Renew Sustain Energy Rev* 62:836–855.

Singh H, Das D (2018) *Biofuels from Microalgae: Biohydrogen.* Springer, Cham, Switzerland, pp. 201–228.

Sorger D, Daum G (2003) Triacylglycerol biosynthesis in yeast. *Appl Microbiol Biotechnol* 61:289–299.

Trchounian A (2015) Critical reviews in biotechnology mechanisms for hydrogen production by different bacteria during mixed-acid and photo-fermentation and perspectives of hydrogen production biotechnology. *Crit Rev Biotechnol* 35:1549–7801.

Yu J, Takahashi P (2007) Biophotolysis-based hydrogen production by Cyanobacteria and Green Microalgae. *Commun Curr Res Educ Top Trends Appl Microbiol* 1:79–89.

# 5 Molecular Biological Approaches for the Improvement of Biofuels Production

## 5.1 INTRODUCTION

Research on biofuel production from renewable biomass has accelerated in recent years to address the rising concerns over the energy security, fossil fuel depletion, and climate change. Biofuels are mainly hydrocarbon fuels that are produced from organic matters. These living materials are generated from a living source. At present, most of the feedstock utilized for biofuel production are grown on arable lands and compete with food crops (Demirbas 2008). However, the yields of biofuels produced from agricultural or industrial wastes are too low to compete with the existing fossil fuel-based economy (Peralta-Yahya and Keasling 2010; Buijs et al. 2013). Molecular biology approaches can aid in improving biofuel production by modification of existing strains or by designing *de novo* microbes with higher titer, yield, and productivities (Torto-Alalibo et al. 2014). Moreover, new bioenergy crops with enhanced desired properties can be developed that can be readily used to extract fuels such as biodiesel (Shih et al. 2016). The use of synthetic biology has paved the way for fabricating or synthesizing new pathways and genetic constructs that would help in reducing the costs of biofuel production processes. These approaches can be used to optimize both microbial hosts and the feedstock for enhancement of biofuel productivities (Peralta-Yahya et al. 2012). Despite some success, several challenges still impede the commercialization of biofuel technologies. In this chapter, various genetic and metabolic engineering tools are described that have been employed so far for the advancement of biofuel production. In addition, the existing and future challenges imposed by the use of such complex biological systems are examined.

## 5.2 METHODS AND TOOLS FOR GENETIC MANIPULATION

The advances in molecular biology techniques have provided several tools for the genetic manipulation of microorganisms. These tools aid in the development of industrial strains which can target specific products of interest with higher yields, titers, and productivities than the unmodified (wild-type) strain of the same species. In this section, various tools available for genetic manipulation are discussed focusing on their application for enhanced biofuel production.

### 5.2.1 Induced Mutation

In induced mutation, mutagenic agents or (mutagens) like X-rays, UV, and other chemicals are used to obtain a mutant with desired traits (Figure 5.1a). These mutagens lead to alteration of gene function either by nitrogenous base replacement, mispairing, or damage. Certain chemicals have similar properties as nitrogenous bases known as base analogs that can be incorporated into the native DNA to produce mutations. For example, 5-bromouracil (5-BU) is an analog of thymine that can pair with both adenine and guanine as opposed to thymine which can only pair with adenine (Holroyd and Van Mourik 2014). This alternate pairing property of 5-BU causes transition in the DNA replication and thus in turn modifies gene function. Alternatively, certain mutagens such as N-methyl-N′-nitro-N-nitrosoguanidine (NTG) does not incorporate into the DNA but causes specific mispairing by alteration of nitrogenous bases (Delić et al. 1970). Other types of mutagens include intercalating agents such acridine orange and ionizing radiation such as UV light, which damage the whole base or the specific base pairing sites. Different mutagenic agents have been tested for improving biofuel production. For example, the UV radiated and ETBR (ethidium bromide) mutant yeast strains, *Pichia stipitis* NCIM 3498 and *Candida shehatae* NCIM 3501 showed higher ethanol production as compared to the wild strains using hemicellulosic hydrolysates of wheat straw as substrates (Koti et al. 2016). In another study, the parent strain *Clostridium acetobutylicum* PTCC-23 was mutagenized for the improved production of butanol by UV exposure, N-methyl-N-nitro-N-nitrosoguanidine, and ethyl methane sulphonate treatments (Syed et al. 2008). These studies suggest positive influence of induced mutations for biofuel production. However, the screening process is tedious and the longevity of the mutant strains is questionable.

### 5.2.2 Gene Deletion or Gene Knockout

Genes are the fundamental codes in DNA by virtue of which specific functions are carried out by the organisms. The elucidation of DNA sequences using recombinant DNA technologies has enabled the identification of the gene patterns of an organism having similarity with the proteins of known functions. Gene deletion refers to the permanent deletion of the gene function from an organism. In this approach, part of DNA or chromosome is deleted using site specific nucleases such as Zinc fingers, TALENS (transcription activator-like effector nucleases), and CRISPR (clustered regularly interspaced short palindromic repeats) (Figure 5.1). On the other hand, gene knockout refers to the suppression or deactivation of a gene function through genetic engineering. Gene knockout is necessary when the deletion of a particular gene is lethal to the organism. Thus, unlike gene deletion where the gene is permanently removed, in gene knockout the genes remain present in an inactive state. There are three main approaches for gene deletion or gene knockout: (1) replacing the gene with other non-functional sequences, (2) introducing an allele whose encoded protein inhibits functioning of the expressed normal protein, and (3) promoting destruction of mRNA expressed from a gene. The first approach leads to the modification of endogenous gene while in the second and third approach the endogenous

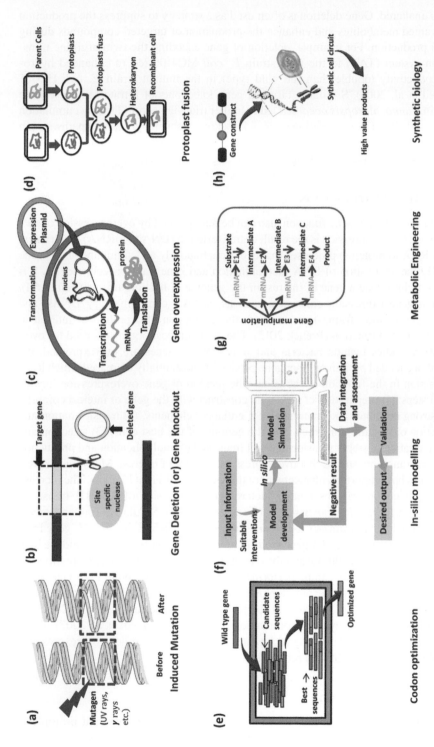

**FIGURE 5.1** Methods of gene manipulation: (a) induced mutation, (b) gene deletion or gene knockout, (c) gene overexpression, (d) protoplast fusion, (e) codon optimization, (f) In silico modeling, (g) metabolic engineering, and (h) synthetic biology.

gene is unaltered. Gene deletion is often used as a strategy to suppress the production of unwanted metabolites and enhance the production of targeted compounds during biofuel production. For example, deletion of gene encoding the twin-arginine translocation system (TAT) in the wild strain *E. coli* MC4100 led to enhanced hydrogen productivity (double than the wild type) in the mutant strain *E. coli* HD701 (Penfold et al. 2003). Similarly, targeted gene knockouts of nitrate reductase gene HygR in *Nannochloropsis oceanica* led to higher triacylglycerol (TAG) accumulation (Wang et al. 2016). Gene deletion or gene knockout of several genes in a pathway can be used to block the entire pathway. The manipulation of entire metabolic pathway will be discussed in the following metabolic engineering section.

### 5.2.3 GENE OVEREXPRESSION

Like in gene deletion, gene function also can be manipulated by overexpression of specific genes. Gene expression is the result of transcription (DNA to mRNA) and translation (mRNA to proteins) processes that occur continuously and simultaneously inside cells. By the application of DNA recombination and gene transfer technologies, it is possible to introduce a gene of interest in the genome of host organisms (Figure 5.1). Different gene expression constructs such as cDNAs (complimentary DNAs) and ORFs (open reading frames) are used for the expression of a protein of interest to enable a targeted function (Prelich 2012). The common host organisms used for overexpression studies include bacteria and yeasts. When genes are overexpressed, the amount of encoded protein or other gene product is substantially increased which leads to alteration in the function of the gene. The method of gene overexpression include several steps: (1) preparation of a plasmid construct with the gene of interest comprising a strong promoter and transcriptional enhancer elements, (2) transformation and integration of plasmid constructs into the genome of the host cells, (3) selection and characterization of stably transformed cell lines, and (4) multiplication and production of mutant transgenic strains that overexpress the protein of interest. Many overexpression studies have reported enhancement of the yields of different biofuels. The increase in lipid production was observed by heterologous expression of AtWRI1 transcription factor in *Nannochloropsis salina* (Kang et al. 2017). Similarly, Tian and others (2017) showed that a strain of *Clostridium thermocellum* expressing the *pdc* gene from *A. pasteurianus* and the *adhA* gene from *T. saccharolyticum* was able to produce high titers of ethanol using cellulose as substrate. It is observed that gene overexpression and gene knockout strategies can be performed simultaneously to obtain high yields of a product. For example, over-expression of transcriptional activator Fhl a, elimination of uptake hydrogenase, and knockout of the formate transporter in *E. coli* BW25113 led to a 4.6-fold increase in hydrogen production from glucose (Maeda et al. 2007). These studies indicate that gene editing techniques (gene knockout or gene overexpression) can coherently improve biofuel production.

### 5.2.4 PROTOPLAST FUSION

Protoplast fusion involves fusion of genetic material of two different protoplasts (cells without cell wall) (Verma and Kumar 2000). It allows transfer of useful genes from one species to another. The mechanism of protoplast fusion is shown in Figure 5.1b. The first step is to remove the cell wall (if present) by the action of specific lytic enzymes. This removal is followed by the fusion of the proto-plasts together either spontaneously by adhering to each other or by the addition of inducing agents such as polyethylene glycol (PEG). This fusion ultimately leads to development of a heterokaryon (hybrid), which comprises of combined genetic material of both the cells and ultimately forms a recombinant cell. The other synkaryons (non-recombinant cells) are discarded. Protoplast fusion has been implemented successfully to improve the characteristics of microbial strains for high-yield biofuel production. For example, rapid starch fermenting strains were developed by fusing haploid strains of *Saccharomyces diastaticus* for enhanced ethanol production (Sakai et al. 1986). In another study by Ge et al., protoplasts of *Saccharomyces cerevisiae W5* and *Candida shehatae 20335* were used to produce fusants that could use xylose and glucose as substrates for ethanol production (Ge et al. 2017). The advantage of protoplast fusion is the ability to mix two genomes, which allows incorporation of positive features of parental cells. However, the disadvantage of this procedure is that only a small number of recom-binant protoplasts are produced, which limits its efficiency.

## 5.2.5 CODON OPTIMIZATION

Although it is expected that protein expression would be enhanced with respect to mRNA transcripts, it may not always be the case. Usually, a group of three bases, known as codons, are required to aid the mRNA-guided translation. These codons are degenerative in nature (i.e., different codons can specify for same amino acids). This degenerative property leads to codon bias (i.e. certain codons are translated more efficiently than others). When expressing proteins across species, codon degen-eracy plays a crucial role as the frequency of codon generation in the organism of origin and in the host organism can differ significantly. Moreover, mutations caused in the codons can alter the post-translational modifications, conformation, stability, and function of the expressed proteins. Therefore, to address these issues, gene or codon optimization is performed to select the ideal sequence for maximum pro-tein expression in any host (Figure 5.1). Different code optimization tools such as CodonW, GeneOptimizer™, and OptimumGene™ have been developed that can be used to obtain the desired sequence of interest that will subsequently favor the functional protein expression. These tools allow alteration of naturally occurring gene sequences and the recombinant gene sequences to attain the highest possible productivity in the chosen expression system. The critical factors involved in codon optimization include codon adaptability, mRNA structure, transcription and transla-tion efficiency, and protein refolding.

## 5.2.6 IN SILICO MODELING

*In silico* modeling refers the establishment of computer-generated models that simu-late the cellular behavior. The primary motive of the modeling is to generate faster simulated growth rates that allow the understanding of particular gene function in a short period of time. In bioprocess development, it is usually used to estimate and optimize the product yields. In addition, these computational models help to elucidate the effects of gene knockouts or gene overexpression of native or engineered genes in the host organisms. Optgene, OptKnock, OptReg, and EMILiO are among the few examples of computational tools used to search the effect of targeted genes in the deleted or overexpressed strains (Khalidi et al. 2016). Despite the successful implementation, it is observed that *in silico* modeling fails to predict the exact model for overall cell behavior due to the limited understanding of the complex mechanisms involved.

### 5.2.7 METABOLIC ENGINEERING

Metabolic engineering allows manipulation and optimization of genetic and regulatory processes inside the cells to increase the yields of desired products. Several subsets of metabolic engineering have emerged over the years with the progress in functional genomics and proteomics. Classical metabolic engineering used the rDNA (recombinant DNA) technology to manipulate the pathway enzymes and alter the cell metabolism in a manner to improve the productivity of the product of interest (Figure 5.1g). However, modern approaches are more inclined towards pathway engineering and targeted metabolic and adaptive cellular processes (Yang et al. 2007).

### 5.2.8 SYNTHETIC BIOLOGY

Synthetic biology involves the design and construction of biological entities such as enzymes, nucleic acids, genetic circuits, or whole biological systems that can be modeled, understood, and tuned to obtain the desired function (MacDonald and Deans 2016). It is based on the principles of synthetic chemistry and integrated circuits in the electronics. The intent of synthetic biology is to develop, design, and build engineered biological systems that can mimic the natural biological processes in a predictable manner. Moreover, it attempts to manipulate the living systems at the molecular level which helps in better understanding the new biological components and systems. Although the application of synthetic biology tools appears to be promising, the scientific and engineering challenges require further exploration.

## 5.3   METABOLIC ENGINEERING FOR THE PRODUCTION OF POTENT BIOENERGY CROPS

### 5.3.1 ENHANCING BIOMASS YIELD

The poor yield of bioenergy crops is one of the major challenges that limit the biofuel production on a commercial scale. Therefore, most of the studies have focused on improving the overall biomass productivity by using simple mutagenic approaches (e.g., random and induced) (Lee et al. 2008). Moreover, it has been observed that improving the photosynthetic efficiency of bioenergy crops can help improve growth and yield. By genetically modifying, the crops by inducing stress resistance against inhibitors (Section 5.3.3), the growth of the bioenergy crop can be prolonged. Some of the successful applications of genetically modified bioenergy crops include sugarcane, corn, and soybean for enhanced ethanol production (Sticklen 2008). Research in this area is fast progressing to obtain high-yielding varieties that are economically more efficient compared to traditional crops.

## 5.3.2 Altering Biomass Composition

The first step for efficient conversion of all biochemicals in feedstock is to identify the major constituents of the biomass. These constituents tend to vary depending upon the growth condition. For an efficient bioenergy production through a fermentation route, it is essential to convert the complex organic components into a simple usable form. This conversion generally termed as pre-treatment of biomass is usually carried out using different physicochemical approaches that are tedious and increase the cost of production of biomass (Baghchehsaraee et al. 2008; Leaño et al. 2012; Ariunbaatar et al. 2014). Alternatively, novel metabolic engineering approaches have enabled the modification or introduction of novel pathways in host organisms that helps higher biofuel production capacities using minimal feedstock pre-treatment.

### 5.3.2.1 Enhancement of Soluble Sugar

Most of the current fermentative technologies are dependent upon the soluble sugar content of biomass (such as sugarcane) that is readily accessible to the microorganisms. By applying suitable metabolic engineering approaches, the amount of soluble sugars of traditional feedstocks can be enhanced. By redirecting the carbon fixation pathway (Calvin cycle) in autotrophic organisms (such as green plants, algae, and cyanobacteria), the photosynthetic capacity can be improved, which in turn increases the sugars productivity (Shih et al. 2016). The key enzyme that plays a vital role in carbon fixation is RuBisCO. The heterologous expression of this enzyme from cyanobacteria to higher crop plants has shown improvement in carbon fixation capabilities in the host organisms (Shih et al. 2016). Alternatively, the metabolic pathway can be channeled towards production of sucrose isomers or other fermentation precursors that do not inhibit the photosynthesis process. It was shown that by altering the cell wall composition with an increase in C6 sugars rather than C5 sugars the ethanol production could be enhanced from maize as feedstock (Torney et al. 2007). This enhancement is because the C6 sugars are more easily fermentable compared to C5 sugars. Some of the strategies used for improving sugar yields are listed in Table 5.1.

### 5.3.2.2 Modification of Lignocellulosic Biomass

Second-generation biofuels are considered superior over first-generation biofuels because they are less expensive and can ameliorate concerns over the food vs. fuel debate. However, their complex composition and high water content lead to lower conversion efficiencies during microbial fermentation and thermochemical processes, respectively. To improve the biofuel production yields and the economic viability of the process, complete use of feedstock (by conversion into fuel and other value-added products) is essential. Due to the abundance of lignocellulosic biomass

**TABLE 5.1**

**Metabolic Engineering Approaches for Producing Bioenergy Crops**

| Biomass | Type of Modification | Metabolic Approach | Targeted Biofuel | References |
|---|---|---|---|---|
| Maize | Altering cell wall composition | Alteration of lignin biosynthetic pathway to naturally reduce lignin content | Ethanol | Torney et al. (2007) |
| Switch grass | Altering cell wall composition | Alteration of lignin biosynthetic pathway to naturally reduce lignin content | NA | Madadi et al. (2017) |
| Poplar | Altering cell wall composition | Downregulation of pectate lyase genes (PtxtPL1-27) for increased pectin and xylan solubility | NA | |
| Alfalfa | Altering cell wall composition | Downregulation of lignin biosynthesis genes | Ethanol | Hisano et al. (2011) |
| *Synechococcus elongatus* PCC 7942 | Increase in soluble sugar (glucose and fructose) | Induced metabolite production via synthetic biology. Expressing "invA" and "glf" genes from *Zymomonas mobilis* | NA | Niederholtmeyer et al. (2010) |
| Sugarcane | Increase in sucrose accumulation | Introduction/ overexpression of sucrose synthase gene | NA | Patrick et al. (2013) |
| Sugarcane | Increase in sucrose accumulation | Sucrose isomerase (SI) expression | Ethanol | Wu and Birch (2007) |
| *Arabidopsis thaliana* | Increasing TAG accumulation | Glyceraldehyde 3-phosphate acyltransferase (GPAT) expression | Biodiesel | Hegde et al. (2015) |
| *Brassica napus* | Increasing fatty acid accumulation | Expression endogenous acetyl-CoA carboxylase (ACC) | | |
| *Nicotiana tabacum* | Increasing fatty acid accumulation | Expression acyl carrier protein (ACP) desaturase | | |
| *Thalassiosira pseudonana* | Increasing lipid accumulation | Knockdown of a multifunctional lipase/ phospholipase/ acyltransferase | | |

*(Continued)*

**TABLE 5.1** (*Continued*)
**Metabolic Engineering Approaches for Producing Bioenergy Crops**

| Biomass | Type of Modification | Metabolic Approach | Targeted Biofuel | References |
|---|---|---|---|---|
| Rice | Inducing heat tolerance | Overexpression of Heat shock protein 101 from *Arabidopsis* | NA | Wu and Birch (2007) |
| Rice | Inducing dehydration tolerance | Overexpression of HVA1 protein a form of Late Embryogenesis Abundant (LEA) proteins from barley | | |

throughout the world, much of the research interest has been focused on its degradation for biofuel production (Somerville et al. 2010). The major constituents of lignocellulosic biomass are cellulose, hemicellulose, and lignin (Nigam and Singh 2011). Other constituents include proteins, fatty acids, sterols, and triglycerides. Lignins are the crucial component of the cell wall of the plants, which pose difficulty for biological degradation. Thus, major molecular approaches are directed towards the modification of lignin content and structure in biomass feedstock to improve its bioconversion (Table 5.1). These approaches mainly focus towards downregulation of genes in lignin biosynthesis pathway and the addition of novel monolignols to remodel lignin structure (Joyce and Stewart 2012).

### 5.3.2.3 Modification of Oil Crops

Oil crops like palm, soy, and jatropha are generally used for the production of biodiesel. As discussed in Chapter 4, the main components for biodiesel production are lipids in the form of triacylglycerols (TAG). So, to enhance biodiesel production from these crops, major metabolic engineering efforts are focused towards enhancement of the lipid accumulation and manipulation of its composition (Shih et al. 2016). Lipid accumulation is brought about by the manipulation of key TAG enzymes (such as diacylglycerol acyltransferases) or by deactivating the competing pathways (such as carbon accumulation pathways) (Table 5.1). Although most of the TAGs are accumulated in the seeds of these crops, efforts are being made to divert the TAG biosynthesis in other major portions of the plants such as vegetative tissues. Similarly, by genetically altering the fatty acid composition, biodiesel characteristics can be significantly improved (Cagliari et al. 2011). Besides oil crops, more recently, advancements in biodiesel production are confined to oleaginous microorganisms such as microalgae and yeasts (Shi et al. 2011). Moreover, different strategies are implemented to express the TAG-producing genes in cyanobacteria, which can be easily manipulated due to its simple cellular structure (Woo and Lee 2017). Such a process, if successfully implemented could provide sustainable means of solar biodiesel production.

### 5.3.3 Inducing Stress Tolerance

Besides enhancing yield and modifying the composition of biomass, producing stress tolerant bioenergy crops could help in biomass growth and development during unfavorable conditions. Different abiotic stresses such as high temperature, salinity, or drought stress lead to denaturation of vital proteins that affect the plant growth. By introducing the necessary genes from the extremophiles to the host of interest, high-yield biomass can be obtained even in extreme environmental conditions. For example, if cold tolerant traits are engineered in the sugarcane biomass, it can be easily grown in cold environmental regions, thereby boosting its availability in such regions (Shih et al. 2016). Similarly, disease resistant strains can help in prolonging the growth of the plant to yield higher biomass. Some examples of metabolic engineering approaches for inducing stress tolerance are summarized in Table 5.1.

## 5.4 MOLECULAR APPROACHES OF STRAIN IMPROVEMENT TO ENHANCE BIOFUEL PRODUCTION

### 5.4.1 Improving Biofuel Yield Using Gene Editing Approaches

Different microorganisms have the ability to generate similar end products using alternative metabolic pathways. Thus, molecular approaches for the enhancement of the product of interest are dependent upon the type of strain and its metabolic machinery. The most common method for strain improvement involves the use of induced mutations or gene editing techniques such as gene knockdown and gene overexpression as described in Section 5.2. For the successful application of gene manipulating techniques, knowledge of the host's genomic structure is of utmost importance. Even for synthetic biology approaches, the role of targeted genes and its impact on the host organism dictates the construction of an optimized biofuel strains (Chubukov et al. 2016). Some of the strategies involved in improving biofuel yields using gene editing techniques are provided in Table 5.2.

---

**TABLE 5.2**

**Metabolic Engineering Approaches for Producing Potent Biofuel Strains**

| Microbe | Strategy | Metabolic Approach | Targeted Biofuel | References |
|---------|----------|--------------------|------------------|-----------|
| *E. coli* | Introducing ethanol producing genes | Inserting genes coding for *alcohol dehydrogenase II* and *pyruvate decarboxylase* from *Z. mobilis* | Ethanol | Ingram et al. (1987) |
| *S. cerevisiae* | Increasing sucrose conversion | Engineering the promoter and 5′ coding sequences of sucrose invertase gene (SUC2) | Ethanol | Basso et al. (2011) |

*(Continued)*

**TABLE 5.2** (*Continued*)
**Metabolic Engineering Approaches for Producing Potent Biofuel Strains**

| Microbe | Strategy | Metabolic Approach | Targeted Biofuel | References |
|---|---|---|---|---|
| *S. cerevisiae* | Improving xylose metabolism | Directed evolution of heterologous transporters GXS1 (from *Candida intermedia*) and XUT3 (from *Scheffersomyces stipitis*) | Ethanol | Young et al. (2012) |
| *E. coli* | Butanol pathway engineering | Construction of modified *Clostridial* 1-butanol pathway and creation of NADH and acetyl-CoA driving forces to direct the flux | Butanol | Shen et al. (2011) |
| *C. acetobutylicum* ATCC 824 | Operon construction | Construction of synthetic isopropanol operon for early isopropanol production and minimize acetate accumulation | Isopropanol, butanol, and ethanol mixture | Dusséaux et al. (2013) |
| *C. acetobutylicum* M5 | Introducing ethanol producing genes | Transformation with plasmid pCAAD carrying the aldehyde alcohol dehydrogenase gene (aad) | Butanol | Nair and Papoutsakis (1994) |
| *E. coli* | Reducing uptake hydrogen activity | Heterologous expression of bidirectional hydrogenase (hoxEFUYH) of *Synechocystis* sp. PCC 6803 | Hydrogen | Maeda et al. (2012) |
| *E. coli* | Introducing hydrogen producing genes | Transformation of [Fe]-hydrogenase gene (hydA) from *Clostridium paraputrificum* M-21 | Hydrogen | Morimoto et al. (2005) |
| *M. thermoautotrophicus* | Introducing methane producing genes | Overexpression of *mcr* and *mrt* genes which encodes methyl coenzyme M reductase II (MRII), the isofunctional enzyme of MRI | Methane | Luo et al. (2002) |
| *C. beijerinckii* DSM 6423 | Improving solvent tolerance | Chemical mutagenesis with N-methyl-N-nitro-N-nitrosoguanidine (NTG) | Isopropanol | Gérando et al. (2016) |

## 5.4.2 Inducing Tolerance Against Toxic Inhibitors and High Biofuel Concentrations

A major limitation for obtaining high yields of biofuels is the inability of fermentative microorganisms to tolerate high titers of the product of interest. The toxicity of most of the biofuel products is due to their hydrophobic nature, which causes disruption in membrane integrity and demolishes the conformation of functional proteins present in the membrane (Tsai et al. 2015). Numerous genetic perturbations have been elicited for improving the tolerance of microorganisms against high concentration of biofuels. However, only marginal increase in the yields has been obtained. Moreover, it is observed that during the hydrolysis of certain biomass, many fermentative inhibitors such as sulfur compounds, furfurals, and organic acids are generated that are detrimental to the microbial growth. Thus, it remains a critical need to improve the microbial tolerance against these hydrolysates. To induce tolerance against several compounds, complex interactions among the different gene targets and optimization of their expression is crucial. In addition, another major factor that inhibits the microbial growth and biofuel production is the physicochemical environment prevailing during the fermentative process. For example, most of the hydrogen-producing organisms are inhibited at acidic pH (below 4). Similarly, certain organisms require mesophilic temperature for their growth and cannot be grown in high temperature regions. By introducing genes specific for tolerance against such physicochemical parameters, the adaptability of microorganisms to the prevailing environmental conditions can be enhanced.

## 5.4.3 Widening the Substrate Spectrum

Another major challenge limiting the biofuel productivity is the poor substrate degradation efficiency. Most of the microorganisms are able to ferment only C6 sugars and not C5 sugars because they lack the necessary enzymes needed to degrade them. Thus, extensive efforts are being made to widen the substrate use capacities of biofuel producing microorganisms (Table 5.2). These approaches mainly focus upon constructing and optimizing gene constructs and promoters for the expression of the necessary enzymes in the host organisms. Moreover, certain protein engineering approaches are used to improve the catalytic efficiency of the rate limiting enzymes (Tsai et al. 2015). By applying such techniques, complete and rapid consumption of complex organic substrates can be feasible.

## 5.4.4 Pathway Engineering

With the success in synthetic biology applications, modern approaches concentrate upon obtaining complete metabolic control by synthesis of novel pathways in the host organisms for improving biofuel production. For example, for obtaining desired concentration of a targeted compound, metabolite sensors are developed with desired input-output relationships (Sekhon and Rahman 2013). A synthetic metabolic pathway is designed by using the "push and pull" principle proposed by

G. Stephanopoulos (Tsai et al. 2015). It involves upstreaming and downstreaming of necessary fluxes in the main metabolic pathway (i.e., overexpression of genes that increase the flux and knock down of genes that divert the flux). Initially, the targeted genes are modeled using computational techniques *in silico* to develop a genetic algorithm through which the desired gene function and expression can be simulated. Once a successful algorithm is generated, attempts are made to engineer the desired pathways *in vivo*.

Several studies have used genome-scale metabolic network modeling to choose strategies to modify yeast strains. Several synthetic biology tools have been developed already that mainly target yeast and cyanobacteria as potential hosts (Sekhon and Rahman 2013; Tsai et al. 2015; Chubukov et al. 2016).

## 5.5 CONCLUSIONS AND PERSPECTIVES

Research over past few years has led to significant improvements in our ability to manipulate gene expression for enhanced biofuel production. To develop economical and sustainable processes for biofuel production, the metabolic pathways of biofuel producers need to be optimally redesigned to achieve high performance. Moreover, new and improved varieties of bioenergy crops must be developed to ensure enough necessary components that can be easily used by fermentative organisms. The major goals of metabolic pathway redesigning for biofuel producer include improved product yield, higher product concentration and productivity, and product tolerance. The production strategy must be designed in a way that the whole process becomes operationally inexpensive by system-wide optimization of midstream and downstream processes.

## REFERENCES

Ariunbaatar J, Panico A, Esposito G, Pirozzi F, Lens PNL (2014) Pretreatment methods to enhance anaerobic digestion of organic solid waste. *Appl Energy* 123:143–156.

Baghchehsaraee B, Nakhla G, Karamanev D, Margaritis A, Reid G (2008) The effect of heat pretreatment temperature on fermentative hydrogen production using mixed cultures. *Int J Hydrogen Energy* 33:4064–4073.

Basso TO, de Kok S, Dario M, do Espirito-Santo JCA, Müller G, Schlölg PS, Silva CP et al. (2011) Engineering topology and kinetics of sucrose metabolism in Saccharomyces cerevisiae for improved ethanol yield. *Metab Eng* 13:694–703.

Buijs NA, Siewers V, Nielsen J (2013) Advanced biofuel production by the yeast saccharomyces cerevisiae. *Curr Opin Chem Biol* 17:480–488.

Cagliari A, Margis R, Dos Santos Maraschin F, Turchetto-Zolet AC, Loss G, Margis-Pinheiro M (2011) Biosynthesis of triacylglycerols (TAGs) in plants and algae. *Int J Plant Biol* 2:40–52.

Chubukov V, Mukhopadhyay A, Petzold CJ, Keasling JD, Martín HG (2016) Synthetic and systems biology for microbial production of commodity chemicals. *NPJ Syst Biol Appl* 2:16009.

Delić V, Hopwood DA, Friend EJ (1970) Mutagenesis by N-methyl-N'-nitro-N-nitrosoguanidine (NTG) in Streptomyces coelicolor. *Mutat Res—Fundam Mol Mech Mutagen* 9:167–182.

Demirbas A (2008) Biofuels sources, biofuel policy, biofuel economy and global biofuel projections. *Energy Convers Manag* 49:2106–2116.

Dusséaux S, Croux C, Soucaille P, Meynial-Salles I (2013) Metabolic engineering of Clostridium acetobutylicum ATCC 824 for the high-yield production of a biofuel composed of an isopropanol/butanol/ethanol mixture. *Metab Eng* 18:1–8.

Ge J, Du R, Song G, Zhang Y, Ping W (2017) Metabolic pathway analysis of the xylose-metabolizing yeast protoplast fusant ZLYRHZ7. *J Biosci Bioeng* 124:386–391.

Gérando HM de, Fayolle-Guichard F, Rudant L, Millah SK, Monot F, Ferreira NL, López-Contreras AM (2016) Improving isopropanol tolerance and production of Clostridium beijerinckii DSM 6423 by random mutagenesis and genome shuffling. *Appl Microbiol Biotechnol* 100:5427–5436.

Hegde K, Chandra N, Sarma SJ, Brar SK, Veeranki VD (2015) Genetic engineering strategies for enhanced biodiesel production. *Mol Biotechnol* 57:606–624.

Hisano H, Nandakumar R, Wang Z-Y (2011) Genetic modification of lignin biosythesis for improved biofuel production. *Biofuels, Glob Impact Renew Energy, Prodcution Agric Tech Adv* 45:223–236.

Holroyd LF, Van Mourik T (2014) Stacking of the mutagenic DNA base analog 5-bromouracil. *Theor Chem Acc* 133:1–13.

Ingram LO, Conway T, Clark DP, Sewell GW, Preston JF (1987) Genetic engineering of ethanol production in *Escherichia coli*. *Appl Environ Microbiol* 53:2420–2425.

Joyce BL, Stewart CN (2012) Designing the perfect plant feedstock for biofuel production: Using the whole buffalo to diversify fuels and products. *Biotechnol Adv* 30:1011–1022.

Kang NK, Kim EK, Kim YU, Lee B, Jeong WJ, Jeong BR, Chang YK (2017) Increased lipid production by heterologous expression of AtWRI1 transcription factor in *Nannochloropsis salina*. *Biotechnol Biofuels* 10:231.

Khalidi O, Guleria S, Koffas MAG (2016) Pathway and strain design for biofuels production. In: *Biotechnology for Biofuel Production and Optimization*. Elsevier, Amsterdam, the Netherlands, pp. 97–116.

Koti S, Govumoni SP, Gentela J, Venkateswar Rao L (2016) Enhanced bioethanol production from wheat straw hemicellulose by mutant strains of pentose fermenting organisms Pichia stipitis and Candida shehatae. *Springerplus* 5:1545.

Leaño EP, Anceno AJ, Babel S (2012) Ultrasonic pretreatment of palm oil mill effluent: Impact on biohydrogen production, bioelectricity generation, and underlying microbial communities. *Int J Hydrogen Energy* 37:12241–12249.

Lee D, Chen A, Nair R (2008) Genetically engineered crops for biofuel production: Regulatory perspectives. *Biotechnol Genet Eng Rev* 25:331–362.

Luo HW, Zhang H, Suzuki T, Hattori S, Kamagata Y (2002) Differential expression of methanogenesis genes of *Methanothermobacter thermoautotrophicus* (formerly *Methanobacterium thermoautotrophicum*) in pure culture and in cocultures with fatty acid-oxidizing syntrophs. *Appl Environ Microbiol* 68:1173–1179.

MacDonald IC, Deans TL (2016) Tools and applications in synthetic biology. *Adv Drug Deliv Rev* 105:20–34.

Madadi M, Penga C, Abbas A (2017) Advances in genetic manipulation of lignocellulose to reduce biomass recalcitrance and enhance biofuel production in bioenergy crops. *J Plant Biochem Physiol* 5:2.

Maeda T, Sanchez-Torres V, Wood TK (2012) Hydrogen production by recombinant Escherichia coli strains. *Microb Biotechnol* 5:214–225.

Maeda T, Vardar G, Self WT, Wood TK (2007) Inhibition of hydrogen uptake in *Escherichia coli* by expressing the hydrogenase from the cyanobacterium *Synechocystis* sp. PCC 6803. *BMC Biotechnol* 7:25.

Morimoto K, Kimura T, Sakka K, Ohmiya K (2005) Overexpression of a hydrogenase gene in *Clostridium paraputrificum* to enhance hydrogen gas production. *FEMS Microbiol Lett* 246:229–234.

Nair R V, Papoutsakis ET (1994) Expression of plasmid-encoded aad in *Clostridium acetobutylicum* M5 restores vigorous butanol production. *J Bacteriol* 176:5843–5846.

Niederholtmeyer H, Wolfstädter BT, Savage DF, Silver PA, Way JC (2010) Engineering cyanobacteria to synthesize and export hydrophilic products. *Appl Environ Microbiol* 76:3462–3466.

Nigam PS, Singh A (2011) Production of liquid biofuels from renewable resources. *Prog Energy Combust Sci* 37:52–68.

Patrick JW, Botha FC, Birch RG (2013) Metabolic engineering of sugars and simple sugar derivatives in plants. *Plant Biotechnol J* 11:142–156.

Penfold DW, Forster CF, Macaskie LE (2003) Increased hydrogen production by Escherichia coli strain HD701 in comparison with the wild-type parent strain MC4100. *Enzyme Microb Technol* 33:185–189.

Peralta-Yahya PP, Keasling JD (2010) Advanced biofuel production in microbes. *Biotechnol J* 5:147–162.

Peralta-Yahya PP, Zhang F, Del Cardayre SB, Keasling JD (2012) Microbial engineering for the production of advanced biofuels. *Nature* 488:320–328.

Prelich G (2012) Gene overexpression: Uses, mechanisms, and interpretation. *Genetics* 190:841–854.

Sakai T, Koo K, Saitoh K, Katsuragi T (1986) Use of protoplast fusion for the development of rapid starch-fermenting strains of *Saccharomyces diastaticus*. *Agric Biol Chem* 50:297–306.

Sekhon KK, Rahman PK (2013) Synthetic biology: A promising technology for biofuel production. *J Pet Environ Biotechnol* 4:1–1.

Shen CR, Lan EI, Dekishima Y, Baez A, Cho KM, Liao JC (2011) Driving forces enable high-titer anaerobic 1-butanol synthesis in *Escherichia coli*. *Appl Environ Microbiol* 77:2905–2915.

Shi S, Valle-Rodríguez JO, Siewers V, Nielsen J (2011) Prospects for microbial biodiesel production. *Biotechnol J* 6:277–285.

Shih PM, Liang Y, Loqué D (2016) Biotechnology and synthetic biology approaches for metabolic engineering of bioenergy crops. *Plant J* 87:103–117.

Somerville C, Youngs H, Taylor C, Davis SC, Long SP (2010) Feedstocks for lignocellulosic biofuels. *Science* 329:790–792.

Sticklen MB (2008) Plant genetic engineering for biofuel production: Towards affordable cellulosic ethanol. *Nat Rev Genet* 9:433–443.

Syed QUA, Nadeem M, Nelofer R (2008) Enhanced butanol production by mutant strains of *Clostridium acetobutylicum* in molasses medium. *Turkish J Biochem Biyokim Derg* 33:25–30.

Tian L, Perot SJ, Hon S, Zhou J, Liang X, Bouvier JT, Guss AM, Olson DG, Lynd LR (2017) Enhanced ethanol formation by *Clostridium thermocellum* via pyruvate decarboxylase. *Microb Cell Fact* 16:171.

Torney F, Moeller L, Scarpa A, Wang K (2007) Genetic engineering approaches to improve bioethanol production from maize. *Curr Opin Biotechnol* 18:193–199.

Torto-Alalibo T, Purwantini E, Lomax J, Setubal JC, Mukhopadhyay B, Tyler BM (2014) Genetic resources for advanced biofuel production described with the gene ontology. *Front Microbiol* 5:528.

Tsai CS, Kwak S, Turner TL, Jin YS (2015) Yeast synthetic biology toolbox and applications for biofuel production. *FEMS Yeast Res* 15(1), 1–15.

Verma N, Kumar V (2000) Protoplast fusion technology and its biotechnological. *International Conference on Industrial Biotechnology.*

Wang Q, Lu Y, Xin Y, Wei L, Huang S, Xu J (2016) Genome editing of model oleaginous microalgae *Nannochloropsis* spp. by CRISPR/Cas9. *Plant J* 88:1071–1081.

Woo HM, Lee HJ (2017) Toward solar biodiesel production from CO2 using engineered cyanobacteria. *FEMS Microbiol Lett* 364:9.

Wu L, Birch RG (2007) Doubled sugar content in sugarcane plants modified to produce a sucrose isomer. *Plant Biotechnol J* 5:109–117.

Yang ST, Liu X, Zhang Y (2007) Metabolic engineering-applications, methods, and challenges. In: *Bioprocessing for Value-Added Products from Renewable Resources*. Elsevier, Amsterdam, the Netherlands, pp. 73–118.

Young EM, Comer AD, Huang H, Alper HS (2012) A molecular transporter engineering approach to improving xylose catabolism in *Saccharomyces cerevisiae*. *Metab Eng* 14:401–411.

# 6 Biohydrogen Production by the Dark Fermentation Process

## 6.1 INTRODUCTION

The need for the development of renewable and sustainable energy fuels has sparked the advent of hydrogen energy research. Hydrogen has several attractive features such as high energy content, carbon neutral, and eco-friendly that makes it a potent fuel for the future (Dunn 2002). Moreover, since the only by-product produced by the combustion of hydrogen is water, it can potentially mitigate global warming and pollution problems. Besides being an attractive fuel source, hydrogen has wider industrial applications in synthesis of products such as ammonia, hydrogenation, metal refining, fuel cells, and portable electronics. It should be noted that hydrogen is the energy carrier and not the energy source and most of the hydrogen produced globally is from fossil energy (i.e., either through reforming natural gas or by coal gasification) (Holladay et al. 2009). However, for renewable energy generation, it is imperative that hydrogen is obtained using renewable energy sources. There are several means of renewable hydrogen production including solar conversion, biomass conversion, and electrolysis using renewable electricity source but each of these suffers from various technical challenges. Biological means of hydrogen production proves to be a less energy intensive and more environmentally friendly process. Figure 6.1 illustrates the various methods for biohydrogen production. Broadly they can be categorized as light dependent and light independent processes. The present chapter focuses on dark fermentative hydrogen production; the photobiological and bioelectrochemical processes will be dealt in subsequent chapters.

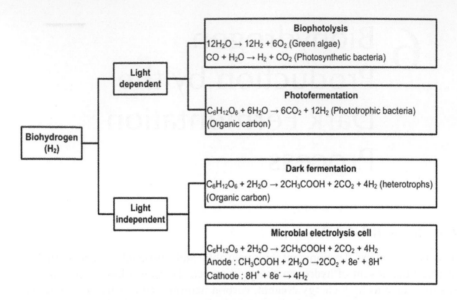

**FIGURE 6.1** Various methods of biological hydrogen production.

## 6.2 PRINCIPLES OF DARK FERMENTATION PROCESS

Owing to the high rate of hydrogen production and additional advantage of waste stabilization, dark fermentative hydrogen production has gained considerable interest in the recent years. The basic principles of dark fermentative hydrogen production involves mixed acid metabolism of facultative (FHL pathway) and obligate anaerobes (PFOR pathway) (Figure 4.5). The maximum theoretical hydrogen yield is 4 moles $H_2$ mole$^{-1}$ glucose consumed when the microorganism follows the acetate pathway (Equation 6.1).

$$C_6H_{12}O_6 + 4H_2O \rightarrow 2CH_3COO^- + 2HCO_3^- + 4H^+ + 4H_2 \qquad (6.1)$$

Similar theoretical yield of 4 moles $H_2$ mole$^{-1}$ glucose consumed is obtained when both acetate and formate are the fermentation end products (Equations 6.2 and 6.3).

$$C_6H_{12}O_6 + 2H_2O \rightarrow 2CH_3COO^- + 2HCOO^- + 4H^+ + 2H_2 \qquad (6.2)$$

$$2HCOOH \rightarrow 2CO_2 + 2H_2 \qquad (6.3)$$

In contrast, if butyrate is the sole end product of glucose fermentation, the maximum theoretical yield of hydrogen is only 2 moles $H_2$ mole$^{-1}$ glucose (Equation 6.4)

$$C_6H_{12}O_6 + 2H_2O \rightarrow CH_3CH_2CH_2COO^- + 2HCO_3^- + 3H^+ + 2H_2 \qquad (6.4)$$

It can be observed from Figure 4.5 that the key intermediate during the dark fermentation process is acetyl CoA that is produced from pyruvate using various enzymatic systems (PFOR or FHL) (Ghimire et al. 2015). The final fate of acetyl CoA determines the total theoretical maximum yield of $H_2$ per mole of sugar consumed. For example, if acetyl CoA is metabolized to acetate, then 4 moles of hydrogen can be obtained as previously described. However, if acetyl CoA is metabolized to butyrate then only 2 moles of hydrogen are produced as 2 NADH molecules are used for butyryl CoA formation. Depending upon the culture conditions and type of microorganisms, simultaneous generation of various end products such as ethanol, acetate, propionate, and butyrate can take place which limits the maximum possible yield of $H_2$.

## 6.3 MICROORGANISMS INVOLVED IN DARK FERMENTATIVE HYDROGEN PRODUCTION

Dark fermentative hydrogen production is carried out by various microorganisms using diverse metabolic pathways as described in Chapters 4 and 5. These organisms can be used either in pure form or can be enriched as co-cultures and mixed cultures, depending upon the type of substrate used for hydrogen production. Various studies have indicated higher hydrogen yields when using mixed cultures due to higher substrate conversion efficiencies (Elshahed 2010). This section provides an insight of the key players involved in dark fermentative hydrogen production.

### 6.3.1 Pure Cultures

There are three main categories of hydrogen-producing bacteria which include (1) spore forming obligate anaerobes, (2) non-spore forming obligate anaerobes, and (3) facultative anaerobes (Cabrol et al. 2017). Most of the obligatory anaerobes belong to the *Clostridium* species and the facultative hydrogen producers are commonly found in the *Enterobacteriaceae* family (Table 6.1). Facultative anaerobes have an advantage over obligate anaerobes because they are less sensitive to oxygen and they require less stringent culture conditions compared to obligate anaerobes (Ghimire et al. 2015). Most of the hydrogen-producing bacteria found in mesophilic conditions are found to be both obligate and facultative in nature, whereas thermophilic hydrogen producers are strictly obligate. The maximum hydrogen production has been reported to be around 2.4–2.8 mol $H_2$ $mol^{-1}$ glucose using obligate anaerobes such as *Clostridium* sp., whereas it ranges from 1 to 1.9 mol $H_2$ $mol^{-1}$ glucose using facultative anaerobes such as *Enterobacter* sp. (Table 6.1).

### 6.3.2 Co-cultures

Collective growth of pre-defined microbial strains under aseptic conditions can be defined as synthetic co-culture. It was suggested previously that hydrogen production could be enhanced by using co-cultures of varying hydrogen-producing bacteria (Pachapur et al. 2015). Co-culturing allows syntrophic interactions among various groups of microorganisms to assist each other to perform more effectively (Mishra et al. 2015). For example, if a complex substrate is used for biohydrogen production, co-culturing of hydrolyzing bacteria (that can convert the complex substrate into simpler monomer) and a hydrogen producing bacteria (that can ferment

**TABLE 6.1**
**Potent Hydrogen-Producing Strains**

| Species | Family | Anaerobe Type | References |
|---|---|---|---|
| *Enterobacter aerogens* | *Enterobacteriaceae* | Facultative | Tanisho (1998) |
| *Clostridium butyricum* | *Clostridiaceae* | Obligate | Kataoka et al. (1997) |
| *Enterobacter cloacae* IIT-BT 08 | *Enterobacteriaceae* | Facultative | Kumar and Das (2000) |
| *Klebsiella pneumoniae* DSM 2026 | *Enterobacteriaceae* | Facultative | Liu and Fang (2007) |
| *Citrobacter* sp. Y19 | *Enterobacteriaceae* | Facultative | Oh et al. (2003) |
| *Clostridium thermocellum* | *Clostridiaceae* | Obligate | Levin et al. (2006) |
| *Clostridium beijerinckii* RZF 1108 | *Clostridiaceae* | Obligate | Zhao et al. (2011) |
| *Enterobacter cloacae* DM11 | *Enterobacteriaceae* | Facultative | Nath et al. (2006) |
| *Clostridium butyricum* KCTC 1875 | *Clostridiaceae* | Obligate | Kim et al. (2008) |

simple monomer to produce hydrogen) can be an effective strategy to obtain higher yields of biohydrogen. Co-cultures have several advantages over single pure cultures such as they can perform complex biological functions, simultaneous substrate consumption, exchange of metabolites, and are less prone to foreign invasion and environmental fluctuations (Pachapur et al. 2015). Another advantage of using co-cultures for hydrogen production is the ability to use facultative and obligate anaerobes simultaneously. Sinha and others (2016) reported that bio-augmenting facultative anaerobe with obligate anaerobe improved $H_2$ yield by 37%. They suggested that by using a facultative anaerobe first, the residual oxygen can be removed, thus creating a complete anaerobic environment which ultimately would promote the growth of the obligate anaerobe. A similar observation was made by Chang and others (2008) who obtained higher process efficiency by using the syntrophic co-culture of aerobic *B. thermoamylovorans* I and anaerobic *C. beijerinckii* L9. Likewise, Qian and others (2011) showed synergistic effect between facultative and obligate anaerobic bacteria to produce stable hydrogen under continuous culture conditions. The biohydrogen yields obtained using various co-culture systems are presented in Table 6.2.

### 6.3.3 MIXED CULTURES

To avoid the upstream process costs such as for media sterilization, various research groups have considered using natural microflora as an inoculum for hydrogen production (Levin 2004; Ren et al. 2007; Kim and Kim 2012; Roy et al. 2012). A mixed culture system is potentially more suitable for transformation of a non-sterile substrate into hydrogen. This system has several advantages over pure or co-cultures such as they are ubiquitous, they can be easily handled, and less chances of contamination (Mishra et al. 2015). Hydrogen-producing mixed microflora are naturally found in various niches such as sewage sludge, municipal sewage sludge, anaerobic digester sludge, and waste-activated sludge (Cabrol et al. 2017). However, in such natural environments, they co-exist with

**TABLE 6.2**

**Hydrogen Production Using Synthetic Co-cultures**

| Co-cultures | Substrate | Approach | References |
|---|---|---|---|
| *Clostridium butyricum* and *Citrobacter freundii* | Starch | Combining facultative and obligate anaerobes | Laurent et al. (2010) |
| *Enterobacter cloacae* IIT-BT 08 and *Citrobacter freundii* IIT-BT L139 | Distillery effluent | Syntrophic association for complete substrate conversion | Mishra et al. (2015) |
| *Klebsiella pneumoniae* and *Clostridium acetobutylicum* | Cane molasses, starchy wastewater, and distillery effluent | Combining facultative and obligate anaerobes | Sinha et al. (2016) |
| *Enterobacter aerogenes* and *Clostridium butyricum* | Starch | Combining facultative and obligate anaerobes | Yokoi et al. (1998) |
| *Clostridium thermocellum* and *Clostridium thermopalmarium* | Cellulose | Combining cellulolytic and non-cellulolytic hydrogen producing bacteria | Geng et al. (2010) |
| *Clostridium thermocellum* and *Thermoanaerobacterium thermosaccharolyticum* | Cellulose | Combining cellulolytic and non-cellulolytic hydrogen producing bacteria | Liu et al. (2008) |
| *Enterobacter cloacae* IIT-BT 08, *Citrobacter freundii* IIT-BT L139, and *Bacillus coagulans* IIT-BT S1 | Glucose | Syntrophic association for complete substrate conversion | Kotay and Das (2009) |

the methanogenic bacteria or other bacteria such as homoacetogens which consume hydrogen (Ahn et al. 2005). Moreover, the presence of sulphate-reducing bacteria and lactate-producing bacteria inhibits hydrogen production because these bacteria compete with hydrogen-producing bacteria for the substrate. Thus, it is essential to pre-treat the sludge to remove the hydrogen-consuming bacteria before using it as an inoculum for biohydrogen production. The pre-treatment can be achieved through heat treatment or acid treatment methods due to the spore-forming nature of hydrogen-producing bacteria under harsh environmental conditions (Das and Veziroglu 2008). Table 6.3 shows the various mixed inoculum sources and their pre-treatment strategies for biohydrogen production. It can be observed that the yields obtained using mixed cultures vary with respect to culture conditions. This variance is due to the fact that no specific metabolic route is defined for mixed cultures and it can be diverted depending upon the favorable conditions (Cabrol et al. 2017).

**TABLE 6.3**

**Hydrogen Production Using Mixed Microbial Culture**

| Inoculum Source | Pre-treatment Strategy | Substrate | $H_2$ Yield | References |
|---|---|---|---|---|
| Sewage sludge and slaughter house sludge | — | Starch effluent | 8.1 mol kg$^{-1}$ COD | Chaitanya et al. (2018) |
| Activated sludge | Heat pre-treatment | Glucose | 1.6 mol mol$^{-1}$ glucose | Baghchehsaraee et al. (2008) |
| Anaerobically digested sludge | | | 2.3 mol mol$^{-1}$ glucose | |
| Anaerobic sludge | Enrichment in anaerobic media prescribed for $H_2$ producers | Distillery effluent | 9.17 mol kg$^{-1}$ COD | Mishra et al. (2015) |
| Cow dung | Heat shock | Glucose | 2.6 mol mol$^{-1}$ glucose | Kumari and Das (2017) |
| Anaerobic sewage sludge | Heat pre-treatment | Sucrose | 3.43 mol mol$^{-1}$ sucrose | Lin and Lay (2005) |
| Anaerobic sewage sludge | Heat treatment | Glucose | 13 mL g$^{-1}$ COD | Kotay and Das (2009) |
| UASB sludge | Chemical treatment with 2-bromoethane sulphonic acid (BESA) | Dairy wastewater | 0.0317 mmol g$^{-1}$ COD | Venkata Mohan et al. (2008) |
| Anaerobic digested sludge | Load shock | sucrose | 1.96 mol mol$^{-1}$ hexose | O-Thong et al. (2009) |
| Bioreactor treating chemical wastewater | Combined heat-shock, chemical, and pH treatments | Dairy waste water | 5.9 mol kg-1 COD | Srikanth et al. (2010) |

## 6.4 FEEDSTOCK FOR DARK FERMENTATIVE BIOHYDROGEN PRODUCTION

For the growth and development of fermentative bacteria, essential nutrients in the form of carbon source, nitrogen source, phosphate, metal ions, and other micronutrients are required. These essential nutrients are obtained from the feedstock used for the dark fermentation process.

Biomass rich in carbohydrate are used mainly as carbon and energy sources for most of the microorganisms (Ntaikou et al. 2010). On the other hand, some essential growth factors (such as vitamins and minerals which act as cofactors) are provided also in the media to promote growth and product formation. In general, the feedstock for the dark fermentation process can be broadly classified into two categories: simple sugars and complex wastes.

## 6.4.1 SIMPLE SUGARS

The main criteria for the selection of a suitable substrate for biohydrogen production include availability, cost, and biodegradability. Initial studies on biohydrogen production focused mainly on carbohydrate rich sources as potential feedstocks. Primarily, $C_6$ (e.g., glucose) and $C_5$ (e.g., xylose) sugars have been used as carbon sources to estimate and compare the maximum theoretical yields. The major sources of these fermentable sugars are first-generation feedstocks as described in Chapter 2. Several studies have shown higher hydrogen production using simple substrates. Mostly glucose has been used as the main carbon source for biohydrogen production since it is the simplest form of carbohydrate (Kataoka et al. 1997; Penfold et al. 2003; Zheng and Yu 2005; Zhao et al. 2011; Penniston and Gueguim Kana 2018). A maximum $H_2$ yield of 6 mol $H_2$ mol$^{-1}$ sucrose was reported by Kumar and Das (2000) using *Enterobacter cloacae* IIT-BT 08. On the other hand, Ho and others (2010) obtained a maximum hydrogen yield of 3.5 mol $H_2$ mol$^{-1}$ cellobiose using *Clostridium* sp. R1. Dessi and others (2018) reported high-yield hydrogen production of 1.2 mol $H_2$ mol$^{-1}$ xylose using a thermophilic mixed culture dominated with *Thermoanaerobacterium* sp. In a similar study by Ren and others (2007), a hydrogen yield of 2.19 mol $H_2$ mol$^{-1}$ xylose was obtained. Although these studies suggest high-rate hydrogen production using simple substrates, these studies are limited mainly to laboratory conditions to estimate the kinetic constants and theoretical yields. For an industrial point of view, hydrogen production using second- and third-generation feedstocks is considered more economical as described in the following section.

## 6.4.2 COMPLEX WASTES

Biohydrogen production has the dual benefit of waste remediation and energy generation when complex biodegradable wastes are considered as substrates for dark fermentative hydrogen production (Das and Veziroglu 2008). Various organic wastes and wastewater ranging from domestic, agricultural, industrial, and forestry wastes have been considered as potential feedstocks for biohydrogen production (Jaseena and Sosamony 2016). Compared to pure substrates, the hydrogen yield obtained using biomass and wastes is much less because of the incomplete conversion of the substrate (Ntaikou et al. 2010). Due to the abundance of lignocellulosic wastes, recent studies have been directed towards using lignocelluloses for biohydrogen production (Saratale et al. 2008). The complex nature of such wastes requires a pre-treatment step to be added before the fermentation step to convert the complex polymers into simple monomeric forms. Since most of these wastes lack essential nutrients required for the growth and development of microorganism, mixed substrates enhance the nutrient composition of wastes. For example, Mishra and others (2017) used groundnut deoiled cake and distillers' dried grain with solubles as co-substrate with distillery effluent to improve the gaseous energy recovery (Mishra et al. 2017). Similarly, Zhu and others (2008) showed enhanced hydrogen production using co-digestion of food waste with primary sludge and waste-activated sludge. These studies suggest that combined nutritional supplements have a positive effect on biohydrogen production. The economic viability of the dark fermentation process can be improved for commercial exploitation by using these non-conventional raw materials.

## 6.5    TYPES OF REACTORS CONFIGURATIONS USED FOR DARK FERMENTATIVE BIOHYDROGEN PRODUCTION

The configuration of the bioreactor is crucial for the long-term performance of dark fermentative biohydrogen production because it influences the prevailing micro-environment for the microbial culture, thereby affecting the population dynamics. Moreover, it establishes the hydrodynamic behavior of the culture media inside the bioreactor and affects the biochemical transformation that is to be carried out in it (Ghimire et al. 2015). The most commonly used reactor type is the batch reactor which comprises of a constant volume vessel in which the biochemical reaction takes place. However, a major drawback is the long down-time and start up. A more widely used configuration for a bioreactor is the continuous stirred-tank reactors (CSTR) that has provision for homogeneous mixing that enhances the contact time between the substrate and the microorganisms. Stirred-tank reactors (STRs) can be operated under varying modes such as batch mode, fed batch mode, or continuous mode (Ntaikou et al. 2010). Batch mode is primarily used for determining cell growth cycle and to determine the cell growth kinetics while continuous mode is considered typically for an industrial application (Das and Veziroglu 2008). Fed batch mode is usually performed to eliminate the substrate inhibition effects. CSTR have several advantages such as simple construction, easy operation, and possibility of temperature and pH control. However, its inherent limitations (i.e., lower solid retention times and cell wash out problems) have prompted the researchers to use alternate configurations such as an anaerobic contact process, packed bed reactors (PBR) (Roy et al. 2014b), membrane bioreactors (MBR) (Bakonyi et al. 2014), upflow anaerobic sludge blanket (UASB) reactors (Zhao et al. 2008), anaerobic sequencing batch reactors (ASBR) (Srikanth et al. 2010), and anaerobic-fluidized bed reactors (AFBR) (Muñoz-Páez et al. 2013). However, a direct comparison of the various configurations of bioreactors for dark fermentative biohydrogen production is not feasible because of the use of various operational conditions. A more detailed description of these reactor configurations is provided in Chapter 13.

## 6.6    EFFECT OF PHYSIOCHEMICAL PARAMETERS ON DARK FERMENTATIVE HYDROGEN PRODUCTION

For the scale up of biohydrogen production, it is essential to estimate the key factors that influence hydrogen productivity. From literature, it can be observed that several physicochemical factors such as the pH, temperature, hydraulic retention time, and hydrogen partial pressure are crucial for effective performance of the dark fermentation process. This section briefly discusses the effect of various factors that influence the biohydrogen production.

## 6.6.1 pH

The operational pH can influence the metabolic pathways to direct either towards hydrogen production or towards hydrogen consumption reactions (Ahn et al. 2005). This influence is because pH can directly affect hydrogenases, the main enzyme responsible for hydrogen production (Ghimire et al. 2015). It was experimentally determined that most of the hydrogen-producing bacteria showed maximum productivity at slightly acidic pH (i.e., ranging from 5.5 to 6) (Ginkel et al. 2001). At this pH, the methanogenic activity is suppressed which aids in proliferation of fermentative organisms. However, certain studies have shown optimum pH ranging from 4.5 to 9 (Ntaikou et al. 2010). The diverse range reported can be attributed to the variation in operational conditions such as inoculum sources, substrate type, and reactor type. During hydrogen production, the pH of the media is drastically reduced due to the formation of volatile fatty acids such as acetic acid, butyric acid, and propionic acid. At a very low pH (<4), the hydrogen production is inhibited because the hydrogenase activity is affected in these ranges. To avoid such inhibition, maintaining the pH throughout the fermentation process or adjusting the pH intermittently has been considered. Both these strategies have been shown to improve the overall hydrogen production (Penniston and Gueguim Kana 2018). The operational pH is dependent upon the type of substrate because it influences the final concentration of volatile fatty acids. Moreover, the pH of the media can influence the formation of solvents from the acids produced. For example, *Clostridium* species can produce butanol and acetone if the pH is below 5 (Ghimire et al. 2015). Since the operation pH can vary according to the type of organism and the substrate used, it is essential to evaluate experimentally the optimal pH to use for the dark fermentation process.

## 6.6.2 TEMPERATURE

Biohydrogen production has been extensively studied at varying operational temperatures (e.g., mesophilic (30°C–45°C), thermophilic (45°C–60°C), and extreme thermophilic (>65°C) ranges) (Ghimire et al. 2015). These temperature ranges are chosen based on the nature and source of the hydrogen-producing microorganisms. Mesophilic bacteria comprise of facultative and obligate anaerobes, whereas thermophilic bacteria are strictly obligate anaerobes. Like pH, temperature can also affect the metabolic pathways by shifting the by-product composition of dark fermentation (Ngoma et al. 2011). Higher hydrogen production rates are reported in the thermophilic range compared to the mesophilic range possibly due to the lower level of contamination at high temperatures. Moreover, by using thermophiles, high-temperature industrial effluents can be used directly without the need of sterilization, thereby lowering the production costs of the process. The higher hydrogen yield during thermophilic dark fermentation can be attributed also to the channeling of the metabolic pathway towards the acetate pathway compared to mesophilic ranges where the butyrate pathway dominates (Ntaikou et al. 2010). However, several reports to the contrary suggest higher butyrate formation with thermophiles (Roy et al. 2014a, 2014c; Varanasi et al. 2015). These differences in final metabolite composition among various studies are not clear and need further elucidation. Although most of the fermentative bacteria are inactive at psychrophilic temperatures, Debowski and others (2014) have successfully demonstrated biohydrogen production at even lower temperatures (<10°C) using cheese whey as substrates (Dębowski et al. 2014). These studies suggest that maintaining optimal temperature is essential for the proper functioning of enzymes that are responsible for channeling metabolic pathways towards hydrogen production.

### 6.6.3 HYDRAULIC RETENTION TIME

Hydraulic retention time (HRT) is a critical parameter that affects continuous biohydrogen production process. HRT can be defined as the time period in which a substrate is available inside the reactor for biochemical conversion. It is calculated by dividing the volume of the reactor with the influent flow rate. The substrate hydrolysis and the subsequent intermediate and product formation are dependent upon the HRT of the process (Elreedy and Tawfik 2015). Furthermore, HRT can be used to control the methanogenic activity due to the varying growth rates of hydrogen-producing bacteria and methanogens (Ghimire et al. 2015). Lower HRT favors hydrogen production because methanogens are washed out of the reactor. Silva-Illanes and others (2017) observed changes in metabolic patterns and the dominant microbial composition while studying the effect of HRT on dark fermentative hydrogen production from glycerol. This observation suggests that HRT plays a key role in determining the microbial community prevailing inside the bioreactor. The microbial diversity changed at varying HRTs during biohydrogen production from palm oil mill effluent (Silva-Illanes et al. 2017). Considering its role in biohydrogen production, several researchers have focused on studying the impact of HRT and the organic loading rate (which determines the HRT) on the dark fermentative process (Shen et al. 2009; Sharma and Li 2010; Juang et al. 2011; Pakarinen et al. 2011; Zhong et al. 2015).

### 6.6.4 HYDROGEN PARTIAL PRESSURE

Hydrogen partial pressure can have a direct influence on the overall gas production during the dark fermentation process. By lowering the hydrogen partial pressure from the headspace of the bioreactor, the mass transfer of dissolved hydrogen from the liquid phase to the gas phase can be improved according to the Le Chatelier's principle (Ghimire et al. 2015). In addition, if the hydrogen partial pressure is increased, then the metabolic activity of the bacteria can shift towards synthesis of more reduced products, thereby reducing the overall hydrogen yield (Jaseena and Sosamony 2016). Thus, it is essential to maintain lower hydrogen partial pressure inside the bioreactor during the dark fermentation process. The most common approach for lowering the hydrogen partial pressure is to continuously remove and collect the gas in a gas collector (Kumar and Das 2000). Alternatively, hydrogen can be stripped from the liquid using nitrogen sparging (Jaseena and Sosamony 2016). Other methods of lowering hydrogen partial pressure include (1) applying vacuum to the headspace that decreases the overall pressure in the system, (2) using a hydrogen permeable membrane to withdraw dissolved $H_2$ from the mixed liquid, and (3) decreasing the operating pressure in the reactor by using a vacuum pump (Ntaikou et al. 2010).

### 6.6.5 NUTRIENT AND TRACE ELEMENTS SUPPLEMENTATION

As described in Section 6.4, the composition of the feedstock is crucial for biohydrogen production. Most of the carbohydrate-rich sources used in the dark fermentation process is deficient of essential nutrients such as nitrogen, phosphate, minerals, and vitamins. Therefore, these are supplied as additional supplements in the media for optimal microbial activity (Jaseena and Sosamony 2016). Nitrogen is an important source for protein synthesis and is the major structural component of enzymes and nucleic acids. Various organic (such as peptone, tryptone, and yeast extract) and inorganic (such as ammonium chloride, and ammonium phosphate) nitrogen sources have been tested for biohydrogen production (Alshiyab et al. 2008b; Alvarado-Cuevas et al. 2013). Depending upon the various carbon and nitrogen sources, often an optimal carbon to nitrogen (C/N) ratio is considered for high-rate hydrogen production. For example, Anzola-Rojas and others (2015) found a hydrogen yield of 3.5 mol $H_2$ $mol^{-1}$ sucrose at a C/N ratio of 137. However, Argun and others (2008) found the highest hydrogen yield of 281 mL $H_2$ $g^{-1}$ starch at a C/N ratio of 200. These studies indicate that the optimal C/N ratio varies with the operational conditions.

Supplementation of various metal ions has shown positive impact on biohydrogen production by dark fermentation (Jaseena and Sosamony 2016). In a study by Karadag and Puhakka (2010), addition of trace amounts iron and nickel led to improvement in hydrogen production by 71%. A similar observation was obtained by Paul and others (2014) while studying biohydrogen produced from dairy effluent. The significant enhancement of hydrogen using these metals can be attributed to their role as cofactors for hydrogenase, the key enzyme for hydrogen production. By supplementing iron, magnesium and copper in the fermentation media, Mishra and Das (2014) showed significant improvement in biohydrogen hydrogen production from distillery effluent. They suggested that trace amounts of magnesium and copper were important for glycolytic enzymes and other kinases and synthetases. However, Alshiyab et al. (2018a) showed that addition of magnesium and calcium was better for biomass growth but not for biohydrogen production. Similarly, Mudhoo and Kumar (2013) showed inhibitory effects of copper, nickel, zinc, cadmium, chromium, and lead for biohydrogen and biomethane production. This variance suggests that the optimal metal ion concentration and its influence on hydrogen production are highly dependent upon the type of microorganisms used for the dark fermentation process.

## 6.7 VALORIZATION OF THE DARK FERMENTATIVE EFFLUENTS

As discussed in Section 6.2, the biohydrogen yield from dark fermentation process is thermodynamically limited to 2 and 4 mol $H_2$ $mol^{-1}$ glucose consumed if the organism follows butyrate and acetate pathways, respectively. So, integration strategies can vary according to the use of dark fermentation effluents for energy or value-added product recovery to improve the overall process efficiency (Figure 6.2).

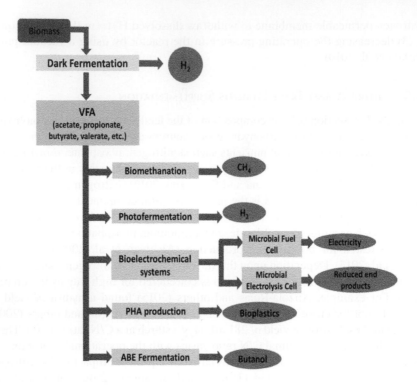

**FIGURE 6.2**  Use of dark fermentation effluents for enhanced energy recovery and value-added product formation.

### 6.7.1  BIOMETHANATION

The subsequent use of the spent medium of the dark fermentation process for methane production has been considered as an ideal integrative technology to harness maximum energy recovery from biomass (Kumari and Das 2015). The two-stage process, known as the biohythane process, has several advantages over the conventional anaerobic digestion process such as higher energy recovery, enhanced waste stabilization efficiency, and separate enrichment of acidogenic and methanogenic microflora (Varanasi et al. 2018). Various studies have reported improvement in the overall energy recovery by using the two-stage biohythane process (Guwy et al. 2011; Kumari and Das 2015; Nualsri et al. 2016; O-Thong et al. 2016). However, technical hurdles remain for the continuous dark fermentation and biomethanation process due to the variation in HRTs and the need of pH optimization before the methanation step, which affects the economy of the process.

### 6.7.2  PHOTOFERMENTATION

Another approach to recover energy from the dark fermentation process is to integrate it with the photofermentation process for enhanced biohydrogen production. Photofermentative bacteria are able to use the volatile fatty acids and convert them to molecular hydrogen in the presence of light (Dasgupta et al. 2010). By combining both dark fermentation and photofermentation, a theoretical maximum yield of 12 mol $H_2$ $mol^{-1}$ glucose can be obtained (Das and Veziroglu 2008). Various researchers have shown improvement in hydrogen yields from the two-stage process compared to individual dark fermentation or photofermentation processes (Ntaikou et al. 2010; Guwy et al. 2011; Xia et al. 2013; Lee et al. 2014). However, several factors limit the practical application of this integrative strategy such as lower photosynthetic efficiency of photofermentative bacteria, complex reactor configurations, require-ment of large surface areas, shading effect, and inhibition of nitrogenase to nitrogen content in the dark fermentation effluent. These limitations must be addressed for the practical scalability of the combined dark fermentation-photofermentation process.

### 6.7.3 BIOELECTROCHEMICAL SYSTEMS

Bioelectrochemical systems (BES) are nascent technologies that use the electrogenic capabilities of certain microorganisms to drive electrochemical reactions. These technologies include electron-producing systems (such as microbial fuel cells) and electron-consuming systems (such as microbial electrolysis cells), which can be used to produce electricity, hydrogen, or other value-added products, respectively. In recent years, various applications of BES have emerged in the areas such as waste reme-diation, water desalination, and nutrient recovery (Kadier et al. 2014; Karluvalı et al. 2015; Choi and Sang 2016). Most studies indicate that short chain carbon compounds such as acetate or butyrate are preferred substrates for electrogenic bacteria (Juang et al. 2011; Velvizhi and Mohan 2012; Karluvalı et al. 2015). Thus, these systems can readily use dark fermentative effluents which comprise of such volatile fatty acids. A more detailed description of BES technologies is provided in Chapter 12.

### 6.7.4 POLYHYDROXYALKANOATES PRODUCTION

Polyhydroxyalkanoates (PHAs) are natural biodegradable polyesters synthesized by microorganisms (Li et al. 2016). They can be a potential substitute for the non-degradable petrochemical plastics and thus can aid in lowering environmental pollu-tion. PHAs are accumulated as internal storage reserves under stress conditions. The most common forms of PHAs include polyhysroxybutyrate and polyhydroxyvaler-ate, which bear properties similar to polypropylene and polyethylene (Poltronieri and Kumar 2017). Using acidified fermentation effluents for PHA production can be attractive because the generation of value-added products would lead to an improved economy of the process. In recent years, several attempts have been made for the concomitant biohydrogen and PHA production using integrated dark fermentation processes (Venkateswar Reddy et al. 2015; Ghimire et al. 2016). However, further investigations are required for its practical application.

### 6.7.5 BUTANOL PRODUCTION

Another value-added product that can be obtained using dark fermentation efflu-ents is butanol, which is naturally produced by several *Clostridium* species

(Syed et al. 2008). Butanol has several industrial applications in products such as paints, solvents, and perfumes. Moreover, it has a high calorific value that makes it an attractive fuel. The properties of butanol are much more similar to gasoline and it can be directly used in internal combustion engines. Although several studies have been conducted over the years for high-rate butanol production, only a few studies are available on biohydrogen and butanol recovery from *Clostridium* species (Tashiro and Sonomoto 2010; Valdez-Vazquez et al. 2015). Since hydrogen is produced in the growth phase of the microorganism and butanol is produced in the stationary phase, by allowing phase separation, both the products can be recovered using a single species. Unlike the previously discussed technologies, this process does not require a separate system and can be continued using the same reactor configuration. The simultaneous production of hydrogen and butanol in the laboratory-scale studies has shown enhanced substrate removal and energy recovery by the integrated process (Ghimire et al. 2015; Mitra et al. 2017). Despite several benefits, the low butanol yield obtained, however, restricts the practical scalability of this technology.

## 6.8    MODELING AND SIMULATION OF DARK
##          FERMENTATION PROCESS

The increasing interest in dark fermentation and its associated technologies necessitates the modeling of the process which would help in better understanding the production pathways and their control. By evaluating the key kinetic parameters for biohydrogen production, the design, analysis, and operation of the fermentation process can be simplified. Several models have been proposed for predicting the biohydrogen production potential, biomass growth, substrate consumption, and by-product formation (Ghimire et al. 2015). The existing mathematical models used for predicting the performance of the dark fermentation process include (1) monod model for cell growth and substrate utilization, (2) logistic model for biomass growth, (3) modified Gompertz model for hydrogen production, (4) Arrhenius model for evaluating the temperature effects, and (5) Luedeking-Piret model for assessing by-product formation (growth, non-growth, or mixed growth associated product formation). The detailed description of these models is not within the scope of this chapter and can be found in other published studies.

## 6.9    CONCLUSION AND PERSPECTIVES

The advances in dark fermentation technology indicate immense potential for practical scalability of biohydrogen production from organic wastes. To find out the industrial feasibility of this process, research on continuous systems is a prerequisite. Moreover, inexpensive and easily available substrates must be explored to reduce the production cost of the process. In addition, the reproducibility of the hydrogen productivity must be established by conducting long-term operational studies and the stability of the microbial consortium must be ensured. By integrating this technology with other renewable systems, the overall yield and the energy efficiency of the process can be enhanced. Thus, impending research in this field must target value-added product recovery using a biorefinery approach. Since several pilot scale studies have successfully demonstrated the feasibility of the dark fermentative hydrogen production process, the commercial application of this technology can be a possibility in the near future.

## REFERENCES

Ahn Y, Park EJ, Oh YK, Park S, Webster G, Weightman AJ (2005) Biofilm microbial community of a thermophilic trickling biofilter used for continuous biohydrogen production. *FEMS Microbiol Lett* 249:31–38.

Alshiyab H, Kalil MS, Hamid AA, Yusoff WMW (2008a) Trace metal effect on hydrogen production using *Cacetobutylicum*. *Online J Biol Sci* 8:1–9.

Alshiyab HS, Kalil MS, Mohtar W, Yusoff W (2008b) Effect of nitrogen source and carbon to nitrogen ratio on hydrogen production using *C. acetobutylicum*. *Am J Biochem Biotechnol* 4:393–401.

Alvarado-Cuevas ZD, Acevedo LGO, Salas JTO, De León-Rodríguez A (2013) Nitrogen sources impact hydrogen production by *Escherichia coli* using cheese whey as substrate. *N Biotechnol* 30:585–590.

Anzola-Rojas M del P, Gonçalves da Fonseca S, Canedo da Silva C, Maia de Oliveira V, Zaiat M (2015) The use of the carbon/nitrogen ratio and specific organic loading rate as tools for improving biohydrogen production in fixed-bed reactors. *Biotechnol Rep* 5:46–54.

Argun H, Kargi F, Kapdan IK, Oztekin R (2008) Biohydrogen production by dark fermentation of wheat powder solution: Effects of C/N and C/P ratio on hydrogen yield and formation rate. *Int J Hydrogen Energy* 33:1813–1819.

Baghchehsaraee B, Nakhla G, Karamanev D, Margaritis A, Reid G (2008) The effect of heat pretreatment temperature on fermentative hydrogen production using mixed cultures. *Int J Hydrogen Energy* 33:4064–4073.

Bakonyi P, Nemestóthy N, Simon V, Bélafi-Bakó K (2014) Fermentative hydrogen production in anaerobic membrane bioreactors: A review. *Bioresour Technol* 156:357–363.

Cabrol L, Marone A, Tapia-Venegas E, Steyer J-P, Ruiz-Filippi G, Trably E (2017) Microbial ecology of fermentative hydrogen producing bioprocesses: Useful insights for driving the ecosystem function. *FEMS Microbiol Rev* 043:158–181.

Chaitanya N, Satish Kumar B, Himabindu V, Lakshminarasu M, Vishwanadham M (2018) Strategies for enhancement of bio-hydrogen production using mixed cultures from starch effluent as substrate. *Biofuels* 9:341–352.

Chang JJ, Chou CH, Ho CY, Chen WE, Lay JJ, Huang CC (2008) Syntrophic co-culture of aerobic *Bacillus* and anaerobic *Clostridium* for bio-fuels and bio-hydrogen production. *Int J Hydrogen Energy* 33:5137–5146.

Choi O, Sang B-I (2016) Extracellular electron transfer from cathode to microbes: Application for biofuel production. *Biotechnol Biofuels* 9:11.

Das D, Veziroglu T (2008) Advances in biological hydrogen production processes. *Int J Hydrogen Energy* 33:6046–6057.

Dasgupta CN, Jose Gilbert J, Lindblad P, Heidorn T, Borgvang SA, Skjanes K, Das D (2010) Recent trends on the development of photobiological processes and photobioreactors for the improvement of hydrogen production. *Int J Hydrogen Energy* 35:10218–10238.

Debowski M, Korzeniewska E, Filipkowska Z, Zieliński M, Kwiatkowski R (2014) Possibility of hydrogen production during cheese whey fermentation process by different strains of psychrophilic bacteria. *Int J Hydrogen Energy* 39:1972–1978.

Dessì P, Porca E, Waters NR, Lakaniemi A-M, Collins G, Lens PNL (2018) Thermophilic versus mesophilic dark fermentation in xylose-fed fluidised bed reactors: Biohydrogen production and active microbial community. *Int J Hydrogen Energy* 43:5473–5485.

Dunn S (2002) Hydrogen futures: Toward a sustainable energy system. *Int J Hydrogen Energy* 27:235–264.

Elreedy A, Tawfik A (2015) Effect of hydraulic retention time on hydrogen production from the dark fermentation of petrochemical effluents contaminated with ethylene glycol. *Energy Procedia* 74:1071–1078.

Elshahed MS (2010) Microbiological aspects of biofuel production: Current status and future directions. *J Adv Res* 1:103–111.

Geng A, He Y, Qian C, Yan X, Zhou Z (2010) Effect of key factors on hydrogen production from cellulose in a co-culture of *Clostridium thermocellum* and *Clostridium thermopalmarium*. *Bioresour Technol* 101:4029–4033.

Ghimire A, Frunzo L, Pirozzi F, Trably E, Escudie R, Lens PNL, Esposito G (2015) A review on dark fermentative biohydrogen production from organic biomass: Process parameters and use of by-products. *Appl Energy* 144:73–95.

Ghimire A, Valentino S, Frunzo L, Pirozzi F, Lens PNL, Esposito G (2016) Concomitant biohydrogen and poly-β-hydroxybutyrate production from dark fermentation effluents by adapted *Rhodobacter sphaeroides* and mixed photofermentative cultures. *Bioresour Technol* 217:157–164.

Ginkel S Van, Sung S, Lay JJ (2001) Biohydrogen production as a function of pH and substrate concentration. *Environ Sci Technol* 35:4726–4730.

Guwy AJ, Dinsdale RM, Kim JR, Massanet-Nicolau J, Premier G (2011) Fermentative biohydrogen production systems integration. *Bioresour Technol* 102:8534–8542.

Ho KL, Chen YY, Lee DJ (2010) Biohydrogen production from cellobiose in phenol and cresol-containing medium using *Clostridium* sp. R1. *Int J Hydrogen Energy* 35:10239–10244.

Holladay JD, Hu J, King DL, Wang Y (2009) An overview of hydrogen production technologies. *Catal Today* 139:244–260.

Jaseena KA, Sosamony KJ (2016) Practical aspects of hydrogen production by dark fermentation—A review. *IRACST—Eng Sci Technol An Int J* 6:2250–3498.

Juang DF, Yang PC, Chou HY, Chiu LJ (2011) Effects of microbial species, organic loading and substrate degradation rate on the power generation capability of microbial fuel cells. *Biotechnol Lett* 33:2147–2160.

Kadier A, Simayi Y, Abdeshahian P, Azman NF, Chandrasekhar K, Kalil MS (2014) A comprehensive review of microbial electrolysis cells (MEC) reactor designs and configurations for sustainable hydrogen gas production. *Alexandria Eng J* 55(1), 427–443.

Karadag D, Puhakka JA (2010) Enhancement of anaerobic hydrogen production by iron and nickel. *Int J Hydrogen Energy* 35:8554–8560.

Karluvalı A, Köroğlu EO, Manav N, Çetinkaya AY, Özkaya B (2015) Electricity generation from organic fraction of municipal solid wastes in tubular microbial fuel cell. *Sep Purif Technol* 156:502–511.

Kataoka N, Miya A, Kiriyama K (1997) Studies on hydrogen production by continuous culture system of hydrogen-producing anaerobic bacteria. *Water Sci Technol* 36(6–7):41–47.

Kim DH, Kim MS (2012) Thermophilic fermentative hydrogen production from various carbon sources by anaerobic mixed cultures. *Int J Hydrogen Energy* 37:2021–2027.

Kim JK, Nhat L, Chun YN, Kim SW (2008) Hydrogen production conditions from food waste by dark fermentation with Clostridium beijerinckii KCTC 1785. *Biotechnol Bioprocess Eng* 13:499–504.

Kotay SM, Das D (2009) Novel dark fermentation involving bioaugmentation with constructed bacterial consortium for enhanced biohydrogen production from pretreated sewage sludge. *Int J Hydrogen Energy* 34:7489–7496.

Kumar N, Das D (2000) Enhancement of hydrogen production by *Enterobacter cloacae* IIT-BT 08. *Process Biochem* 35:589–593.

Kumari S, Das D (2015) Improvement of gaseous energy recovery from sugarcane bagasse by dark fermentation followed by biomethanation process. *Bioresour Technol* 194:354–363.

Kumari S, Das D (2017) Improvement of biohydrogen production using acidogenic culture. *Int J Hydrogen Energy* 42:4083–4094.

Laurent B, Hiligsmann S, Hamilton C, Masset J, Thonart P (2010) Fermentative hydrogen production by *Clostridium butyricum* CWBI1009 and *Citrobacter freundii* CWBI952 in pure and mixed cultures. *Biotechnologie, Agronomie, Société et Environnement Biotechnology, Agronomy, Society and Environment [BASE]*, 14(S2), 541–548.

Lee WS, Seak A, Chua M, Yeoh HK, Ngoh GC (2014) A review of the production and applications of waste-derived volatile fatty acids. *Chem Eng J* 235:83–99.

Levin D (2004) Biohydrogen production: Prospects and limitations to practical application. *Int J Hydrogen Energy* 29:173–185.

Levin D, Islam R, Cicek N, Sparling R (2006) Hydrogen production by *Clostridium thermocellum* 27405 from cellulosic biomass substrates. *Int J Hydrogen Energy* 31:1496–1503.

Li Z, Yang J, Loh XJ (2016) Polyhydroxyalkanoates: Opening doors for a sustainable future. *NPG Asia Mater* 8:e265–e265.

Lin CY, Lay CH (2005) A nutrient formulation for fermentative hydrogen production using anaerobic sewage sludge microflora. *Int J Hydrogen Energy* 30:285–292.

Liu F, Fang B (2007) Optimization of bio-hydrogen production from biodiesel wastes by *Klebsiella pneumoniae*. *Biotechnol J* 2:374–380.

Liu Y, Yu P, Song X, Qu Y (2008) Hydrogen production from cellulose by co-culture of *Clostridium thermocellum* JN4 and *Thermoanaerobacterium thermosaccharolyticum* GD17. *Int J Hydrogen Energy* 33:2927–2933.

Mishra P, Balachandar G, Das D (2017) Improvement in biohythane production using organic solid waste and distillery effluent. *Waste Manag* 66:70–78.

Mishra P, Das D (2014) Biohydrogen production from *Enterobacter cloacae* IIT-BT 08 using distillery effluent. *Int J Hydrogen Energy* 39:7496–7507.

Mishra P, Roy S, Das D (2015) Comparative evaluation of the hydrogen production by mixed consortium, synthetic co-culture and pure culture using distillery effluent. *Bioresour Technol* 198:593–602.

Mitra R, Balachandar G, Singh V, Sinha P, Das D (2017) Improvement in energy recovery by dark fermentative biohydrogen followed by biobutanol production process using obligate anaerobes. *Int J Hydrogen Energy* 42:4880–4892.

Mudhoo A, Kumar S (2013) Effects of heavy metals as stress factors on anaerobic digestion processes and biogas production from biomass. *Int J Environ Sci Technol* 10:1383–1398.

Muñoz-Páez KM, Ruiz-Ordáz N, García-Mena J, Ponce-Noyola MT, Ramos-Valdivia AC, Robles-González IV, Villa-Tanaca L, Barrera-Cortés J, Rinderknecht-Seijas N, Poggi-Varaldo HM (2013) Comparison of biohydrogen production in fluidized bed bioreactors at room temperature and 35°C. In: *International Journal of Hydrogen Energy*. Pergamon, New York, pp. 12570–12579.

Nath K, Kumar A, Das D (2006) Effect of some environmental parameters on fermentative hydrogen production by *Enterobacter cloacae* DM11. *Can J Microbiol* 52:525–532.

Ngoma L, Masilela P, Obazu F, Gray VM (2011) The effect of temperature and effluent recycle rate on hydrogen production by undefined bacterial granules. *Bioresour Technol* 102:8986–8991.

Ntaikou I, Antonopoulou G, Lyberatos G (2010) Biohydrogen production from biomass and wastes via dark fermentation: A review. *Waste and Biomass Valorization* 1:21–39.

Nualsri C, Kongjan P, Reungsang A (2016) Direct integration of CSTR-UASB reactors for two-stage hydrogen and methane production from sugarcane syrup. *Int J Hydrogen Energy* 41:17884–17895.

Oh YK, Seol EH, Kim JR, Park S (2003) Fermentative biohydrogen production by a new chemoheterotrophic bacterium Citrobacter sp. Y19. *Int J Hydrogen Energy* 28:1353–1359.

O-Thong S, Prasertsan P, Birkeland N-K (2009) Evaluation of methods for preparing hydrogen-producing seed inocula under thermophilic condition by process performance and microbial community analysis. *Bioresour Technol* 100:909–918.

O-Thong S, Suksong W, Promnuan K, Thipmunee M, Mamimin C, Prasertsan P (2016) Two-stage thermophilic fermentation and mesophilic methanogenic process for biohythane production from palm oil mill effluent with methanogenic effluent recirculation for pH control. *Int J Hydrogen Energy* 41:21702–21712.

Pachapur VL, Sarma SJ, Brar SK, Le Bihan Y, Buelna G, Verma M (2015) Biological hydrogen production using co-culture versus mono-culture system. *Environ Technol Rev* 4:55–70.

Pakarinen O, Kaparaju P, Rintala J (2011) The effect of organic loading rate and retention time on hydrogen production from a methanogenic CSTR. *Bioresour Technol* 102:8952–8957.

Paul JS, Quraishi A, Thakur V, Jadhav SK (2014) Effect of ferrous and nitrate ions on biological hydrogen production from dairy effluent with anaerobic wastewater treatment process. *Asian J Biol Sci* 7:165–171.

Penfold DW, Forster CF, Macaskie LE (2003) Increased hydrogen production by *Escherichia coli* strain HD701 in comparison with the wild-type parent strain MC4100. *Enzyme Microb Technol* 33:185–189.

Penniston J, Gueguim Kana EB (2018) Impact of medium pH regulation on biohydrogen production in dark fermentation process using suspended and immobilized microbial cells. *Biotechnol Biotechnol Equip* 32:204–212.

Poltronieri P, Kumar P (2017) Polyhydroxyalkanoates (PHAs) in industrial applications BT. In: Martínez LMT, Kharissova OV, Kharisov BI (Eds.) *Handbook of Ecomaterials*. Springer International Publishing, Cham, Switzerland, pp. 1–30.

Qian CX, Chen LY, Rong H, Yuan XM (2011) Hydrogen production by mixed culture of several facultative bacteria and: Anaerobic bacteria. *Prog Nat Sci Mater Int* 21:506–511.

Ren N, Xing D, Rittmann BE, Zhao L, Xie T, Zhao X (2007) Microbial community structure of ethanol type fermentation in bio-hydrogen production. *Environ Microbiol* 9:1112–1125.

Roy S, Ghosh S, Das D (2012) Improvement of hydrogen production with thermophilic mixed culture from rice spent wash of distillery industry. *Int J Hydrogen Energy* 37:15867–15874.

Roy S, Kumar K, Ghosh S, Das D (2014a) Thermophilic biohydrogen production using pretreated algal biomass as substrate. *Biomass Bioenergy* 61:157–166.

Roy S, Vishnuvardhan M, Das D (2014b) Continuous thermophilic biohydrogen production in packed bed reactor. *Appl Energy* 136:51–58.

Roy S, Vishnuvardhan M, Das D (2014c) Improvement of hydrogen production by newly isolated *Thermoanaerobacterium thermosaccharolyticum* IIT BT-ST1. *Int J Hydrogen Energy* 39:7541–7552.

Saratale GD, Chen S Der, Lo YC, Saratale RG, Chang JS (2008) Outlook of biohydrogen production from lignocellulosic feedstock using dark fermentation—A review. *J Sci Ind Res. (India)* 67:962–979.

Sharma Y, Li B (2010) Optimizing energy harvest in wastewater treatment by combining anaerobic hydrogen producing biofermentor (HPB) and microbial fuel cell (MFC). *Int J Hydrogen Energy* 35:3789–3797.

Shen L, Bagley DM, Liss SN (2009) Effect of organic loading rate on fermentative hydrogen production from continuous stirred tank and membrane bioreactors. *Int J Hydrogen Energy* 34:3689–3696.

Silva-Illanes F, Tapia-Venegas E, Schiappacasse MC, Trably E, Ruiz-Filippi G (2017) Impact of hydraulic retention time (HRT) and pH on dark fermentative hydrogen production from glycerol. *Energy* 141:358–367.

Sinha P, Gaurav K, Roy S, Balachandar G, Das D (2016) Improvement of biohydrogen production with novel augmentation strategy using different organic residues. *Int J Hydrogen Energy* 41:14015–14025.

Srikanth S, Venkata Mohan S, Lalit Babu V, Sarma PN (2010) Metabolic shift and electron discharge pattern of anaerobic consortia as a function of pretreatment method applied during fermentative hydrogen production. *Int J Hydrogen Energy* 35:10693–10700.

Syed QUA, Nadeem M, Nelofer R (2008) Enhanced butanol production by mutant strains of *Clostridium acetobutylicum* in molasses medium. *Turkish J Biochem Biyokim Derg* 33:25–30.

Tanisho S (1998) Hydrogen production by facultative anaerobe *Enterobacter aerogenes*. In: *BioHydrogen*. Springer US, Boston, MA, pp. 273–279.

Tashiro Y, Sonomoto K (2010) Advances in butanol production by *Clostridia*. *Technol Educ Top App* 1383–1394.

Valdez-Vazquez I, Pérez-Rangel M, Tapia A, Buitrón G, Molina C, Hernández G, Amaya-Delgado L (2015) Hydrogen and butanol production from native wheat straw by synthetic microbial consortia integrated by species of *Enterococcus* and *Clostridium*. *Fuel* 159:214–222.

Varanasi JL, Kumari S, Das D (2018) Improvement of energy recovery from water hyacinth by using integrated system. *Int J Hydrogen Energy* 43(3):1308–1318.

Varanasi JL, Roy S, Pandit S, Das D (2015) Improvement of energy recovery from cellobiose by thermophillic dark fermentative hydrogen production followed by microbial fuel cell. *Int J Hydrogen Energy* 40(26):8311–8321.

Velvizhi G, Mohan SV (2012) Electrogenic activity and electron losses under increasing organic load of recalcitrant pharmaceutical wastewater. *Int J Hydrogen Energy* 1–10.

Venkata Mohan S, Lalit Babu V, Sarma PN (2008) Effect of various pretreatment methods on anaerobic mixed microflora to enhance biohydrogen production utilizing dairy wastewater as substrate. *Bioresour Technol* 99:59–67.

Venkateswar Reddy M, Kotamraju A, Venkata Mohan S (2015) Bacterial synthesis of polyhydroxyalkanoates using dark fermentation effluents: Comparison between pure and enriched mixed cultures. *Eng Life Sci* 15:646–654.

Xia A, Cheng J, Lin R, Lu H, Zhou J, Cen K (2013) Comparison in dark hydrogen fermentation followed by photo hydrogen fermentation and methanogenesis between protein and carbohydrate compositions in Nannochloropsis oceanica biomass. *Bioresour Technol* 138:204–213.

Yokoi H, Tokushige T, Hirose J, Hayashi S, Takasaki Y (1998) $H_2$ production from starch by a mixed culture of *Clostridium butyricum* and *Enterobacter aerogenes*. *Biotechnol Lett* 20:143–147.

Zhao B, Yue Z, Zhao Q, Mu Y, Yu H, Harada H, Li Y (2008) Optimization of hydrogen production in a granule-based UASB reactor. *Int J Hydrogen Energy* 33:2454–2461.

Zhao X, Xing D, Fu N, Liu B, Ren N (2011) Hydrogen production by the newly isolated *Clostridium beijerinckii* RZF-1108. *Bioresour Technol* 102:8432–8436.

Zheng X-J, Yu H-Q (2005) Inhibitory effects of butyrate on biological hydrogen production with mixed anaerobic cultures. *J Environ Manage* 74:65–70.

Zhong J, Stevens DK, Hansen CL (2015) Optimization of anaerobic hydrogen and methane production from dairy processing waste using a two-stage digestion in induced bed reactors (IBR). *Int J Hydrogen Energy* 40:15470–15476.

Zhu H, Parker W, Basnar R, Proracki A, Falletta P, Béland M, Seto P (2008) Biohydrogen production by anaerobic co-digestion of municipal food waste and sewage sludges. *Int J Hydrogen Energy* 33:3651–3659.

# 7 Biohydrogen Production by Photobiological Processes

## 7.1 INTRODUCTION

Biohydrogen can be directly harnessed from solar energy by the action of certain oxygenic and anoxygenic photosynthetic microorganisms (McKinlay and Harwood 2010). These organisms produce hydrogen to remove the excess reducing equivalents produced during the photosynthetic electron transport (Skizim et al. 2012). The mechanisms of photobiological hydrogen production depend on the presence or absence of oxygen in the surrounding environment (Figure 7.1). Oxygenic photosynthetic microbes such as microalgae and cyanobacteria produce hydrogen though the biophotolysis process. On the other hand, anoxygenic photoheterotrophs such as green and purple bacteria produce hydrogen by the photofermentation process. Both mechanisms have been extensively studied in literature. However, the fundamental basis of the triggering of hydrogen production pathways in these organisms is still not clearly understood (Dasgupta et al. 2010). Photobiological hydrogen production can serve a dual purpose: (1) lower the greenhouse gas emissions by using the atmospheric $CO_2$ (biophotolysis) and (2) reducing environmental pollution by using wastes and wastewaters (photofermentation). However, low hydrogen yields obtained from these processes limit their practical feasibility. A major limitation of the biophotolysis process is the activities of the key enzymes nitrogenases and hydrogenases, which are inhibited during oxygen evolution (Dasgupta et al. 2010). The presence of uptake hydrogenases reduces the overall hydrogen yields of the photofermentation process. Currently, various attempts are being made to understand the metabolic and regulatory aspects of photobiological hydrogen production to enhance the hydrogen-producing capacities of these organisms. In addition, the advances in reactor engineering, genetic engineering, and other molecular tools have shown promising potential to improve the conversion efficiencies of these processes. In this chapter, the fundamental aspects of photobiological hydrogen production are described with a key emphasis on the recent advancements with respect to photobioreactor designs, alteration in media components, new isolates, and synthetic biology applications. In addition, the various parameters influencing the hydrogen productivity are discussed.

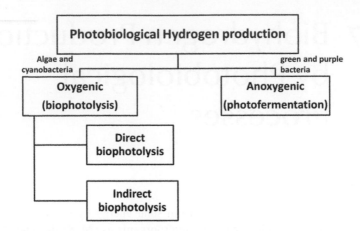

**FIGURE 7.1** Photobiological hydrogen production processes.

## 7.2 OXYGENIC PHOTOBIOLOGICAL HYDROGEN PRODUCTION

In oxygenic environments, the electrons needed for $H_2$ production are derived from water by the light-induced water oxidation reaction of the photosynthetic organisms, which leads to oxygen evolution (Equation 7.1):

$$2H_2O \rightarrow O_2 + 4H^+ + 4e^-$$  (7.1)

The reaction in Equation 7.1 is carried out by the PSII reaction center of the light-harvesting complex (LHC II) present in the thylakoid membrane of the photosynthetic organisms. Usually, these electrons are further used in the Calvin cycle to fix $CO_2$ into storage carbohydrates. However, under favorable conditions and in the absence of oxygen, these electrons are used up by the hydrogenases or nitrogenases to produce molecular hydrogen through the proton reduction reaction (Equation 7.2):

$$4H^+ + 4e^- \rightarrow 2H_2$$  (7.2)

The overall biophotolysis reaction for splitting water into molecular oxygen and molecular hydrogen can be represented as in Equation 7.3:

$$2H_2O \rightarrow O_2 + 2H_2$$  (7.3)

Depending upon the fate of electrons derived from water, oxygenic photobiological biohydrogen production can be classified into direct and indirect biophotolysis.

### 7.2.1 DIRECT BIOPHOTOLYSIS

When the electrons obtained from water are directly used up for hydrogen production, it is termed as direct biophotolysis. Hydrogenase is the key enzyme responsible for hydrogen production during direct biophotolysis. It is postulated that the partial inhibition of PSII creates an anaerobic environment that ultimately leads to the expression of the hydrogenase enzyme (Bandyopadhyay et al. 2010). However, even in the presence of a minimal amount oxygen, the activity of hydrogenase is inhibited. This inhibition is due to the fact that oxygen reacts with the H cluster (a complex iron-sulfur cofactor) of the hydrogenase enzyme resulting in its rapid inactivation and eventual degradation (Swanson et al. 2015). Since the $O_2$ evolution is inevitable during the photosynthesis process, the hydrogen yields are drastically affected during direct photolysis process. Various attempts have been made to decrease the oxygen sensitivity of hydrogenase enzymes. Melis and others (2000) showed that hydrogen production was enhanced under sulfur deprivation conditions due to the reversible inactivation of PSII and oxygen evolution reaction. They suggested operation of a two-stage direct biophotolysis set up so that the carbon accumulation phase (stage 1) can be separated from the hydrogen production phase (stage 2) and the hydrogen production phase is operated under a sulfur deprivation mode to improve the biohydrogen yields. Bingham and others (2012) developed mutant [FeFe] hydrogenases that had decreased oxygen sensitivity. They reported that saturation and site-directed mutagenesis were useful techniques to develop mutants with less oxygen sensitivity. Some of the other strategies to improve efficiency of direct biophotolysis in green microalgae and cyanobacteria are listed in Table 7.1.

## 7.2.2 INDIRECT BIOPHOTOLYSIS

Indirect biophotolysis is when the electrons derived for water are used up for carbon assimilation and the newly synthesized storage carbohydrates serve as substrates for

## TABLE 7.1
## Strategies to Improve Direct Photolysis in Green Microalgae and Cyanobacteria

| Organism | Strategy | References |
|---|---|---|
| Chlamydomonas. reinhardtii CC125 | Truncating the chlorophyll antenna size of PSII to improve the photon capturing efficiency | Berberoglu et al. (2008) |
| Chlamydomonas sp. | Random and directed mutagenesis to produce less oxygen-sensitive hydrogenases | Mathews and Wang (2009) |
| Chlamydomonas. reinhardtii | Codon optimization to insure a very low partial pressure of oxygen within the cells | Wu et al. (2011) |
| Chlamydomonas reinhardtii D1 | Amino acid substitution leading to mutant with lower chlorophyll content, higher photosynthetic capacity, higher respiration rate, and prolonged period of $H_2$ production | Torzillo et al. (2009) |
| Synechococcus elongatus | Expression of Clostridial [FeFe]-hydrogenase by direct electroporation technique | Miyake and Asada (1997) |
| Anabaena sp. PCC 7120 | Disruption of uptake hydrogenase gene | Masukawa et al. (2002) |
| Synechocystis PCC6803 | Genetic manipulation for overproduction of hydrogenase | Veaudor et al. (2018) |

biohydrogen production. The overall mechanism for indirect biophotolysis can be represented by Equations 7.4 and 7.5:

$$12H_2O + 6CO_2 \rightarrow C_6H_{12}O_6 + 6O_2 \tag{7.4}$$

$$C_6H_{12}O_6 + 12H_2O \rightarrow 12H_2 + 6CO_2 \tag{7.5}$$

Nitrogenases and bidirectional hydrogenases play vital roles during indirect bio-photolysis (Azwar et al. 2014). Indirect biophotolysis is advantageous over direct biophotolysis because the oxygen evolution is separated partially from the hydrogen evolution reaction. The partial separation is brought about by two main approaches (i.e., spatial separation and temporal separation) (Figure 4.3). Spatial separation is observed in certain heterocystous cyanobacteria in which a specialized cell lacking a PSII reaction center is developed (heterocyst). In these cells, due to the lack of PSII, oxygen evolution does not occur. The electrons derived from the sugars from the vegetative cells are used for the ferredoxin reduction in the photosynthetic electron transport chain. Finally, through the catalytic action of nitrogenase, hydrogen is produced (Figure 4.3). In contrast, during temporal separation, the oxygen evolution and the hydrogen evolution are separated by the day and night cycles, as such the oxygenic photosynthesis and carbon fixation occur during the day while the storage polysaccharides are used for the hydrogen production during the night. Temporal separation usually is observed in algae and non-heterocystous cyanobacteria (McKinlay and Harwood 2010). A higher rate of hydrogen production is observed from indirect biophotolysis compared to direct biophotolysis (Manish and Banerjee 2008). However, the reported yields are not attractive for commercial exploitation of this process for biohydrogen production. The major disadvantage is the low solar energy conversion efficiency of the photosynthetic organisms that reduce the total number of electrons available for hydrogen production. It is estimated that about 10% solar energy conversion efficiency is required for the indirect biophotolysis to be competitive with the alternative sources of renewable hydrogen production (Pandu and Joseph 2012). However, the current observed solar conversion efficiencies are below 5% (Kosourov et al. 2017). The conversion efficiencies of cyanobacteria and green algae can be improved by genetically altering the light-harvesting complexes within these organisms. Some of the alternative approaches to improve the biohydrogen yield through the indirect biophotolysis process are presented in Table 7.2.

## 7.3   ANOXYGENIC PHOTOBIOLOGICAL HYDROGEN PRODUCTION

Certain photoheterotrophic bacteria have the capability to produce biohydrogen under anoxic conditions by a process known as photofermentation. These bacteria comprise of light-harvesting photosystems (bacterial photosystems) similar to photosynthetic organisms. However, they are incapable of water oxidation and thus in turn oxygen evolution due to the lack of a PSII reaction center (Levin 2004). Two classes of photoheterotrophic bacteria have been observed that possess photofermentative

**TABLE 7.2**

**Strategies to Improve Indirect Biophotolysis in Green Microalgae and Cyanobacteria**

| Organism | Strategy | References |
|---|---|---|
| *Chlamydomonas reinhardtii* | Random gene insertion for enhancing starch reserves and reducing dissolved oxygen concentration | Kruse et al. (2005) |
| *Chlamydomonas reinhardtii* | Introduction of HUP1 (hexose uptake protein) from *Chlorella kessleri* for enhanced hexose utilization from internal (storage carbohydrate) as well as external medium | Doebbe et al. (2007) |
| *Synechocystis* sp. PCC 6803 | Silica sol-gel encapsulation to alter the cell metabolism and potentially increase reductant availability for $H_2$ production | Dickson et al. (2009) |
| *Synechococcus* PCC7942 | Heterologous expression of [FeFe] hydrogenase in the heterocysts | Asada et al. (2000) |
| *Anabaena variabilis* | Nutrient supplementation and optimization of physicochemical parameters | Shah et al. (2001) |
| *Anabaena siamensis* | Cell immobilization and nutrient deprivation to enhance heterocyst formation | Taikhao and Phunpruch (2017) |

capabilities: green bacteria and purple bacteria (McKinlay and Harwood 2010). They evolve hydrogen with the help of nitrogenase enzyme under nitrogen-limiting conditions by using light energy and organic compounds as substrates (Manish and Banerjee 2008). The most widely studied organisms among the photofermentative bacteria are the purple non-sulfur (PNS) bacteria such as *Rhodobacter* sp., *Rhodopseudomonas* sp., and *Rhodospirillum* sp. (Basak and Das 2007). Compared to biophotolysis, photofermentation has several advantages which include (1) higher yields, (2) no oxygen evolution that facilitates the activity of key $H_2$ producing enzymes, (3) use of broad spectrum of light, and (4) capability for using exogenous organic substrates thereby aiding in bioremediation. Compared to biophotolysis, PNS bacteria require less free energy (~8.5 KJ mol$^{-1}$ $H_2$ for lactate) to produce hydrogen from organic substrates (Basak and Das 2007). Moreover, several PNS bacteria are reported to produce hydrogen using inorganic substrates such as $S_2O_3^{2-}$, $H_2S$, and $Fe^{2+}$ (McKinlay and Harwood 2010). Another major advantage of photofermentation is the ability to drive thermodynamically unfavorable reactions by using solar energy (i.e., it can potentially convert an entire substrate to hydrogen) (Azwar et al. 2014). Since these bacteria can metabolize short chain organic acids such as acetate, propionate, and butyrate, the photofermentation process can be integrated with the dark fermentation process to obtain higher substrate conversion efficiency and overall hydrogen yields (Guwy et al. 2011). In spite of several advantages, the production volume of hydrogen is still too low. One of the possible reasons is the presence of uptake hydrogenases (which consume hydrogen during nitrogen fixation) that lower the maximum possible biohydrogen yields. By eliminating the uptake hydrogenases from the PNS bacteria using genetic tools, the hydrogen yields can be significantly enhanced. Moreover, as in case of biophotolysis, several attempts have been made

**TABLE 7.3**

**Strategies to Improve Biohydrogen Production by Photofermentation**

| Organism | Strategy | References |
|---|---|---|
| *Rhodobacter sphaeroides* | Elimination of polyhydroxyalkanoate pathway and knockout of uptake hydrogenase | Franchi et al. (2004) |
| *Rhodobacter capsulatus* | Genetic modification of electron transport chain and elimination of uptake hydrogenases | Öztürk et al. (2006) |
| *Rhodobacter sphaeroides* | Disruption of PII like proteins GlnB and GlnK | Kim et al. |
| *Rhodospirillum rubrum* | Genetic recombination and cloning of gene encoding Fe-only hydrogenase of *Clostridium acetobutylicum* | (2008) |
| *Rhodopseudomonas faecalis* RLD-53 | Separation of $CO_2$ from reaction system | Liu et al. (2009) |
| *Rhodobacter sphaeroides* | Supplementation and optimization of $Fe^{2+}$ | Zhu et al. (2007) |
| *Rhodopseudomonas faecalis* RLD-53 | Bioflocculation induced by $Ca^{2+}$ | Xie et al. (2013) |

to improve the light conversion efficiencies of photosynthetic bacteria to further improve the biohydrogen yields (Table 7.3).

## 7.4 RECTORS CONFIGURATION

Due to the requirement of continuous light irradiation during photobiological hydrogen production, the design of the photobioreactor is crucial for obtaining maximum photosynthetic efficiencies and hydrogen yields. Compared to conventional bioreactors used for chemoheterotrophs (batch or continuously stirred-tank reactor), the photobioreactors for biohydrogen production must comprise several additional features such as (1) transparent structure to allow maximum penetration of light, (2) strict anaerobic environment to avoid hydrogen inhibition, (3) novel mixing strategies, (4) controlled temperature conditions (due to excessive heat generated by light irradiation and self-shading of cells), (5) higher area/volume ratio, (6) provisions for nutrient supply and pH control, and (7) flexible material of construction (low cost, durable, non-toxic, and resistant to weathering). Various photobioreactor configurations have been utilized for hydrogen production which include shaking flasks, stirred tank, tubular (vertical, horizontal, or helical), flat panel, and vertical-column type of photobioreactors (Figure 7.2). Each of these reactor configurations is associated with several advantages and several disadvantages. For example, a tubular photobioreactor has a higher illumination area but has high concentrations of dissolved oxygen. Similarly, a vertical-column bioreactor has several advantages such as low cost, ease of operation, and compatibility but has poor illumination area. The flat panel photobioreactor is suitable for achieving high photosynthetic efficiencies and effective control over gas pressure but temperature maintenance and inefficient mixing patterns limit the practical scalability of this type of configuration. The various operational features of different configurations of photobioreactors along with their advantages and disadvantages is not within the scope of this chapter and can be found elsewhere (Dasgupta et al. 2010).

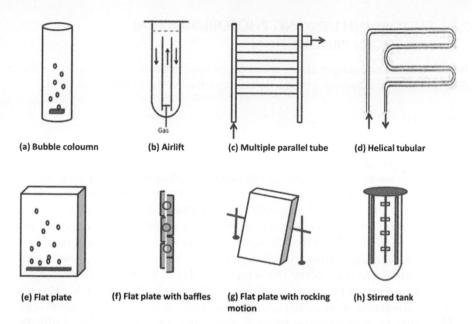

(a) Bubble coloumn    (b) Airlift    (c) Multiple parallel tube    (d) Helical tubular

(e) Flat plate    (f) Flat plate with baffles    (g) Flat plate with rocking motion    (h) Stirred tank

**FIGURE 7.2** Various configurations of photobioreactors for biohydrogen production: (a) bubble column, (b) airlift, (c) multiple parallel tube, (d) helical tubular, (e) flat plate, (f) flat plate with baffles, (g) flat plate with rocking motion, and (h) stirred tank.

Over the past few decades, several attempts have been made to develop suitable photobioreactors for hydrogen production (Kunjapur and Eldridge 2010). Tamburic and others (2011) developed a novel flat-plate bioreactor system comprising of transparent polymethyl methacrylate sheets. Their system was composed of two main compartments: (1) a primary compartment to hold the culture in which mixing is achieved using a recirculating gas-lift and (2) a secondary compartment to control the temperature and wavelength of radiation. Using this configuration, they reported a hydrogen yield of 105 mL $L^{-1}$ with a maximum hydrogen production rate of 1.1 mL $L^{-1}$ $h^{-1}$ and a photochemical conversion efficiency of 0.24% (Tamburic et al. 2011). A similar configuration was proposed by Skjånes and others (2016) who developed a flat panel photobioreactor that could be operated separately for biomass growth and hydrogen production. In their study, a rocking motion was applied to the flat panel to ascertain mixing, gas exchange, and large surface area between the culture and the headspace (Skjånes et al. 2016). Chen and others (2006) developed an innovative internal optical-fiber illuminating system to improve the efficiency of photobioreactor. In their study, different strategies such as use of solid carriers and optimization of acetate concentration were employed to improve phototrophic hydrogen production. By using these strategies, a maximum hydrogen yield of 3.63 mol $H_2$ $mol^{-1}$ acetate was obtained, which was higher than most of the studies reported for phototrophic hydrogen production (Chen and Chang 2006). These studies indicate that, although several advancements have been made with respect to photobioreactor designs, the development of suitable configuration with effective engineering innovation still remains a challenge.

## 7.5   FACTORS INFLUENCING PHOTOBIOLOGICAL HYDROGEN PRODUCTION

Besides the reactor configuration, various operational physicochemical factors affect the photobiological hydrogen production as described in the following sections.

### 7.5.1   pH

pH is the most influential factor that controls the phototrophic hydrogen production. It not only affects the biomass growth but also is crucial for the activity of key enzymes involved in hydrogen production (i.e., hydrogenases and nitrogenases). The optimal pH is dependent upon the type of organism involved. An initial pH of $7.0 \pm 0.2$ has been suggested as suitable for cell growth and hydrogen production for *Rhodobacter sphaeroides* (Akroum-Amrouche et al. 2011). In these species, pH follows the trend of the growth such that the pH increases as the growth increases. However, it decreases during hydrogen production when bacterium growth stops (Akroum-Amrouche et al. 2011). The decrease in pH can be attributed to hydrogen-consuming and proton-producing reactions. On the other hand, an optimal pH range of 6–10 was reported for microalgal growth and hydrogen production (Alalayah et al. 2014). Unlike photofermentative bacteria, algae are observed to produce hydrogen at slightly alkaline pH (Alalayah et al. 2014). This observation could be due to the prevailing favorable conditions for the biophotolytic hydrogenases that lead to high hydrogen production rates at these pH conditions. Thus, by optimizing and controlling pH in photobioreactors, biomass growth and biohydrogen production could be substantially improved.

### 7.5.2   TEMPERATURE

Like pH, temperature also affects the growth rate and activity of the hydrogen-producing enzymes. An optimal temperature range of 25°C–35°C is reported for the growth of most algal species (Dasgupta et al. 2010) while for the cyanobacteria it is in the range of 30°C–40°C (Tiwari and Pandey 2012). For the thermophilic species, the optimal temperature for growth and hydrogen production can exceed 60°C. During the photobioreactor operation, the temperature deviates significantly due to the self-shading of cells. This self-shading causes loss of extra absorbed energy through the fluorescence and heat, thus leading to a rise in temperature (Dasgupta et al. 2010). Although higher temperatures offer an advantage of lower oxygen evolution, it can denature the functional proteins of the cell, thereby retarding the microbial growth. Miyamoto and others (1979) demonstrated the possibility of controlling the evolved $H_2/O_2$ ratio during biophotolysis by changing the operational temperatures. Similarly, Özgür and others (2010) showed that outdoor photobiological hydrogen production using *Rhodobacter capsulatus* was strongly affected by the fluctuations in temperature. They suggested that change in temperature caused the bacteria to expend their energy for adaptation to the varying temperature for their survival and in turn decreased the hydrogen yields obtained (Özgür et al. 2010). These studies implicate the necessity of maintaining controlled temperature conditions for obtaining maximum hydrogen production rates.

### 7.5.3 LIGHT INTENSITY

Light is the source of energy for most of the photosynthetic organisms. Therefore, the intensity and dispersion of light during photobiological hydrogen production directly affects the photosynthetic efficiency and in turn hydrogen yield. It was previously reported that increase in light intensity increased the biomass growth and hydrogen production (Akroum-Amrouche et al. 2011). They concluded that under the influence of light intensity, *Rhodobacter sphaeroides* produced hydrogen in an exponential phase as well as in stationary phase (Akroum-Amrouche et al. 2011). An optimal light intensity range of 50–200 µE m$^{-2}$ s$^{-1}$ was found to be suitable for microalgal biomass growth and hydrogen production (Dasgupta et al. 2010). However, intensities greater than 200 µE m$^{-2}$ s$^{-1}$ can lead to photoinhibition. Moreover, high light intensities caused oxidative stress in the cells (Kosourov et al. 2017). On the contrary, certain organisms produce hydrogen in the absence of light (i.e., dark conditions). It was reported by Tiwari and Pandey (2012) that certain cyanobacteria favored dark conditions for hydrogen production. In such organisms, stable dark/night cycles are used such that during the day the growth occurs at the optimal light intensity and during the night hydrogen production takes place. This phenomenon is observed mainly in non-heterocystous cyanobacteria such as *A. variabilis* SPU 003, which produces hydrogen only during darkness (Tiwari and Pandey 2012).

### 7.5.4 MIXING

As discussed in Section 7.4, different types of agitation systems have been developed for microalgal growth and biohydrogen production depending upon the type of photobioreactor configuration. It is speculated the mechanical agitation can hamper the biomass growth by causing shear stress on the cells (Dasgupta et al. 2010). In aerobic growth, aeration and bubble re-circulation mainly provide adequate mixing in the photobioreactor. However, in anaerobic conditions, such as during hydrogen production, mixing is a challenge. Various studies have reported the use of diaphragm or centrifugal pumps for providing mixing in the photobioreactors (Dasgupta et al. 2010). In addition, the produced hydrogen gas can provide localized mixing inside the reactors. However, the efficiency of such a system is too low. Depending upon the agitation system, different mixing regimes (such as laminar flow or turbulent flow) can be developed inside the photobioreactor. The unfavorable regimes can cause damage to the microbial cells and thus needs to be carefully monitored. Certain species are able to tolerate high levels of turbulence while other are easy susceptible to damaging (Sobczuk et al. 2006). Therefore, the study of various hydrodynamic stresses incurred by microalgae and cyanobacteria during photobiological hydrogen production is essential for obtaining high biomass and hydrogen productivities.

### 7.5.5 NUTRIENT MEDIA

The media composition plays a vital role in the growth and development of microorganisms. Various kind of nutrient stresses influence the hydrogen production (Gonzalez-Ballester et al. 2015). For example, the presence of different nitrogen sources such as nitrate, nitrite, or ammonium inhibit hydrogen production in *Anabaena variabilis* SPU 003 and *A. cylindrica* (Tiwari and Pandey 2012). For photoheterotrophic growth, the addition of a carbon source is essential. Short carbon organic acids such as acetate or butyrate are more favorable during photofermentative hydrogen production compared to the presence of hexoses or pentoses (Özgür et al. 2010). It was reported that in heterocystous cyanobacteria, the addition of various sugars stimulated hydrogen production as the electron donation by the simple organic compounds enhanced the nitrogenase activity (Tiwari and Pandey 2012). Besides carbon and nitrogen, sulphur and phosphorous deprivation has significant effect on photobiological hydrogen production. It has been reported by various studies that sulphur deprivation leads to enhanced biohydrogen yields (McKinlay and Harwood 2010). This result is because during sulphur deficiency, the PSII reaction center is degraded and oxygen evolution is inhibited. In contrast, PSI is unaffected by the sulphur deprivation. Thus, the absence of oxygen evolution favors hydrogen production in sulphur-deprived cells. Similarly, phosphorus deficiency also leads to PSII inactivation, which leads to the establishment of anaerobic environment favorable for hydrogen production (Gonzalez-Ballester et al. 2015). However, in case of phosphorus deprivation, the inactivation of PSII is much slower compared to sulphur-deprived cells. Although these nutrient deficiencies enhance hydrogen production in certain species, they can have a detrimental effect on the metabolic patterns of cell cultures leading to loss of cell viability. Thus, careful examination must be performed to ensure a positive effect on biomass growth and hydrogen productivity.

### 7.5.6 Other Factors

Besides the parameters discussed previously, several other factors such as dissolved oxygen, the mode of operation (batch, continuous, fed batch), presence of metal ions as cofactors, and $CO_2$ concentration influence photobiological hydrogen production. The effect of these parameters varies depending upon the metabolic potential of the organism used. It was reported that addition of trace elements such as copper, zinc, and nickel show enhanced hydrogen production by cyanobacteria because they act as co-factors for nitrogenase (Tiwari and Pandey 2012). However, Kumar and Polasa (1991) showed the inhibitory effect of nickel and copper during hydrogen production by *Oscillatoria subbrevis*. Similarly, low $CO_2$ concentration in the headspace (4%–8% v/v) enhanced hydrogen productivity of *Anabaena variabilis* ATCC 29413 (Salleh et al. 2014). However, high $CO_2$ concentration (15%) was observed to be beneficial for hydrogen production by marine algae *Chlamydomonas* sp. MGA 161 (Miura et al. 1993). These results signify the need for optimal operating conditions with respect to the microbial species used for hydrogen production.

### 7.6　CONCLUSION AND FUTURE PROSPECTS

Biohydrogen production using photobiological systems is an effective means to obtain renewable and clean energy generation. Through these processes, solar energy can be directly converted to molecular hydrogen without the expense of high-cost carbohydrate-rich substrates. However, the major disadvantage is the low photosynthetic efficiencies that results in low hydrogen yields compared to the dark fermentation process. To make the process commercially viable, various technical challenges must be overcome. Different innovative photobioreactor designs are needed that allow uniform light dispersion inside the reactor. Moreover, efficient mixing strategies must be used to facilitate homogeneous distribution of nutrients and to avoid temperature and light gradients. In addition to the bioreactor designs, the major influencing parameters must be optimized with respect to the type of organism used. New strains with higher photosynthetic capacities must be developed that can tolerate harsh environmental conditions. To use the maximum amount of the solar spectrum, co-culture or mixed cultures strategy can be used. Furthermore, the metabolic pathway engineering for enhanced solar conversion efficiencies and higher hydrogen yields can be explored.

## REFERENCES

Akroumṅ-Amrouche D, Abdi N, Lounici H, Mameri N (2011) Effect of physico-chemical parameters on biohydrogen production and growth characteristics by batch culture of *Rhodobacter sphaeroides* CIP 60.6. *Appl Energy* 88:2130–2135.

Alalayah WM, Alhamed YA, Al-zahrani A, Edris G (2014) Experimental investigation parameters of hydrogen production by Algae *Chlorella Vulgaris*. In: *International Conference on Chemical, Environment & Biological Sciences*. pp. 41–43.

Asada Y, Koike Y, Schnackenberg J, Miyake M, Uemura I, Miyake J (2000) Heterologous expression of clostridial hydrogenase in the cyanobacterium *Synechococcus* PCC7942. *Biochim Biophys Acta—Gene Struct Expr* 1490:269–278.

Azwar MY, Hussain MA, Abdul-wahab AK (2014) Development of biohydrogen production by photobiological, fermentation and electrochemical processes: A review. *Renew Sustain Energy Rev* 31:158–173.

Bandyopadhyay A, Stöckel J, Min H, Sherman LA, Pakrasi HB (2010) High rates of photobiological $H_2$ production by a cyanobacterium under aerobic conditions. *Nat Commun* 1:139.

Basak N, Das D (2007) The prospect of purple non-sulfur (PNS) photosynthetic bacteria for hydrogen production: The present state of the art. *World J Microbiol Biotechnol* 23:31–42.

Berberoglu H, Pilon L, Melis A (2008) Radiation characteristics of *Chlamydomonas reinhardtii* CC125 and its truncated chlorophyll antenna transformants tla1, tlaX and tla1-CW+. *Int J Hydrog Energy* 33:6467–6483.

Bingham AS, Smith PR, Swartz JR (2012) Evolution of an [FeFe] hydrogenase with decreased oxygen sensitivity. In: *International Journal of Hydrogen Energy*. Pergamon, New York, pp. 2965–2976.

Chen CY, Chang JS (2006) Enhancing phototropic hydrogen production by solid-carrier assisted fermentation and internal optical-fiber illumination. *Process Biochem* 41:2041–2049.

Dasgupta CN, Jose Gilbert J, Lindblad P, Heidorn T, Borgvang SA, Skjanes K, Das D (2010) Recent trends on the development of photobiological processes and photobioreactors for the improvement of hydrogen production. *Int J Hydrog Energy* 35:10218–10238.

Dickson DJ, Page CJ, Ely RL (2009) Photobiological hydrogen production from *Synechocystis* sp. PCC 6803 encapsulated in silica sol-gel. *Int J Hydrog Energy* 34:204–215.

Doebbe A, Rupprecht J, Beckmann J, Mussgnug JH, Hallmann A, Hankamer B, Kruse O (2007) Functional integration of the HUP1 hexose symporter gene into the genome of *C. reinhardtii*: Impacts on biological $H_2$ production. *J Biotechnol* 131:27–33.

Franchi E, Tosi C, Scolla G, Della Penna G, Rodriguez F, Pedroni PM (2004) Metabolically engineered *Rhodobacter sphaeroides* RV strains for improved biohydrogen photoproduction combined with disposal of food wastes. *Mar Biotechnol* 6:552–565.

Gonzalez-Ballester D, Jurado-Oller JL, Fernandez E (2015) Relevance of nutrient media composition for hydrogen production in *Chlamydomonas*. *Photosynth Res* 125:395–406.

Kim EJ, Lee MK, Kim MS, Lee JK (2008) Molecular hydrogen production by nitrogenase of *Rhodobacter sphaeroides* and by Fe-only hydrogenase of *Rhodospirillum rubrum*. *Int J Hydrog Energy* 33:1516–1521.

Kumar S, Polasa H (1991) Influence of nickel and copper on photobiological hydrogen production and uptake in Oscillatoria subbrevis strain 111. *Proc Indian Nat Sci Acad* 576:281–285.

Kosourov S, Murukesan G, Seibert M, Allahverdiyeva Y (2017) Evaluation of light energy to $H_2$ energy conversion efficiency in thin films of cyanobacteria and green alga under photoautotrophic conditions. *Algal Res* 28:253–263.

Kruse O, Rupprecht J, Bader KP, Thomas-Hall S, Schenk PM, Finazzi G, Hankamer B (2005) Improved photobiological $H_2$ production in engineered green algal cells. *J Biol Chem* 280:34170–34177.

Kunjapur AM, Eldridge RB (2010) Photobioreactor design for commercial biofuel production from Microalgae. *Ind Eng Chem Res* 49:3516–3526.

Levin D (2004) Biohydrogen production: Prospects and limitations to practical application. *Int J Hydrog Energy* 29:173–185.

Liu BF, Ren NQ, Ding J, Xie GJ, Cao GL (2009) Enhanced photo-$H_2$ production of *R. faecalis* RLD-53 by separation of $CO_2$ from reaction system. *Bioresour Technol* 100:1501–1504.

Manish S, Banerjee R (2008) Comparison of biohydrogen production processes. *Int J Hydrog Energy* 33:279–286.

Masukawa H, Mochimaru M, Sakurai H (2002) Disruption of the uptake hydrogenase gene, but not of the bidirectional hydrogenase gene, leads to enhanced photobiological hydrogen production by the nitrogen-fixing cyanobacterium *Anabaena* sp. PCC 7120. *Appl Microbiol Biotechnol* 58:618–624.

Mathews J, Wang G (2009) Metabolic pathway engineering for enhanced biohydrogen production. *Int J Hydrog Energy* 34:7404–7416.

McKinlay JB, Harwood CS (2010) Photobiological production of hydrogen gas as a biofuel. *Curr Opin Biotechnol* 21:244–251.

Melis A, Zhang L, Forestier M, Ghirardi ML, Seibert M (2000) Sustained photobiological hydrogen gas production upon reversible inactivation of oxygen evolution in the green alga *Chlamydomonas reinhardtii*. *Plant Physiol* 122:127–136.

Miura Y, Yamada W, Hirata K, Miyamoto K, Kiyohara M (1993) Stimulation of hydrogen production in algal cells grown under high $CO_2$ concentration and low temperature. *Appl Biochem Biotechnol* 39–40:753–761.

Miyake M, Asada Y (1997) Direct electroporation of clostridial hydrogenase into cyanobacterial cells. *Biotechnol Tech* 11:787–790.

Miyamoto K, Hallenbeck PC, Benemann JR (1979) Hydrogen production by the thermophilic alga mastigocladus laminosus: Effects of nitrogen, temperature, and inhibition of photosynthesis. *Appl Environ Microbiol* 38(3), 440–446.

Özgür E, Uyar B, Öztürk Y, Yücel M, Gündüz U, Eroğlu I (2010) Biohydrogen production by *Rhodobacter capsulatus* on acetate at fluctuating temperatures. *Resour Conserv Recycl* 54:310–314.

Öztürk Y, Yücel M, Daldal F, Mandaci S, Gündüz U, Türker L, Eroğlu I (2006) Hydrogen production by using *Rhodobacter capsulatus* mutants with genetically modified electron transfer chains. *Int J Hydrog Energy* 31:1545–1552.

Pandu K, Joseph S (2012) Comparisons and limitations of biohydrogen production processes: A review. *Int J Adv Eng Technol* 2:2231–1963.

Salleh SF, Kamaruddin A, Uzir MH, Mohamed AR (2014) Effects of cell density, carbon dioxide and molybdenum concentration on biohydrogen production by *Anabaena variabilis* ATCC 29413. *Energy Convers Manag* 87:599–605.

Shah V, Garg N, Madamwar D (2001) Ultrastructure of the fresh water cyanobacterium *Anabaena variabilis* SPU 003 and its application for oxygen-free hydrogen production. *FEMS Microbiol Lett* 194:71–75.

Skizim NJ, Ananyev GM, Krishnan A, Dismukes GC (2012) Metabolic pathways for photobiological hydrogen production by nitrogenase- and hydrogenase-containing unicellular cyanobacteria *Cyanothece*. *J Biol Chem* 287:2777–2786.

Skjånes K, Andersen U, Heidorn T, Borgvang SA (2016) Design and construction of a photobioreactor for hydrogen production, including status in the field. *J Appl Phycol* 28:2205–2223.

Sobczuk TM, Camacho FG, Grima EM, Chisti Y (2006) Effects of agitation on the microalgae *Phaeodactylum tricornutum* and *Porphyridium cruentum*. *Bioprocess Biosyst Eng* 28:243–250.

Swanson KD, Ratzloff MW, Mulder DW, Artz JH, Ghose S, Hoffman A, White S et al., (2015) [FeFe]-hydrogenase oxygen inactivation is initiated at the H cluster $^{2}$Fe subcluster. *J Am Chem Soc* 137:1809–1816.

Taikhao S, Phunpruch S (2017) Increasing hydrogen production efficiency of $N_2$-Fixing cyanobacterium *Anabaena siamensis* TISTR 8012 by Cell Immobilization. In: *Energy Procedia*. Elsevier, Amsterdam, the Netherlands, pp. 366–371.

Tamburic B, Zemichael FW, Crudge P, Maitland GC, Hellgardt K (2011) Design of a novel flat-plate photobioreactor system for green algal hydrogen production. *Int J Hydrog Energy* 36:6578–65911.

Tiwari A, Pandey A (2012) Cyanobacterial hydrogen production—A step towards clean environment. *Int J Hydrog Energy* 37:139–150.

Torzillo G, Scoma A, Faraloni C, Ena A, Johanningmeier U (2009) Increased hydrogen photoproduction by means of a sulfur-deprived *Chlamydomonas reinhardtii* D1 protein mutant. *Int J Hydrog Energy* 34:4529–4536.

Veaudor T, Ortega-Ramos M, Jittawuttipoka T, Bottin H, Cassier-Chauvat C, Chauvat F (2018) Overproduction of the cyanobacterial hydrogenase and selection of a mutant thriving on urea, as a possible step towards the future production of hydrogen coupled with water treatment. *PLoS One* 13:e0198836.

Wu S, Xu L, Huang R, Wang Q (2011) Improved biohydrogen production with an expression of codon-optimized hemH and lba genes in the chloroplast of *Chlamydomonas reinhardtii*. *Bioresour Technol* 102:2610–2616.

Xie GJ, Liu BF, Wen HQ, Li Q, Yang CY, Han WL, Nan J, Ren NQ (2013) Bioflocculation of photo-fermentative bacteria induced by calcium ion for enhancing hydrogen production. *Int J Hydrog Energy* 38:7780–7788.

Zhu H, Fang HHP, Zhang T, Beaudette LA (2007) Effect of ferrous ion on photo heterotrophic hydrogen production by *Rhodobacter sphaeroides*. *Int J Hydrog Energy* 32:4112–4118.

# 8 Biomethanation

## 8.1 INTRODUCTION

Anaerobic digestion, or biomethanation, is a process by which organic wastes are converted to methane and $CO_2$. Besides the gaseous energy recovery, this process produces digested organic residues that can be used as an organic fertilizer due to the presence of ammonia and humus. This residue also increases the water retention properties of the soil. Besides methane and $CO_2$, the biogas also is comprised of $N_2$, $H_2$, and $H_2S$ in minor quantities. Various industrial and domestic organic wastes can be treated using the biomethanation process. The anaerobic digestion process is favored over other waste treatment processes because it produces renewable energy in the form of methane and thus can improve the overall economy of the process (Ward et al. 2008; Sárvári Horváth et al. 2016). The organic materials remain either in solid or in liquid forms and may be used as substrate for the biomethanation process. The chemical composition of each of these materials varies. In the last several decades, emphasis has been given to the study of different raw materials individually and in their combinations.

## 8.2 MICROBIOLOGY

Various studies have reported the microbiological aspects of the biomethanation process (Smith and Hungate 1958; Pohland and Ghosh 1971; Zeikus 1977; Balch et al. 1979; Hobson et al. 1981; Hungate and Stack 1982; Wuhrmann 1982; Jablonski et al. 2015). Barker (1956) was the first to be able to obtain highly enriched cultures of several methane-producing strains that were not axenic in nature (Wuhrmann 1982). However, in the later studies, the isolation of a pure strain of methane producers was reported by several researchers (Smith and Hungate 1958; Jablonski et al. 2015). It was suggested that methanogens usually thrive on low-carbon substrates such as acetic acid and formic acid for methane production, while hydrogen is the electron donor. An in-depth analysis of biomethanation process by Pohland and Ghosh (1971) revealed that anaerobic digestion was a four-step process which comprises hydrolysis, acidogenesis (acid forming), acetogenesis (acetate forming), and methanogenesis (methane forming) stages. There are two key players in anaerobic digestion process: acidogens and methanogens. Certain acidogens can convert the organics to volatile fatty acids by simultaneous hydrolysis and fermentation (Hobson et al. 1981). Examples of such bacteria include *Ruminococcus albus*, *R. flavefaciens*, *Butvrivibrio fibrisolvens*, *Bacteriodes succinogens*, *Cellobacterium cellulosolvens*, *Eubacterium ruminantium*, *Bacteriodes*, *Amylophilus*, *B. ruminicola*, and *Anaerovibrio lipolytica* (Bryant 1963; Zeikus 1977; Balch et al. 1979; Hungate and Stack 1982). Usually, the mesophilic sewage sludge comprises a population of $10^8$–$10^9$ mL$^{-1}$ hydrolytic bacteria (Zeikus 1977; Hobson et al. 1981). Studies on hydrolytic bacteria that utilize specific carbon substrates demonstrated $10^7$ proteolytic and $10^5$ celluloytic bacteria per milliliter of sewage sludge (Smith and Hungate 1958; Sárvári Horváth et al. 2016).

Methanogens are obligate anaerobes that convert the volatile fatty acids (produced during the acidogenic phase) into methane and $CO_2$. Various species of methane-producing bacteria include *Methanobacterium omelianskii, M. bryanti, M. formicicum, M. thermoautotrophicum, Methanobrevibacter arboriphilus, M. ruminantium, M, smithi, M. vannielli, Methanomicrobium mobile, Methanoqenium cariaci, Methanospirillum hungatei, Methanosarcina barkeri, Metnanococcus mazei, Methanobacterium sohuqenii, M. suboxydans, M. propionicum,* and *Methanococcus thermolithotrophicus* (Bryant et al. 1967; Siebert and Hattingh 1967; Zeikus and Wolee 1972; Das and Roy 2017).

The characteristics of the methanogenic bacteria are given in Table 8.1. *Methanosarcina barkeri* and *Methanococcus mazei* are of special interest due to their versatility with respect to various substrates that can be used (Sahm 1984).

## TABLE 8.1
## Characteristics of Methanogenic Bacteria

| Species | Morphology | Substrate | Cell Wall Component |
|---|---|---|---|
| *M. formicicum* | Rods and filaments | $H_2$, formate | Pseudomurein |
| thermoautotrophicum | | $H_2$ | |
| *Methanobrevibacter* | Lancet-shaped cocci short | $H_2$, formate | Pseudomurein |
| ruminantium | rods | $H_2$, formate | |
| smithii | | $H_2$ | |
| arboriphilus | | | |
| *Methanococcus* | Motile irregular small | $H_2$, formate | Polypeptide subunits |
| vanniellii | cocci pseudosarcina | $H_2$, formate | |
| voltae | | $H_2$, formate | |
| thermolithotrophics | | $H_2$, methanol, | |
| mazei | | methylamines, | |
| | | acetate | |
| *M. mobile* | Motile short rods | $H_2$, formate | Polypeptide subunits |
| *M. cariaci* | Motile irregular small cocci | $H_2$, formate | Polypeptide subunits |
| *Methanospirillum hungatei* | Motile regular curved rods | $H_2$, formate | Polypeptide |
| *Methanosarcina barkeri* | Irregular cocci as single cells methanol packets, pseudoparenchyma | $H_2$, acetate | Heteropolysaccharide |
| *Methanothrix soehngenii* | Rods to long filaments | Acetate | No muramic acid |
| *Methanothermus fervidus* | Non-motile | $H_2$ | Pseudomurein |

*Source:* Sahm, H., *Advances in Biochemical Engineering/Biotechnology*, Vol. 29, Fiechter A (Ed.), Springer-Verlag, Berlin, Germany, New York, 83–115, 1984; Zinder, S.H. and Mah, R.A., *Appl. Environ. Microbiol.*, 38, 996–1008, 1979; Taylor, G.T., *Prog. Indust. Microbiol.*, 16, 231–329, 1982.

The methanogenic bacteria are strict anaerobes. They require lower redox potential (–330 mV which corresponds to a concentration of 1 molecule of oxygen in about $10^{56}$ L of water) for growth compared to most other anaerobic bacteria. Oxygen is a potent inhibitor of methanogenesis. These bacteria can be Gram positive or Gram negative and they have quite different cell shapes (Table 8.1). Balch et al. (1979) proposed texonomy for 13 species of methanogens. After that several new strains were isolated. The more interesting thermophilic strains are *Methanosarcina* species, *Methanobacterium soehngenii*, *Methanococcus mazei*, and *Methanothermus fervidus* (Zeikus 1977; Zinder and Mah 1979). Among these methanogens, *M. fervidus* is able to grow near the boiling point of water. Methanogenesis also occurs in nature at 0°C, but most pure strains of methanogens have their growth optimum around 40°C (mesophiles) and at 65C°–75°C (thermophiles) (Zeikus 1977).

## 8.3 BIOCHEMISTRY

The biochemical pathways involved in methane production have been carefully examined by using synthetic media, which mimics the constituents present in the sewage or other animal wastes (Hobson et al. 1981). It is observed that the type of carbohydrate present in the organic waste and its breakdown are the crucial rate-limiting steps of the anaerobic digestion process (Ghosh and Klass 1977). For complex wastes such as lignocellulose, the major sugar components include cellulose, hemicellulose, and lignin. The hydrolysis of cellulose is performed by the action of the cellulase enzyme released by cellulolytic bacteria. These bacteria differ from each other in the relative amounts of the cellulase enzymes (endoglucanase and exo-glucanase) and the ability to attack various forms of cellulose (Hobson et al. 1981; Chesson et al. 1982). Various wastes contain cellulose in varying forms that makes it inaccessible for degradation. For example, the cellulose in domestic sewage is in a free form that can be easily degraded while cellulose in agricultural residues is associated with the lignin complex that slows down the degradation rate (Ghosh et al. 1980). Lignin is the non-biodegradable component of the cell wall that prevents the cellolytic bacteria from adhering to the plant fibers (Cowling and Kirk 1976; Das 1985). This adherence is a prerequisite for optimum bacterial attack. However, a number of compounds of the same types such as lignin were found to be degraded to methane and carbon dioxide by the mixture of bacteria in a domestic sewage sludge digester (Healy et al. 1980; Chesson et al. 1982). Besides cellulose, hemicellulose also can be degraded by the action of hemicellulases, which are less complex compared to the cellulolytic enzymes (Williams and Withers 1981). Cellulose and hemicellulose are first digested to short chain carboxylic acids that are then rapidly converted to acetic acid and propionic acid (Hungate 1950; Wood 1961; Horvath 1974; Scharer and Moo-Young 1979). The preliminary step in the carbohydrate breakdown follows the glycolysis pathway ending by pyruvate formation. The fermentation step then liberates the hydrogen required for methane generation (Wood 1961; Taylor 1975).

In addition to carbohydrate, the breakdown of protein and lipid also leads to methane production. Protein digestion is initiated by the extracellular hydrolysis of proteins into peptides and amino acids through the action of proteases (Ward et al. 2008; Sárvári Horváth et al. 2016). Amino acids are then further degraded to ammonia and organic acids (Hobson et al. 1981). Lipid biodegradation occurs through the action of lipases that convert the lipid molecules to long and short chain fatty acids. These fatty acids are further degraded by the β-oxidation process (McCarty et al. 1962).

The biochemical characteristics of methanogenic bacteria are different from other bacteria in the following manners:

1. Limited substrate flexibility for methane production
2. Complex cellular composition
3. Synthesis of unconventional chemicals such as glycerol ethers
4. rRNA composition dissimilar to the normal rRNA of bacteria
5. Small genome size (Taylor 1975; Zinder and Mah 1979; Sahm 1984)

Barker and his coworkers (Barker 1956; Taylor and Wolfe 1974; Jetten et al. 1992) established the mechanism of methane formation. They showed that carbon dioxide, methanol, or acetate can be used as precursor of methane by methanogenic bacteria (Figure 8.1). It was further proved that the methyl group of acetate or methanol can be transferred as it is and can be reduced thereafter (Wolfe 1982). Of the two major substrate systems of methanogens—(1) the hydrogen-carbon dioxide (and formate) system and (2) the acetate system—the former has been studied extensively compared to latter (Wolfe 1982).

McBride and Wolfe (1971) fractionated the acidic molecule formed during the reduction of carbon dioxide to methane. He named this compound coenzyme M since it was involved in methyl transfer (Wolfe 1982). Later, Taylor and Wolfe (1974) identified coenzyme M as 2-mercaptoethane sulphonic acid. The coenzyme M was required as a growth factor by *Methanobrevibacter ruminanthium* and actively transported into the organism (Wolfe 1982). The methanogenesis by *M. ruminanthium* and *Methanosarcina barkeri* was specifically inhibited by the structural analogs of coenzyme M (e.g., 2-bromoethane sulfonic acid and chloroethane sulfonic acid).

Significant progress has been made in the study of methyl reductase after the discovery of methyl-coenzyme M (Wolfe 1982). It was observed that 1 mole of methane was produced from 1 mole of methyl-coenzyme M added, when hydrogen and ATP

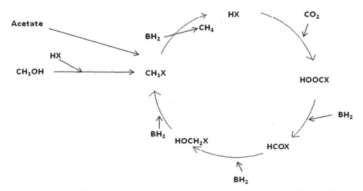

X = methyl-coenzyme M (CH₃SCH₂CH₂SO⁻₃) or 2-mercaptoethanesulfonate (HS–CoM)

**FIGURE 8.1**   Barker's scheme for methanogenesis.

were present in excess in the medium. However, when equal amounts of hydrogen and $CO_2$ (1: 1) were added to the same medium, the methane yield increased by 12 fold. This effect, known as the RPG effect, was consistent with a new substrate added to the media (Gunsalus and Wolfe 1977). It can be deduced that for each mole of methyl-coenzyme M, 11 moles of $CO_2$ are reduced to methane. Most of the recent studies are focused on determining the factor required for conversion of $CO_2$ to methane (Wolfe 1982).

In addition to coenzyme M, the coenzyme $F_{420}$ has been detected in all methanogens, which participates as an electron carrier in the NADP-linked hydrogenase and formate dehydrogenase system in the methanogens (Cheeseman et al. 1972). Two additional co-factors ($F_{430}$ and $F_{342}$) were found in which co-factor F430 from *Methanosarcina barkeri* contained nickel (Thauer 1982). The stabilization of different acids is reported by several researchers (Wood 1961; Lawrence 1971).

Acetic acid

$$CH_3COOH + H_2O \rightarrow CH_4 + H_2CO_3$$

Propionic acid

First step: $CH_3CH_2COOH + \frac{1}{2}H_2O \rightarrow CH_3COOH + \frac{3}{4}CH_4 + \frac{1}{4}CO_2$

Second step: $CH_3COOH + H_2O \rightarrow CH_4 + H_2CO_3$

Overall: $CH_3CH_2COOH + 3/2 H_2O \rightarrow 7/4 CH_4 + \frac{1}{4}CO_2 + H_2CO_3$

Butyric acid

First step: $CH_3CH_2CH_2COOH + H_2CO_3 \rightarrow 2CH_3COOH + \frac{1}{2}CH_4 + \frac{1}{2}CO_2$

Second step: $CH_3COOH + 2 H_2O \rightarrow 2CH_4 + 2H_2CO_3$

Overall: $CH_3CH_2CH_2COOH + 2H_2O \rightarrow 5/2CH_4 + \frac{1}{2}CO_2 + H_2CO_3$

## 8.4 THERMODYNAMICS AND KINETICS

### 8.4.1 THERMODYNAMICS

In the biomethanation process, 1 mole of glucose produces 3 moles of acetic acid (Ljungdahl and Andreesen 1976). The change of free energy is −74.3 kCals according to the equation

$$C_6H_{12}O_6 + 4 H_2O \rightarrow 2 CH_3COO^- + HCO_3^- + 4 H_2 + 4 H^+, \Delta F^\circ = -49.2 \text{ kCals}$$

$$2HCO_3^- + 4H_2 + H^+ \rightarrow CH_3COO^- + 4H_2, \Delta F^\circ = -25.1 \text{ kCals}$$

$$C_6H_{12}O_6 \rightarrow 3 CH_3COO^- \ \Delta F^\circ = -74.3 \text{ kCals}$$

In the preceding process, the formation of acetic acid from glucose is favorable because the standard free energy change is found to be negative (Wuhrmann 1982).

The following reactions indicate that methane formation is comparatively favorable compared to acetic acid formation from carbon dioxide and hydrogen.

$$2 \, CO_2 + 4 \, H_2 \rightarrow CH_3COOH + 2 \, H_2O, \, \Delta F^\circ = -18.7 \text{ kCals}$$

$$CO_2 + 4 \, H_2 \rightarrow CH_4 + 2 \, H_2O, \, \Delta F^\circ = -33.2 \text{ KCals}$$

The degradation of volatile fatty acids to methane, as occurred in this process, has a different change of standard free energy.

$$CH_3COO^- + H_2O \rightarrow CH_4 + HCO_3^-, \, \Delta F^\circ = -7.4 \text{ KCals}$$

$$CH_3CH_2COO^- + 2H_2O \rightarrow CH_3COO^- + 3H_2 + CO_2, \, \Delta F^\circ = +19.5 \text{ KCals}$$

$$CH_3CH_2CH_2COO^- + 2H_2O \rightarrow 2CH_3COO^- + 2H_2 + H^+ \, \Delta F^\circ = +9.95 \text{ KCals}$$

Degradation of butyric and propionic acids are not favored in a thermodynamic sense. Free energy change involved in the reduction of carbon dioxide by NAD(P)H is shown in the following equations.

$$4 \, NAD(P)H + 4H^+ \rightarrow 4 \, NAD(P)^+ + 4H_2, \, \Delta F^\circ = +18.4 \text{ KCals}$$

$$CO_2 + 4 \, H_2 \rightarrow CH_4 + 2 \, H_2O, \, \Delta F^\circ = -33.2 \text{ KCals}$$

SUM: $4H^+ + 4NAD(P)H + CO_2 \rightarrow 4NAD(P)^+ + CH_4 + 2H_2O, \, \Delta F^\circ = -14.8 \text{ KCals}$

The change of standard free energy is very much negative in the case of the reduction of carbon dioxide to methane compared to other methanation reactions mentioned previously.

## 8.4.2  KINETICS

The degradation of solid biomass containing cellulose, hemicellulose, proteins, and lipids in anaerobic digestion involves the sequential step of hydrolysis, acidification, and methanation. Only the last two steps are involved in the digestion of the simple compounds such as monosaccharides or amino acids. Study of the kinetics of these three digestion steps requires detailed information. A few kinetic information, however, are available for these steps. This information is applicable mainly for hydrolysis and acidification.

Pretorius (1969) reviewed the published kinetic data applicable to the overall anaerobic digestion process. Lawrence and McCarty (1969) developed the rate constants for biomethanation of various volatile fatty acids (e.g., acetic, propionic, and butyric acid). It was suggested that methanation was the rate-limiting step. This suggestion was still questionable because faster rates of reactions were assumed for hydrolysis and acidification compared to the methanation reaction. However, several researchers reported some indirect evidence that the hydrolysis could be the rate-limiting step in the overall digestion of cellulosic feeds (Ghosh and Pohland 1977; Kotze et al. 1968). In the case of biomethanation of simple carbohydrates (e.g., glucose), acetic acid degradation to methane was the rate-limiting step (Ghosh and Pohland 1977; Van den Berg et al. 1981).

Biomethanation is a two-step process, the first being generally known as the acidogenic phase, which leads to the hydrolytic cleavage of higher polymeric compounds into intermediary volatile fatty acids. The second step, known as the methanogenic phase comprises of reduction of these acids into mainly methane and carbon dioxide. This process may be represented by the following reaction steps (Equation 8.1):

Organic residue (A) $\rightarrow$ volatile fatty acids (B) $\rightarrow$ (methane + carbon dioxide)    (8.1)

Das (1985) did the kinetic analysis based on the carbon balance using mixed organic residues comprised of water hyacinth, wastewater grown algae, cow dung, and rice husk and observed that the first step reaction is a first order and the second has a fractional order of reaction of 0.70. The kinetic model gives the concentration profile of individual components at any time, $t$ (Equation 8.2):

$$C_A = 13.3 e^{-0.lt}$$

$$\text{and } \frac{dC_B}{dt} = 1.064 e^{-0.lt} - 0.57 C_B^{0.7}$$

(8.2)

The concentration of B was estimated by solving the equation numerically. The results showed the maxima at the end of 4 days that also was found from experimental observation.

The factors influencing the anaerobic digestion process include (1) rate of product formation, (2) acids spectrum, and (3) rate of substrate use. Different operating conditions are important (e.g., retention time, loading rate, temperature, pH, and biokinetic parameters such as $\mu$, $\mu_{max}$, and $K_s$). Ghosh and Pohland (1977) studied the characteristics of acid formers using wastewater sludge and Lozana and others (1980) studied the same with a pure substrate such as sucrose. Andrews and Pearson (1965) also showed the kinetics and characteristics of volatile fatty acids formation in this process. Lozano and others (1980) showed the correlation between the specific rate of acid formation and the cell growth rate using sucrose as a substrate.

$$\frac{1}{X} r_p = 1.7 \mu + 1.6$$

(8.3)

where $X$ is the cell mass per unit volume, g $L^{-1}$, $\mu$ is the specific growth rate of microorganism, $h^{-1}$, $r_p$ is the rate of acid formation, g $L^{-1}h^{-1}$

The maximum volatile fatty acid productivity reported was 1.7 g $L^{-1}h^{-1}$ as acetic acid (Lozano et al. 1980). Several researchers have reported the kinetic constants for methanogenic bacteria (Ghosh et al. 1980; Andrews and Graff 1971). At mesophilic temperature, Ghosh and others (1980) reported the maximum specific growth rate ($\mu_{max}$) of 0.49 $d^{-1}$ and saturation constant ($K_s$) of 4.2 g $L^{-1}$ in a batch system of mixed methanogenic population with acetic acid as the substrate. Chen and Hashimoto (1978) studied different kinetic parameters in the case of biomethanation of different wastes (e.g., sewage sludge, municipal refuse, and livestock residue). Ghosh and others (1980) considered a kinetic relationship for the anaerobic digestion of kelp and sewage sludge at steady state without recycling (Equation 8.4):

$$\theta = \frac{K}{\mu_{max}} \frac{1}{\left(1-VS_R\right)L\theta} + \frac{1}{\mu_{max}} \tag{8.4}$$

where $\theta$ = hydraulic detention time, d; $k$ = half velocity constant, g volatile solids $L^{-1}$ $d^{-1}$; $\mu_{max}$ = maximum specific growth rate constant, $d^{-1}$; VSR = volatile solid reduction, %; L = loading rate, g volatile solids $L^{-1}$ $d^{-1}$.

$\mu_{max}$ and $k$ values using kelp and sewage sludge were found to be equal to 0.09 $d^{-1}$, 0.38 $d^{-1}$ and 6.8 g $L^{-1}$, 63 g $L^{-1}$, respectively (Ghosh et al. 1980). Buswell and others (1939) studied the anaerobic digestion of different organic residues and presented a general formula for the conversion of complex material to carbon dioxide and methane neglecting the fraction of substrate converted to microorganisms (Equation 8.5):

$$C_aH_bC_c + \left(a - \frac{b}{4} - \frac{c}{2}\right)H_2O \rightarrow \left(\frac{a}{2} - \frac{b}{8} - \frac{c}{4}\right)CO_2 + \left(\frac{a}{2} + \frac{b}{8} - \frac{c}{4}\right)CH_4 \tag{8.5}$$

Klass and Ghosh (1977) reported an empirical formula for kelp $C_{2.61}H_{4.63}O_{2.23}$ $N_{0.1}S_{0.01}$, Ash. The theoretical methane yield, neglecting the effect of nitrogen and sulphur, can be written as in Equation 8.6:

$$C_{2.61}H_{4.63}O_{2.23} + 0.34\ H_2O \rightarrow 1.285\ CO_2 + 1.32\ CH_4 \tag{8.6}$$

From Equation 8.6, it is found that 1 kg of kelp is considered with a minerals content of 267 g. The maximum methane and carbon dioxide yields are about 297 L and 288 L, respectively.

## 8.5  ORGANIC WASTES AS SUBSTRATE

### 8.5.1  SOLID BIOMASS

Agro-residues are renewable feedstock for energy production. In the last few years, the research in the field of biogas production has increased considerably. Hobson (1982) tested various agricultural residues for methane production and concluded that the type of substrate affected the methane yield. Other parameters that contribute to significant methane production include operational temperature, hydraulic retention time, solid content, and ash content (Hobson 1982). Several researchers used various agro-residues as substrate for the biomethanation process at laboratory and pilot scales (Hobson 1982; Hayes 1980). These agro-residues consist of animal wastes and agricultural biomass. Hobson and others (1980) found that the digestibility of dairy cattle wastes was less compared to other wastes. Jewell (1976) investigated the energy flow on a dairy farm with 100 cows. The effect of a multireactor system in anaerobic digestion using cow dung as substrate showed higher volatile solids reduction, methane yield, and nitrogen-phosphorus-potassium (NPK) value of about 1.5 times compared to a batch system. The acid productivity also was much higher in the case of the multireactor system ($1.9 \text{ g L}^{-1}\text{d}^{-1}$ compared to the batch measurement of $0.6 \text{ g L}^{-1}\text{d}^{-1}$). The energy recovery as methane was double compared to batch system. Calzada and Rolz (1983) used various agricultural residues such as wheat straw and rice straw for biomethane production (Calzada and Rolz 1983). Various studies have suggested high-rate methane production using seaweeds such as water hyacinth, duck weed, and water pennywort as feedstocks (Wolverton and Donald 1981; Ghosh et al. 1980; Das 1985). Moreover, algae were found to be a suitable substrate for high-rate methane production (Samson and LeDuy 1982). Besides individual feedstocks, few studies have reported higher methane yields using mixed biomass (i.e., co-digestion of different feedstocks) (Ghosh et al. 1980; Das 1985).

## 8.5.2 INDUSTRIAL WASTES

The industrial wastes are mostly liquid in nature. Several researchers reported the characteristics of the various industrial wastes (Ghosh et al. 1982; Das et al. 1983). A two-phase biomethanation process was used for the treatment of industrial wastes. Several pilot and full-scale plants have been demonstrated using distillery, beet sugar, citric acid, and soft drinks wastes. The gas yield among these wastes was considerably varied (Ghosh et al. 1982). Maximum yield was observed for distillery effluent ($0.73 \text{ m}^3 \text{ Kg}^{-1}$ COD added). In India, there are a few reports available on the treatment of distillery wastes by this process (Das et al. 1983). Ghose and Das (1982) studied the effect of enriched methane culture and pH on COD reduction. Sen and others (1962) established the effect of various types of mixing and various nitrogen additives on this process.

## 8.5.3 PROCESS PARAMETERS

Several researchers reported the effect of temperature on the biomethanation process using different residues (Das 1985; Chen et al. 1980; Pfeffer and Liebman 1975). Chen and others (1980) observed that the kinetic parameters changed at different temperatures while the methane yield remained constant. Different optimum mesophilic temperatures were reported (Pfeffer and Liebman 1975). The gas yield and production rate at 60°C were higher compared to mesophilic temperature (Pfeffer 1974). However, Schellenback (1980) observed that the mesophilic culture had a higher methane yield per kilogram of volatile solid added and a higher destruction efficiency (Schellenbach 1980). The pH required for the growth of many bacterial species inherent in anaerobic digestion is not known. The pH of the fermentation broth changes significantly during the anaerobic digestion process (Zeikus 1980). It is reported that the volatile fatty acid production increases when the pH is controlled at 6.0 rather than in a pH uncontrolled system using sucrose as substrate (Lozano et al. 1980). The optimum pH for the growth of methanogenic bacteria is between 6.8 and 8.0 (Cohen et al. 1979). However, Cohen and others (1979) noted the optimum pH for acid and methane formers as 6.0 and 7.8, respectively. The effect of various types of mixing (e.g., mechanical agitation, digester content recirculation, and gas recirculation) have already been reported by several researchers (Stafford et al. 1980). For anaerobic digestion of solid residues, agitation reduced the change of floating scum formation, which affected the digester performance (Stafford et al. 1980). Hills (1979) reported the optimum C:N ratio of 25.1 using dairy cow manure as a substrate (Wate et al. 1983). The optimum C:N ratio was found to depend on the substrate. It was observed that the optimum C:N ratio was dependent on temperature (Ghosh et al. 1980; Das 1985). The effect of different heavy metal ions was studied by several researchers (Das 1985; Buswell 1939). The inhibition of methane production by the combined metal ions was found to be effective at a lower metal concentration compared to each metal separately (Wate et al. 1983). However, few metal ions have a stimulatory effect on the biomethanation process.

## 8.6  STRATEGIES FOR THE IMPROVEMENT OF BIOGAS PRODUCTION

Improvement of biogas production in anaerobic digestion processes can be achieved in several ways (Buswell 1939; Sanders et al. 2000; Jantsch et al. 2002; Schmidt et al. 2013; Moestedt 2015; Schwarz 2001; Sousa et al. 2007; Stroot et al. 2001; Shakeri Yekta et al. 2014; Sierra-Alvarez et al. 1994; Song et al. 2005; Dapelo and Bridgeman 2018):

- Optimizing the organic loading rates and hydraulic retention time
- Determining the rate-limitation steps to increase the biogas production
- Balancing nutrients and trace elements to increase the microbial activity
- Using mixed organic residues
- Improving the process operating parameters such as a two-stage process, recycling the digested materials, and mixing.
- Pre-treating the raw materials
- Maintaining proper temperature and pH

## 8.7  SUMMARY AND CONCLUSION

Biogas production or biomethanation by anaerobic digestion process is well established for the beneficial use of organic wastes that otherwise pose severe environmental pollution problems.

The main advantage of this process is the applicability at small scale and commercial scale. China and India have successfully implemented this process at the small scale in rural areas. Several industries use their industrial effluent for the biogas generation at the large scale successfully. A significant amount energy required in the distillation of alcohol can be used from the biogas generated in the anaerobic digestion process. In addition, it helps to meet the disposal standard set for the effluent and the digested materials are used successfully as biofertilizer, which not only improves the nutritional value of the soil but also the water retention capacity of the soil. The two-stage biomethanation process and use of mixed residues are suitable for the improvement of the efficiency of the process by increasing the microbial activities to a great extent.

## REFERENCES

Andrews JF, Graff SP (1971) Application of process kinetics to design of anaerobic processes. In *Anaerobic Biological Treatment Processes*, American Chemical Society Advances in Chemistry Series 105, pp. 126–162.

Andrews JF, Pearson EA (1965) Kinetics and characteristics of volatile acid production in anaerobic fermentation processes. *Int J Air Water Pollut* 9:439–461.

Balch WE, Fox GE, Magrum LJ, Woese CR, Wolfe RS (1979) Methanogens: Reevaluation of a unique biological group. *Microbial Rev* 43:260–296.

Barker HA (1956) *Bacteriol Fermentations*, John Wiley & Sons, New York.

Bryant MP (1963) Biosynthesis of branched-chain amino acids from branched-chain fatty acids by rumen bacteria. *Arch Biochem Biophys* 101:269–277.

Bryant MP, Wolin EA, Wolin MJ, Wolf RS (1967) Methanobacillus omelianskii, a symbiotic association of two species of bacteria. *Arch Microbiol* 59:20–31.

Buswell AM (1939) *Anaerobic Fermentation Ill.* State Water Survey Bulletin no. 32, Urbana, IL.

Calzada JF, Rolz C (1983) Anaerobic digestion in the integrated utilization of coffee wastes. *Proceedings of Third International Symposium on Anaerobic Digestion.* Boston, MA, pp. 315–324.

Cheeseman P, Toms-Wood A, Wolfe RS (1972) Isolation and properties of a fluorescent compound, factor 420, from *Methanobacterium* strain MOH. *J Bacteriol* 112:527–531.

Chen YR, Hashimoto AG (1978) Kinetics of methane fermentation. *Biotechnol Bioeng Symp* 8:269–282.

Chen YR, Varel VH, Hashimoto AG (1980) Effect of temperature on methane fermentation kinetics of beef-cattle manure. *Biotechnol Bioeng Symp* 10:325–339.

Chesson A, Stewart CS, Wallace RJ (1982) Influence of plant phenolic acids on growth and cellulolytic activity of rumen bacteria. *Appl Environ Microbiol* 44(3):597–603.

Cohen A, Zoetemeyer RJ, Deursen A van, Andel J G van (1979) Anaerobic digestion of glucose with separated acid production and methane formation. *Water Res* 13:571–580.

Cowling EB, Kirk TK (1976) Properties of cellulose and lignocellulosic materials as substrates for enzymatic conversion processes. *Biotechnol Bioeng Symp* 6:95–123.

Dapelo D, Bridgeman J (2018) Assessment of mixing quality in full-scale, biogas-mixed anaerobic digestion using CFD. *Bioresour Technol* 265:480–489.

Das D (1985) Optimization of Methane Production From Agricultural Residues. PhD dissertation, BERC, IIT Delhi.

Das D, Ghose TK, Gopalakrisnan KS, Joshi AP (1983) *Treatment of Distillery Wastes by a Two Phase Biomethanation Process. Symposium papers* "Energy from Biomass and Wastes VII" Florida, pp. 601–626.

Das D, Roy S (2017) *Biohythane: Fuel for the Future*, Pan Stanford Publishing, Singapore.

Ghose TK, Das D (1982) Maximization of energy recovery in biomethanation Processes: Part-II Use of mixed residue in batch system. *Process Bioch* 17:39–42.

Ghosh S, Henry MP, Klass DL (1980) Bioconversion of waterhyacinth-coastal bermuda grass-MSW-sludge blends to methane. *Biotechnol Bioeng Symp* 10:163–187.

Ghosh S, Klass D (1977) Two-phase anaerobic digestion. Symposium papers "Clean Fuels from Biomass and Wastes," IGT, Chicago, IL, pp. 373–415.

Ghosh S, Ombregt JP, DeProost VH, Pipyn P (1982). Symposium papers "Energy from Biomass and Wastes," Klass DL (Ed.), IGT, Chicago, IL.

Gunsalus RP, Wolfe RS (1977) Stimulation of $CO_2$ reduction to methane by methyl-coenzyme M in extracts of *Methanobacterium*. *Biochem Biophys Res Commun* 76:790–795.

Hayes TD, Jewell WJ, Kabrick RM (1980) Heavy metals removal from sludges using combined biological/chemical treatment. *Proceeding 34th Industrial Waste Conference.* Purdue University, West Lafayette, 529–543.

Healy JB, Young LY, Reinhardt M (1980) Methanogenic decomposition of ferulic acid, a model lignin derivative. *Appl Environ Microbiol* 39:436–444.

Hills DJ (1979) Effects of carbon: Nitrogen ratio on anaerobic digestion of dairy manure. *Agric Waste* 1:267–278.

Hobson PN (1982) Biogas production from agricultural wastes. *Experimentia* 38:206–209.

Hobson PN, Bousfield S, Summers R (1981) *Methane Production from Agricultural and Domestic Wastes*. ASP, London, UK.

Horvath RS (1974) Evolution of anaerobic-energy- yielding metabolic pathwaysof the pro-caryotes. *J Theor Biol* 48:361–371.

Hungate RE (1950) The anaerobic mesophilic cellulolytic bacteria. *Bacteriol Rev* 14:1–45.

Hungate RE, Stack RJ (1982) Phenylpropionic acid: A growth factor for *Ruminococcus albus*. *Appl Environ Microbiol* 44:79–83.

Jablonski S, Rodowicz P, Łukaszewicz M (2015) Methanogenic archaea database containing physiological and biochemical characteristics. *Int J Syst Evolution Microbiol* 65:1360–1368.

Jantsch TG, Angelidaki I, Schmidt JE, Braña de Hvidsten BE, Ahring BK (2002) Anaerobic biodegradation of spent sulphite liquor in a UASB reactor. *Bioresour Technol* 84:15–20.

Jetten MSM, Stams AJM, Zehnder AJB (1992) Methanogenesis from acetate: A comparison of the acetate metabolism in Methanothrix soehngenii and Methanosarcina spp. *FEMS Microbiol Letts* 88:181–197.

Jewell WJ et al. (1976) US ERDA Report no. TID- 27164.

Kotze PG, Thiel PG, Toerien DF, Hattingh WHJ, Siebert ML (1968) A biological and chemical study of several anaerobic digesters. *Water Res* 2:195–213.

Lawrence AW (1971) Application of process kinetics to design of anaerobic processes. In *Anaerobic Biological Treatment Processes*. American Chemical Society Advances in Chemistry Series 105, pp. 163–189.

Lawrence AW, McCarty PL (1969) Kinetics of methane fermentation in anaerobic treatment. *J Water Pollut Cont Fed* 41:R1–R17.

Ljungdahl LG, Andreesen JR (1976) *Microbial Production and Utilization of Gases*. Schlegel HG (Ed.), Gottingen, Germany.

Lozano IT, Cornok I, Goma G (1980) *Symposium on Bioconversion and Biochemical Engineering II*, Ghose TK (Ed.), IIT Delhi, New Delhi, India, 113–132.

McBride BC Wolfe RS (1971) A new coenzyme of methyl transfer, coenzyme M. *Biochemistry* 8:2317–2324.

McCarty PL et al. (1962) Microbiology of anaerobic digestion. Report No. R 62–69, Sedgwick Laboratories of Sanitary Sciences, MIT, Cambridge, MA.

Moestedt J (2015) Biogas Production From Thin Stillage: Exploring the Microbial Response to Sulphate and Ammonia. PhD Thesis, Swedish University of Agricultural Sciences, Uppsala, Sweden.

Pfeffer JT (1974) On-line monitoring of the methanogenic fermentation. *Biotechnol Bioeng* 16:771–787.

Pfeffer JT, Liebman JC (1975) Biological conversion of organic refuse to methane. *NSF*, PB-245 795.

Pohland FG, Ghosh S (1971) Developments in anaerobic stabilization of organic wastes—The two-phase concept. *Environ Lett* 1:255–266.

Pretorius WA (1969) Anaerobic digestion III: Kinetics of anaerobic fermentation. *Water Res* 3:545–558.

Sahm H (1984) *Advances in Biochemical Engineering/Biotechnology*, Vol. 29, Fiechter A (Ed.), Springer-Verlag, Berlin, Germany, 83–115.

Samson R, LeDuy A (1982) Biogas production from anaerobic digestion of *Spirulina maxima* algal biomass. *Biotechnol Bioeng* 24:1919–1924.

Sanders WTM, Geerink M, Zeeman G, Lettinga G (2000) Anaerobic hydrolysis kinetics of particulate substrates. *Water Sci Technol* 41:17–24.

Sárvári Horváth I, Tabatabaei M, Karimi K, Kumar R (2016) Recent updates on biogas production—A review. *Biofuel Res J* 10:394–402.

Scharer JM, Moo-Young M (1979) *Advances in Biochemical Engineering Vol. II.* Ghose TK, Fiechter A, Blakebrough N (Eds.) Springer-Verlang, New York, 85–101.

Schellenbach S (1980) Microbial conversion: Gasification. Symposium papers *Energy from Biomass and Wastes*, Klass DL, Weatherley JW (Eds.), IGT, Chicago, IL, 445–494.

Schmidt T, Pröter J, Scholwin F, Nelles M (2013) Anaerobic digestion of grain stillage at high organic loading rates in three different reactor systems. *Biomass Bioenerg* 55:285–290.

Schwarz WH (2001) The cellulosome and cellulose degradation by anaerobic bacteria. *Appl Microbiol Biotechnol* 56:634–649.

Sen EP, Bhaskaran TR (1962) Anaerobic digestion of liquid molasses distillery wastes. *J Water Pollut Cont Fed* 54:1170–1175.

Shakeri YS, Lindmark A, Skyllberg U, Danielsson A, Svensson BH (2014) Importance of reduced sulfur for the equilibrium chemistry and kinetics of Fe(II), Co(II) and Ni(II) supplemented to semi-continuous stirred tank biogas reactors fed with stillage. *J Hazard Mater* 269:83–88.

Siebert ML, Hattingh WHJ (1967) Estimation of methane producing bacterial number by the most probable number (MNP) technique. *Water Res* 1:13–19.

Sierra-Alvarez R, Field JA, Kortekaas S, Lettinga G (1994) Overview of the anaerobic toxicity caused by organic forest industry wastewater pollutants. *Water Sci Technol* 29:353–363.

Smith PH, Hungate RE (1958) Isolation and characterization of *Methanobacterium ruminantium* sp. *J Bacteriol* 75:713–718.

Song H, Clarke WP, Blackall LL (2005) Concurrent microscopic observations and activity measurements of cellulose hydrolyzing and methanogenic populations during the batch anaerobic digestion of crystalline cellulose. *Biotechnol Bioeng* 91:369–378.

Sousa DZ, Pereira MA, Smidt H, Stams AJM, Alves MM (2007) Molecular assessment of complex microbial communities degrading long chain fatty acids in methanogenic bio-reactors. *FEMS Microbiol Ecol* 60:252–265.

Stafford DA, Hawkes DL, Horton R (1980) *Methane Production from Waste Organic Matter*, CRC Press, Boca Raton, FL.

Stroot PG, McMahon KD, Mackie RI, Raskin L (2001) Anaerobic codigestion of municipal solid waste and biosolids under various mixing conditions—I: Digester performance. *Water Res* 35:1804–1816.

Taylor CD, Wolfe RS (1974) Structure and Methylation of Coenzyme M (HSCH, CH, SO3). *J Biol Chem* 249:4879–4885.

Taylor GT (1975) The biology of methanogenic bacteria. *Process Biochem* 10:29–31.

Taylor GT (1982), The methanogenic bacteria. *Prog Indust Microbiol* 16:231–329.

Thauer RK (1982) *Anaerobic Digestion 1981*, Hughes DE et al. (Eds.), Elsevier Biomedical Press, Amsterdam, the Netherlands, 37–44.

Van den Berg C, Lentz CP, Armstrong DW (1981) *Advance Biotechnology* Vol. II, Moo-Young M, Robinson CW (Eds.), Pergamon Press, Canada, 251.

Ward AJ, Hobbs PJ, Holliman PJ, Jones DL (2008) Optimisation of the anaerobic digestion of agricultural resources. *Bioresour Technol* 99(17):7928–7940.

Wate SR, Chakraborti T, Subrahmanyam PVR (1983) Effects of cobalt on biogas production from cattle dung. *Indian J Environ Health* 5:179–190.

Williams AG, Withers SE (1981) Hemicellulose degrading enzymes synthesized by rumen bacteria. *J Appl Bacteriol* 51:375–385.

Wolfe RS (1982) Biochemistry of methanogenesis. *Experimentia* 38:198–201.

Wolverton BC, McDonald RC (1981) Energy from vascular plant wastewater treatment Systems, *Economic Botany* 35:224–232.

Wood WA (1961) *The Bacteria Metabolism Vol. II*, Gunsalus IC, Stanier RY (Eds.), AP Press, New York, 59–149.

Wuhrmann K (1982) Ecology of methanogenic system in nature. *Experientia* 38:193–198.

Zeikus JG (1977) The biology of methanogenic bacteria. *Bacteriol Rev* 41:514–541.

Zeikus JG (1980) *Anaerobic Digestion*. Stafford DA, Wheatley BI, Hughes DE (Eds.), Applied Science Publishers, London, UK, 61–87.

Zeikus JG, Wolee RS (1972) Methanobacterium thermoautotrophicus sp. n., an anaerobic, autotrophic, extreme thermophile. *J Bacteriol* 109:707–713.

Zinder SH, Mah RA (1979) Isolation and characterization of a thermophilic Strain of *Methanosarcina* unable to use $H_2$–$CO_2$ for methanogenesis. *Appl Environ Microbiol* 38:996–1008.

# 9 Bioethanol

## 9.1 INTRODUCTION

Renewable biofuel production using biomass is considered sustainable and environmentally friendly. Bioethanol is among the first biofuels obtained from the sugar and starch feedstocks using microbial fermentation technology (Vohra et al. 2014). Industrial bioethanol production is usually carried out by the yeast *Saccharomyces cerevisiae*. Certain bacteria such as *Zymomonas mobilis* also have been reported to produce high yields of ethanol (Stewart et al. 1983). Bioethanol already has been exploited commercially in many countries such as the United States, Brazil, Canada, and Europe (Kim and Dale 2004). The characteristic properties of ethanol make it an attractive source for liquid transportation fuel (Table 9.1). Due to the high octane number and high flammability, it can be directly used in internal combustion engines with high compression ratios. However, due to the risk of a cold starting problem and the corrosive nature of ethanol, it is usually blended with gasoline. Besides its fuel characteristics, ethanol has several other advantages over conventional fossil fuels such as low toxicity, less greenhouse gas emission, carbon neutral, and production of less toxic by-products. The world statistics for ethanol production show that in 2016 a total of 117.7 million cubic meters of ethanol was produced of which 98.8 million cubic meters was used as fuel (www.statista.com). Despite this success, bioethanol production using food-based feedstocks raises serious concerns over its potential interference with the food market. This concern has necessitated the use of non-food-based feedstocks such as industrial, agricultural, or forestry wastes (second-generation feedstocks). However, bioethanol production from the second-generation feedstocks still suffers from various challenges such as handling and transport of the feedstock, inefficient biomass pre-treatment methods, and lower energy efficiency. In addition, the current estimated costs of bioethanol production are not that attractive compared to crude oil prices because most expenses incurred are the feedstock costs. In this chapter, the current status of ethanol production from various feedstocks and the conversion technologies involved are highlighted. In addition, the economics and the environmental impact of the bioethanol production technologies are discussed in detail.

## 9.2 BIOETHANOL PRODUCTION PROCESSES

Fermentative microorganisms (such as yeasts and bacteria) use $C_5$ and $C_6$ sugars for bioethanol production (Mohd Azhar et al. 2017). Ethanol is a by-product of fermentative metabolism, which is used to derive energy under anaerobic conditions (Figure 4.6). The process for ethanol production depends upon the type of feedstock used for bioethanol production. Figure 9.1 shows the various steps involved in

## TABLE 9.1
## Characteristic Properties of Bioethanol as Liquid Transportation Fuel

| Parameter | Property |
|---|---|
| Molecular formula | $C_2H_5OH$ |
| Molecular mass | 46.07 g mol$^{-1}$ |
| Density | 0.789 kg L$^{-1}$ |
| Boiling temperature | 78.5°C |
| Freezing point | −117°C |
| Flash point | 12.8°C |
| Vapour pressure | 50 mm Hg |
| Higher heating value | 29800 KJ kg$^{-1}$ |
| Lower heating value | 21090 KJ kg$^{-1}$ |
| Specific heat capacity (l) | 2.46 J/g °C |
| Viscosity | 1.2 mPa s |
| Refractive index | 1.36 |
| Octane number | 99 |

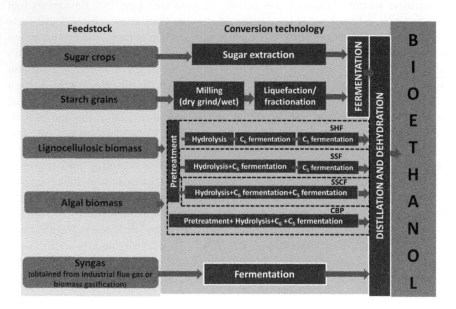

**FIGURE 9.1**    Bioconversion of various types of feedstocks for bioethanol production.

bioethanol production using various raw materials. In general, the overall ethanol production can be categorized into three major steps: (1) extraction of soluble sugars, (2) conversion of sugar to ethanol (i.e., fermentation), and (3) separation and purification of ethanol (Vohra et al. 2014). Depending upon the complexity of the raw material, additional steps can be combined to obtain the soluble sugars prior to the fermentation step.

## 9.2.1 Sugar Platform

Sugar crops such as sugarcane, sugar beet, and sweet sorghum are comprised mainly of glucose, fructose and sucrose as their major component (Devarapalli and Atiyeh 2015). By using the established techniques such as grinding and crushing, these sugars can be extracted for fermentative conversion to ethanol (Singh et al. 2017). After fermentation, the ethanol is concentrated and purified using distillation and dehydration processes for ethanol separation and recovery (Figure 9.1). Sugar crops provide the simplest means of sugar transformation to ethanol. These crops produce higher yields of sugar (and thus in turn higher yields of ethanol) with low conversion costs. However, the fermentation is interrupted at the ethanol concentration of 10%–15% (v/v) because the growth of fermentative organisms is inhibited beyond this concentration. In addition, high temperature, osmotic stress, acidity, and bacterial contamination are a few other problems that affect the growth of the yeast during ethanol fermentation (Vohra et al. 2014). The continuing debate over food vs. fuel has drastically affected the use of sucrose-based feedstocks for ethanol. Efforts are being made to improve the yields of the existing sugar feedstocks to meet the need of food and fuel by introducing new feedstock varieties that do not compete with the existing food crops. Furthermore, recently reusing the sugar-processed residues (such as bagasse and tops) as the biomass in a biorefinery approach for second-generation biofuel has gained interest (see Section 9.2.3).

## 9.2.2 Starch Platform

Different grains such as corn, wheat, and barley are comprised of starch as their storage polysaccharide. Starch is a polymer of glucose comprised of amylose and amylopectin. This polymer cannot be directly used by yeast and requires a pre-processing step prior to fermentation. Once the harvesting is done, milling is performed using either a dry-grind process or wet-milling process (Figure 9.1). The extraction method for glucose differs in both these processes. There are three major steps in the dry-grind process: (1) grinding, (2) cooking and liquefaction, and (3) saccharification (Singh et al. 1999). Grinding involves breakage of large particles into smaller particles that can be mixed with water to form a slurry. In the cooking and liquefaction step, partial hydrolysis of starch occurs through the enzyme $\alpha$-amylase, which acts on internal $\alpha$-(1,4)-glycosidic bonds to form glucose dimers such as dextrin. During the next saccharification step, glucoamylase is used, which acts on both $\alpha$-(1,4)- and $\alpha$-(1,6)-glycosidic bonds for further hydrolysis of dimers to glucose monomers. The obtained glucose is then subjected to yeast fermentation, concentration, separation, and recovery steps similar to those used for obtaining ethanol from sugar feedstocks. The by-product of the dry-grind process is DDGS (distiller's dried grains with solubles), which is of high commercial value (Kim and Dale 2004).

The wet-milling process is more energy intensive compared to the dry-grind process. It allows the extraction and use of each component of the plants for value-added product recovery. The first step of the wet-milling process is steeping in which grains are soaked in dilute aqueous sulphur dioxide solution. Steeping is followed by removal of germ, fiber, and starch. The separated starch is washed, dried, and subjected to saccharification and fermentation to yield ethanol. Compared to the dry-grind process, the wet-milling process results in slightly lower ethanol yields due to the formation of other value-added co-products such as artificial sweeteners, oil, and gluten meal. Most of the corn ethanol production plants in the United States use the dry-grind mills to maximize capital returns per gallon of ethanol (Vohra et al. 2014). However, similar to sugar-based feedstocks, the use of starch-based feedstocks impact the costs of food crops and land usage, thus negating its use for bioethanol production. These disadvantages have given rise to the need for second-generation ethanol production using non-food-based feedstocks.

### 9.2.3 CELLULOSE PLATFORM

The sustainability of the first-generation feedstocks is questionable because of food security and land usage issues. Therefore, major interest has been targeted towards the development of second-generation biofuels using non-food-based biomass sources. Due to the global abundance of lignocellulosic biomass, various research groups have focused on cellulose-based biofuel production (Kim and Dale 2004). Lignocellulosic biomass sources have several advantages such as they are abundantly available, virtually inexhaustible, have a low net carbon emission, and are energy efficient. Lignocellulosic wastes include agricultural and forest residues, industrial wastes, dedicated herbaceous, softwood and hardwood, and municipal solid wastes (Limayem and Ricke 2012).

Table 9.2 shows lower bioethanol yields using various lignocellulosic biomass sources. The major components of these biomass sources are cellulose (30%–50%),

---

**TABLE 9.2**
**Bioethanol Production Using Lignocellulosic Biomass**

| Feedstock | Process | Pre-treatment | Fermenting Microorganism | Ethanol Yield/Conc. | References |
|---|---|---|---|---|---|
| Corn cob | SHF | Enzymatic hydrolysis | *Saccharomyces cerevisiae* | 2.82 L kg$^{-1}$ | Syawala et al. (2013) |
| Sugarcane bagasse | | | | 2.99 L kg$^{-1}$ | |
| Duckweed | SSF | Steam explosion | *Saccharomyces cerevisiae* | n.a. | Zhao et al. (2015) |
| Wheat straw | SSF | Steam explosion | *Saccharomyces cerevisiae* | 51.5 g L$^{-1}$ | Chen et al. (2008) |
| Rice straw | SSF | Alkali + enzymatic hydrolysis | *Mucor circinelloides* J | 30.5 g L$^{-1}$ | Takano and Hoshino (2018) |
| Wheat straw | SSCF | Acetic acid + steam | Recombinant *Saccharomyces cerevisiae* | 0.36 g g$^{-1}$ | Bondesson and Galbe (2016) |
| Kraft paper mill sludge | SSCF | Enzymatic hydrolysis | Recombinant *Escherichia coli* (ATCC-55124) | 42 g L$^{-1}$ | Kang et al. (2010) |
| Switchgrass | CBP | – | Engineered *Caldicellulosiruptor bescii* | 12.8 mM | Chung et al. (2014) |

hemicellulose (20%–35%), lignin (20%–30%), and in some cases pectins (2%–20%) (Vohra et al. 2014). The composition of these constituents varies depending upon the type of biomass (Table 2.2). There are four main stages in lignocellulosic ethanol production: (1) pre-treatment, (2) hydrolysis, (3) fermentation, and (4) distillation (Figure 9.1). The pre-treatment step is needed to make the celluloses and hemicelluloses accessible to the hydrolyzing enzymes. Figure 2.3 shows the various methods used for pre-treatment of lignocellulosic biomass. The type of method used depends upon the biomass type and composition. Due to the rigid structure of lignin, often one or more pre-treatment methods are combined to extract the maximum amount of sugars possible from the lignocellulosic biomass. However, the need for the additional pre-treatment steps imposes extra labor and processing costs, thus reducing the economic viability of the process.

After pre-treatment, one of the following steps is performed depending upon the process conditions: (1) separate hydrolysis and fermentation (SHF), (2) simultaneous saccharification and fermentation (SSF), (3) simultaneous saccharification and co-fermentation (SSCF), and (4) consolidated bioprocessing (CBP) (Devarapalli and Atiyeh 2015). In SHF, enzymatic hydrolysis and fermentation are carried out in separate reactors, while in SSF both the hydrolysis and fermentation occur in same reactor. A major disadvantage of both these processes is the inability to use xylose. In SSCF, xylose and glucose are co-fermented in the same reactor using genetically modified strains that can ferment $C_6$ and $C_5$ sugars. However, in SSCF, the hydrolysis step is separated from the co-fermentation step. CBP, on the other hand, is an integrated technology in which both hydrolysis and co-fermentation are carried out using a single organism (Figure 9.1). CBP technology is still in its early developmental stages and requires further scrutiny for use in industrial plants. Depending upon the culture conditions, one of the previously listed processes is used for ethanol. For purification and recovery, similar processes such as distillation and dehydration, considered for sugar- and starch-based feedstocks, are used. Ethanol production from lignocellulosic feedstock suffers from various disadvantages that limit their industrial application. The harsh pre-treatment techniques, lack of high-yielding strains, ethanol toxicity, and co-production of inhibitory compounds are some of the major challenges that need utmost consideration. Although substantial advancements have been made in genetic and enzymatic technologies to improve the sugar-extracting capabilities of fermenting microorganisms, a technological breakthrough is needed for successful implementation in commercial-scale systems.

### 9.2.4 ALGAL BIOMASS PLATFORM

Bioethanol production using third-generation feedstock such as microalgae has gained considerable research interest in recent years. Algae are photosynthetic microorganisms that derive energy either phototrophically (i.e., using light as energy), heterotrophically (i.e., using organic carbon as source of energy), or mixotrophically (i.e., using booth light and organic carbon as energy source). They are composed mainly of carbohydrates, proteins, and lipids that can be used for biofuel production (Chew et al. 2017). The accumulation of algal polysaccharides (mostly sugar, starch, and cellulose) can be enhanced by using appropriate nutrient starvation conditions or other environmental stresses during cultivation (Markou et al. 2012). It was previously reported that by adjusting the carbon/nitrogen (C/N) ratio, the biomass and carbohydrate productivity of *Chlorella vulgaris* UTEX 1803 was significantly enhanced (Jerez-Mogollón et al. 2012). Similarly, Cheng and others (2017) improved the carbohydrate and starch accumulation in *Chlorella* sp. AE10 by optimizing the $CO_2$ concentration, light intensity, and initial nitrogen concentration. They showed the highest starch accumulation of 77.6% w/w (dry basis) when a two-stage cultivation process with cell dilution was used (Cheng et al. 2017). These accumulated starch and other carbohydrates can be readily used for ethanol production. Compared to lignocellulosic feedstock, algae have several advantages such as high growth rate, limited land usage, simple structure, year-round production, adaptability to changing environmental conditions, and $CO_2$ sequestration ability, which make it a potential biofuel feedstock (Chew et al. 2017). The bioethanol production strategy is similar to those followed for lignocellulosic feedstock (Figure 9.1). However, mild pre-treatment is sufficient for extraction of carbohydrates from algae (Özçimen and Inan 2015). Algae can be used for ethanol production either directly, or after the extraction of biodiesel and other value-added compounds (such as pharmaceuticals and nutraceuticals) in a biorefinery approach (Demirbas 2010). Table 9.3 lists the various studies performed on ethanol production using micro or macroalgae as substrates. However, the major disadvantage is the harvesting process which accounts for 20%–30% of the total production costs of microalgae production (Barros et al. 2015).

## TABLE 9.3
### Ethanol Production Using Micro and Macroalgae

| Algae | Pre-treatment | Fermenting Microbe | Ethanol Yield/Conc. | References |
|---|---|---|---|---|
| **Macroalgae** | | | | |
| *Sargassum* sp. | Acid + enzymatic hydrolysis | *Saccharomyces cerevisiae* | 2.79 g L$^{-1}$ | Borines et al. (2013) |
| *Laminaria digitata* | Dilute acid + enzymatic hydrolysis | *Kluyveromyces marxianus* | 6 g L$^{-1}$ | Obata et al. (2016) |
| *Saccharina latissima* | — | *Saccharomyces cerevisiae* | 0.45% (v/v) | Adams et al. (2009) |
| *Kappaphycus alvarezii* | Acid hydrolysis | *Saccharomyces cerevisiae* | 0.21 g g$^{-1}$ | Meinita et al. (2012) |
| **Microalgae** | | | | |
| *Chlorococcum humicola* | Acid hydrolysis | *Saccharomyces cerevisiae* | 7.2 g L$^{-1}$ | Harun and Danquah (2011) |
| *Chlorella vulgaris* FSP-E | Acid hydrolysis | *Zymomonas mobilis* ATCC 29191 | 11.7 g L$^{-1}$ | Ho et al. (2013) |
| *Chlamydomonas reinhardtii* | Dilute acid hydrothermal | *Saccharomyces cerevisiae* S288C | 14.6 g L$^{-1}$ | Nguyen et al. (2009) |
| *Scenedesmus obliquus* CNW-N | Acid hydrolysis | *Zymomonas mobilis* ATCC 29191 | 0.202 g g$^{-1}$ | Miranda et al. (2012) |

## 9.2.5 Syngas Platform

A recent approach for obtaining ethanol from biomass is the syngas platform (Figure 9.1). In syngas fermentation, non-food-based feedstocks such as second and third-generation feedstocks are first converted to syngas using gasification technology and this gas is further fermented to ethanol using acetogenic bacteria (Devarapalli and Atiyeh 2015). Syngas is mainly a mixture of CO, $CO_2$, and $H_2$. However, depending upon the type of feedstock from which it is produced, it can contain trace amounts of $CH_4$, $NH_3$, $H_2S$, NO, and some other hydrocarbons (Munasinghe and Khanal 2010). In addition to syngas, the industrial flue gases which are comprised of similar gas composition as syngas also can be converted to ethanol by acetogenic bacteria (Devarapalli and Atiyeh 2015). The syngas-fermenting organisms mainly include *Clostridium autoethanogenum*, *Clostridium ljungdahlii*, *Acetobacterium woodii*, *Butyribacterium methylotrophicum*, and *Clostridium carboxidivorans* (Henstra et al. 2007). These organisms metabolize syngas using the Wood-Ljungdahl pathway and convert it to organic acids and alcohols (Daniell et al. 2012). The overall biochemical reactions to convert syngas to ethanol and acetic acid can be represented by following equations:

$$6CO + 3H_2O \rightarrow C_2H_5OH + 4CO_2 \qquad (9.1)$$

$$2CO_2 + 6H_2 \rightarrow C_2H_5OH + 3H_2O \qquad (9.2)$$

$$3CO + 3H_2 \rightarrow C_2H_5OH + CO_2 \qquad (9.3)$$

$$4CO + 2H_2O \rightarrow CH_3COOH + 2CO_2 \qquad (9.4)$$

$$2CO_2 + 4H_2 \rightarrow CH_3COOH + 2H_2O \qquad (9.5)$$

$$2CO + 2H_2 \rightarrow CH_3COOH \qquad (9.6)$$

The acetate obtained in Equations 9.4 through 9.6 can be further metabolized to ethanol using the acetone–butanol–ethanol (ABE) pathway (Figure 4.6). Compared to conventional hydrolysis and saccharification processes, syngas fermentation has several advantages such as utilization of complete biomass, high conversion efficiencies, no need of an additional pre-treatment step, and limited toxic products formation. However, limitations in gas-liquid mass transfer, syngas impurities, and microorganism sensitivity to surrounding environment are few challenges that need to be overcome. The syngas platform for bioethanol production is relatively new and needs further exploration.

## 9.3 STRATEGIES USED FOR PURIFICATION AND RECOVERY OF BIOETHANOL

The final concentration of ethanol after fermentation is usually around 3%–12% w/v. The fermentation broth usually is comprised of various by-products along with ethanol and, therefore, its separation is an energy-intensive process. Distillation is the common method to concentrate and purify ethanol up to 95% (Jesús Mendoza-Pedroza and Gabriel Segovia-Hernandez 2018). Besides distillation, various alternate strategies have been used to reduce the energy requirement and to obtain ethanol with desired concentrations as shown in Figure 9.2.

### 9.3.1 DISTILLATION

Distillation is based upon the principle of the difference in boiling points of components in a mixture. By heating the liquid mixture, the low boiling point components are concentrated in the vapor phase, which can be condensed to obtain the pure form of the desired component. Since ethanol is more volatile than water (boiling points: ethanol 78°C, water 100°C), distillation provides effective separation of ethanol from the fermentation broths. After distillation, an azeotropic mixture of ethanol (95%) and water (5%) is obtained, which is known as hydrated ethanol (Nigam and Singh 2011). This azeotropic mixture can be dehydrated further using molecular sieves such as zeolites to obtain anhydrous ethanol containing up to 99.6% ethanol. Alternatively, a non-polar solvent (such as benzene or cyclohexane) can be added externally to break the ethanol-water azeotrope. When the solvent added is more volatile than water, it is called azeotropic distillation and if the solvent added is less volatile than water, it is called extractive distillation (Jesús Mendoza-Pedroza and Gabriel Segovia-Hernandez 2018). The by-product of the distillation process is vinnase or stillage, which can be used as fertilizer (Figure 9.2a). Although the process is well established, the high energy input increases the product recovery costs.

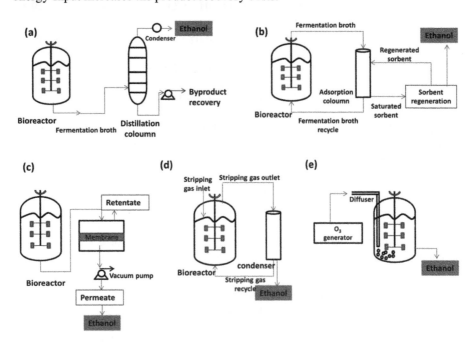

**FIGURE 9.2** Bioethanol purification strategies: (a) distillation, (b) adsorption, (c) pervaporation (d) gas stripping, and (e) ozonation.

### 9.3.2 Adsorption

Adsorption is a process by which components in a mixture are simply adsorbed on the adsorbent depending on their physical and chemical properties. It can be used for *in situ* product recovery (i.e., the separation is concomitantly performed during ethanol production). By using such a process, the effects of product inhibition can be overcome. Due to the low energy requirement, adsorption has been most actively investigated for ethanol purification (Chandel et al. 2007). Adsorption of ethanol from the aqueous media proceeds through the development of hydrophobic interactions between the surface of the adsorbent matrix and the alkyl chain of alcohol (Figure 9.2b). High surface area adsorbents are required to effectively separate the ethanol-water mixture. Activated carbon, alumina, and hydrophobic zeolites are the most common adsorbents used for purification of ethanol (Hashi et al. 2010). After adsorption, thermal or pressure swing regeneration methods are used to separate the adsorbate molecules through vaporization and to recover the adsorbents.

### 9.3.3 Pervaporation

Pervaporation is another *in situ* recovery method that can be used for ethanol separation from fermentation broths. It is basically a combination of membrane permeation and evaporation techniques (Aroujalian and Raisi 2009). In this method, a porous hydrophobic membrane is used such that ethanol is recovered by adsorption on the membrane followed by diffusion to the permeate side (Figure 9.2c). Ethanol is then recovered through evaporation by applying a vacuum at the permeate side (Jesús Mendoza-Pedroza and Gabriel Segovia-Hernandez 2018). The most common membranes used for pervaporation include organic polymeric membrane (such as polydimethylsiloxane) and inorganic membranes (such as zeolites) (Peng et al. 2010). It is possible to concentrate up to 80%–99% bioethanol using pervaporation technique, however. The limitation of this process includes membrane fouling and low selectivity, which limits its effective application in fermentation industries.

### 9.3.4 Gas Stripping

Gas stripping is a simple approach in which oxygen-free gas ($N_2$, $CO_2$, or $H_2$) is circulated in the fermentation media that selectively adsorbs the target compounds (Strods and Mezule 2017). Similar to the pervaporation method, gas stripping involves transport of target molecules from bulk liquid to gas phase. The absence of a membrane helps in avoiding the risk of gas clogging or fouling. Once the targeted product is removed by the stripped gas, it is subsequently collected using a condenser and the aqueous phase gas is recycled back to the fermenter for further product removal (Figure 9.2d). During ethanol recovery, gas stripping is performed to remove the impurities present in the fermentation media such as acetaldehyde. Previously, it was reported that ethanol yields were significantly enhanced when carbon dioxide was sparged through the fermentation broth (Hashi et al. 2010). The most dominant variable in gas stripping is gas flow rate which significantly affects the product recovery process (Strods and Mezule 2017). Rapid gas flow rate results in the removal of undesirable components and can cause foaming in the bioreactor. Various research activities are underway to improve the selectivity of the gas stripping techniques (Onuki et al. 2015).

### 9.3.5 OZONATION

Ozone or trioxygen is an inorganic molecule that can decompose various kinds of compounds due to its strong oxidation potential (~2.07 V) (Fernandes et al. 2018). Ozone decomposes rapidly to form radicals which can instantly react with organic compounds. By ozonation, the physical and chemical properties of the components in a mixture can be varied to improve the biodegradability, volatility, and decrease toxicity (Onuki et al. 2015). In ethanol fermentation, ozone is used to remove the impurities without affecting the ethanol yields (Figure 9.2e). However, a major disadvantage is the presence of non-oxidizable substances in the fermentation broth that cannot be removed by ozonation. Moreover, incomplete oxidation may lead to production of unwanted by-products (Onuki et al. 2015).

## 9.4 ECONOMICS AND ENVIRONMENTAL IMPACT OF BIOETHANOL PRODUCTION

Although bioethanol production has been successfully commercialized using first-generation feedstock, the sustainability of the process can only be achieved using non-food-based energy crops and wastes. To make the second and third-generation bioethanol production economically competitive, the bioconversion of biomass must be made cost effective (Chandel et al. 2007). It has been suggested that a biomass biorefinery approach by using each biomass component for producing value-added products can greatly improve the bioethanol production economics (Schlosser and Blahusiak 2011). Besides the conversion processes, the biomass harvesting and yield, cost of enzymes, design of bioreactor, and biocatalyst regeneration and contamination are major contributing factors that affect the economic viability of the ethanol production process. The cost of ethanol production using the corn dry-grind process was modeled by using SuperPro Designer® software (Kwiatkowski et al. 2006). They predicted a rise in ethanol production cost from US $0.89 gal$^{-1}$ to US $1.38 gal$^{-1}$ due to the increase in corn costs from US $0.071 kg$^{-1}$ to US $0.125 kg$^{-1}$ (Kwiatkowski et al. 2006). Similarly, a preliminary study was performed by NREL (National Renewable Energy Laboratory, USA) to estimate the costs of lignocellulosic ethanol production using Aspen Plus software (Mcaloon et al. 2000). They suggested the possibility of ethanol production costs of 20 cents L$^{-1}$ using lignocellulosic biomass by using synthetic enzymes and by consolidating bioprocessing (Mcaloon et al. 2000). The current bioethanol selling price is estimated to be US $1.57 gal$^{-1}$ using corn-based ethanol production in the United States (https://grains.org), while it is approximated to US $1.84 gal$^{-1}$ using sugarcane fermentation in Brazil (https://www.reuters.com). These costs can be improved further by the improvements in enzyme application and by innovative fermentation technologies.

Ethanol production is considered a carbon–neutral process because the $CO_2$ released during ethanol production is recycled back to the plants, which use the $CO_2$ to carry out photosynthesis. In addition, the toxicity of exhaust gases released during ethanol production is much less compared to its crude oil counterpart (Chandel et al. 2007). Ethanol comprises of 35% oxygen which helps in complete combustion of fuel, thereby reducing the particulate emissions and the associated health risks to human. It is estimated that by using ethanol-blended fuels such as E85 (85% ethanol, 15% gasoline), approximately 37% net greenhouse gas emissions can be reduced (Manzetti and Andersen 2015). Moreover, bioethanol blending with petrodiesel helps in decreasing the cetane number, heating value, aromatic fractions, and viscosity which leads to complete burning of blended fuels and less emissions (Lapuerta et al. 2008). The net effect of ethanol use results in overall decrease in ozone formation, thereby reducing the potential hazards to the natural vegetation and ecosystem. Currently, different countries such as the United States, Brazil, China, India, Japan, Russia, and Canada have made mandatory government policies for blending ethanol with gasoline or diesel. Moreover, efforts are underway to develop new commercialization plants and strategies to increase the ethanol percentage in the blend mixture.

## 9.5   CONCLUSION AND FUTURE OUTLOOK

Bioethanol offers a potential renewable and sustainable solution to the global energy security problems. Due to the worldwide use of ethanol-blended transportation fuels, the demand for the bioethanol production is rapidly increasing, thus triggering the commencement of the second-generation bioethanol market. The future of cellulosic-based bioethanol production is dependent upon various technical, social, and political challenges. In addition, ethical issues relating to the use of genetically modified organisms, affordability, land usage issues, and the potential monopolization of bioresources need to be addressed. The major technical barrier for the cellulose-based bioethanol production is the pre-treatment or hydrolysis step, which needs utmost research attention. Use of newly designed cocktail of enzymes for complete hydrolysis and improving the overall process design can contribute to increasing the fermentation efficiency. Moreover, developing a technology for recovering and reusing hydrolyzing enzymes can aid in lowering the costs of the process. Another major challenge is to reduce the inhibitor concentration that drastically lowers the ethanol yield. Competent microorganisms that are able to thrive in toxic environmental conditions are needed that can produce ethanol even in high inhibitor concentrations. Metabolic engineering approaches also can be used to direct the microorganism's pathways for high-yield ethanol production. Furthermore, development of suitable strains capable of metabolizing $C_6$ and $C_5$ sugars is imperative for obtaining maximum substrate conversion efficiencies. Various pilot and demonstration bioethanol plants are needed to make the cellulosic-based ethanol production commercially feasible. In addition, integration strategies for the complete use of feedstock are necessary for improving the overall economics of the process.

## REFERENCES

Adams JM, Gallagher JA, Donnison IS (2009) Fermentation study on saccharina latissima for bioethanol production considering variable pre-treatments. *J Appl Phycol* 21:569–574.

Aroujalian A, Raisi A (2009) Pervaporation as a means of recovering ethanol from lignocellulosic bioconversions. *Desalination* 247:509–517.

Barros AI, Gonçalves AL, Simões M, Pires JCM (2015) Harvesting techniques applied to microalgae: A review. *Renew Sustain Energy Rev* 41:1489–1500.

Bondesson PM, Galbe M (2016) Process design of SSCF for ethanol production from steam-pretreated, acetic-acid-impregnated wheat straw. *Biotechnol Biofuels* 9:222.

Borines MG, de Leon RL, Cuello JL (2013) Bioethanol production from the macroalgae *Sargassum* spp. *Bioresour Technol* 138:22–29.

Chandel AK, Es C, Rudravaram R, Narasu ML, Rao V, Ravindra P (2007) Economics and environmental impact of bioethanol production technologies: An appraisal. *Biotechnol Mol Biol Rev* 2:14–32.

Chen H, Han Y, Xu J (2008) Simultaneous saccharification and fermentation of steam exploded wheat straw pretreated with alkaline peroxide. *Process Biochem* 43:1462–1466.

Cheng D, Li D, Yuan Y, Zhou L, Li X, Wu T, Wang L, Zhao Q, Wei W, Sun Y (2017) Improving carbohydrate and starch accumulation in *Chlorella* sp. AE10 by a novel two-stage process with cell dilution. *Biotechnol Biofuels* 10:75.

Chew KW, Yap JY, Show PL, Suan NH, Juan JC, Ling TC, Lee DJ, Chang JS (2017) Microalgae biorefinery: High value products perspectives. *Bioresour Technol* 229:53–62.

Chung D, Cha M, Guss AM, Westpheling J (2014) Direct conversion of plant biomass to ethanol by engineered *Caldicellulosiruptor bescii*. *Proc Natl Acad Sci* 111:8931–8936.

Daniell J, Köpke M, Simpson SD (2012) Commercial biomass syngas fermentation. *Energies* 5:5372–5417.

Demirbas A (2010) Use of algae as biofuel sources. *Energy Convers Manag* 51:2738–2749.

Devarapalli M, Atiyeh HK (2015) A review of conversion processes for bioethanol production with a focus on syngas fermentation. *Biofuel Res J* 2:268–280.

Fernandes A, Boczkaj G, Głazowska J, Tomczak-Wandzel R, Kamiński M (2018) Comparison of ozonation and evaporation as treatment methods of recycled water for bioethanol fermentation process. *Waste Biomass Valori* 9:1141–1149.

Harun R, Danquah MK (2011) Influence of acid pre-treatment on microalgal biomass for bioethanol production. *Process Biochem* 46:304–309.

Hashi M, Tezel FH, Thibault J (2010) Ethanol recovery from fermentation broth via carbon dioxide stripping and adsorption. In: *Energy and Fuels*. American Chemical Society, pp. 4628–4637.

Henstra AM, Sipma J, Rinzema A, Stams AJ (2007) Microbiology of synthesis gas fermentation for biofuel production. *Curr Opin Biotechnol* 18:200–206.

Ho SH, Huang SW, Chen CY, Hasunuma T, Kondo A, Chang JS (2013) Bioethanol production using carbohydrate-rich microalgae biomass as feedstock. *Bioresour Technol* 135:191–198.

https://grains.org Ethanol Market and Pricing Data—February 20, 2018—U.S. GRAINS COUNCIL. https://grains.org/ethanol_report/ethanol-market-and-pricing-data-february-20-2018/. Accessed October 3, 2018.

https://www.reuters.com Brazil ethanol prices plunge in the first week of new cane crop | Reuters. https://www.reuters.com/article/us-brazil-ethanol/brazil-ethanol-prices-plunge-in-the-first-week-of-new-cane-crop-idUSKBN1HG32R. Accessed October 3, 2018.

Jerez-Mogollón SJ, Rueda-Quiñonez LV, Alfonso-Velazco LY, Barajas-Solano AF, Barajas-Ferreira C, Kafarov V, Jaimes-duarte D, Resumen A (2012) Improvement of lab-scale production of microalgal carbohydrates for biofuel production. *CT&F-Ciencia, Tecnología y Futuro* 5(1):103–116.

Jesús Mendoza-Pedroza J, Gabriel Segovia-Hernandez J (2018) Alternative schemes for the purification of bioethanol: A comparative study. *Recent Adv Petrochem Sci* 4.

Kang L, Wang W, Lee YY (2010) Bioconversion of Kraft paper mill sludges to ethanol by SSF and SSCF. *Appl Biochem Biotechnol* 161:53–66.

Kim S, Dale BE (2004) Global potential bioethanol production from wasted crops and crop residues. *Biomass and Bioenergy* 26:361–375.

Kwiatkowski JR, McAloon AJ, Taylor F, Johnston DB (2006) Modeling the process and costs of fuel ethanol production by the corn dry-grind process. *Ind Crops Prod* 23:288–296.

Lapuerta M, Armas O, Herreros JM (2008) Emissions from a diesel–bioethanol blend in an automotive diesel engine. *Fuel* 87:25–31.

Limayem A, Ricke SC (2012) Lignocellulosic biomass for bioethanol production: Current perspectives, potential issues and future prospects. *Prog Energy Combust Sci* 38:449–467.

Manzetti S, Andersen O (2015) A review of emission products from bioethanol and its blends with gasoline. Background for new guidelines for emission control. *Fuel* 140:293–301.

Markou G, Angelidaki I, Georgakakis D (2012) Microalgal carbohydrates: An overview of the factors influencing carbohydrates production, and of main bioconversion technologies for production of biofuels. *Appl Microbiol Biotechnol* 96:631–645.

McAloon A, Taylor F, Yee W, Ibsen K, Wooley R (2000) Determining the cost of producing Ethanol from corn starch and lignocellulosic feedstocks. *National Renewable Energy Laboratory Report* 2000.

Meinita MDN, Kang JY, Jeong GT, Koo HM, Park SM, Hong YK (2012) Bioethanol production from the acid hydrolysate of the carrageenophyte *Kappaphycus alvarezii* (cottonii). *J Appl Phycol* 24:857–862.

Miranda JR, Passarinho PC, Gouveia L (2012) Pre-treatment optimization of *Scenedesmus obliquus* microalga for bioethanol production. *Bioresour Technol* 104:342–348.

Mohd Azhar SH, Abdulla R, Jambo SA, Marbawi H, Gansau JA, Mohd Faik AA, Rodrigues KF (2017) Yeasts in sustainable bioethanol production: A review. *Biochem Biophys Reports* 10:52–61.

Munasinghe PC, Khanal SK (2010) Biomass-derived syngas fermentation into biofuels: Opportunities and challenges. *Bioresour Technol* 101:5013–5022.

Nguyen MT, Choi SP, Lee J, Lee JH, Sim SJ (2009) Hydrothermal acid pretreatment of *Chlamydomonas reinhardtii* biomass for ethanol production. *J Microbiol Biotechnol* 19:161–166.

Nigam PS, Singh A (2011) Production of liquid biofuels from renewable resources. *Prog Energy Combust Sci* 37:52–68.

Obata O, Akunna J, Bockhorn H, Walker G (2016) Ethanol production from brown seaweed using non-conventional yeasts. *Bioethanol* 2:134–145.

Onuki S, Koziel JA, Jenks WS, Cai L, Rice S, van Leeuwen J (Hans) (2015) Ethanol purification with ozonation, activated carbon adsorption, and gas stripping. *Sep Purif Technol* 151:165–171.

Özçimen D, Inan B (2015) An overview of bioethanol production from algae. In: *Biofuels— Status and Perspective*. InTech.

Peng P, Shi B, Lan Y (2010) A review of membrane materials for Ethanol recovery by pervaporation. *Sep Sci Technol* 46:234–246.

Schlosser S, Blahusiak M (2011) Biorefinery for production of chemicals, energy and fuels. *Elektroenergetika* 4:8–15.

Singh S, Adak A, Saritha M, Sharma S, Tiwari R, Rana S, Arora A, Nain L (2017) Bioethanol production scenario in India: Potential and policy perspective. In: *Sustainable Biofuels Development in India*. Springer International Publishing, Cham, Switzerland, pp. 21–37.

Singh V, Moreau RA, Doner LW, Eckhoff SR, Hicks KB (1999) Recovery of fiber in the corn dry-grind ethanol process: A feedstock for valuable coproducts. *Cereal Chem J* 76:868–872.

Stewart GG, Panchal CJ, Russell I, Sills AM (1983) Biology of ethanol-producing microorganisms. *Crit Rev Biotechnol* 1:161–188.

Strods M, Mezule L (2017) Alcohol recovery from fermentation broth with gas stripping: System experimental and optimisation. *Agronomy Res* 15(3), 897–904.

Syawala DS, Wardiyati T, Maghfoer MD (2013) Production of bioethanol from corncob and sugarcane bagasse with hydrolysis process using *Aspergillus niger* and *Trichoderma viride*. *J Environ Sci Toxicol Food Technol* 5:49–56.

Takano M, Hoshino K (2018) Bioethanol production from rice straw by simultaneous saccharification and fermentation with statistical optimized cellulase cocktail and fermenting fungus. *Bioresour Bioprocess* 5:16.

Vohra M, Manwar J, Manmode R, Padgilwar S, Patil S (2014) Bioethanol production: Feedstock and current technologies. *J Environ Chem Eng* 2:573–584.

www.statista.com • Global ethanol production for fuel use 2017|Statistic. https://www.statista.com/statistics/274142/global-ethanol-production-since-2000/. Accessed September 26, 2018.

Zhao X, Moates GK, Elliston A, Wilson DR, Coleman MJ, Waldron KW (2015) Simultaneous saccharification and fermentation of steam exploded duckweed: Improvement of the ethanol yield by increasing yeast titre. *Bioresour Technol* 194:263–269.

# 10 Biobutanol

## 10.1 INTRODUCTION

The global demand for renewable energy generation is increasing rapidly due to the looming scarcity of conventional fossil fuel-based resources. In this regard, major focus has been put on bioethanol as an alternative fuel source considering its sustainability, ability to lower the net greenhouse gas (GHG) emissions, and economic viability. Although bioethanol production from renewable biomass sources has seen commercial success, the associated disadvantages such as low energy content and high corrosive aggressiveness have boosted the research focus towards alternate bioalcohols as potentials biofuels (Minteer 2011). Butyl alcohol or butanol is an attractive energy source since it has higher calorific value and is non-corrosive and non-hygroscopic in nature (Kaminski et al. 2011). Butanol is a colorless flammable alcohol with a four carbon, straight-chain structure. It has wide industrial applications in the manufacture of paints, cosmetics, adhesives, perfumes, solvents, pharmaceuticals, and polymers (Zhang et al. 2016). The fact has been established that butanol is a better fuel additive than ethanol and can be used directly in internal combustion engines without any modification (Bharathiraja et al. 2017). Table 10.1 provides the major fuel properties of butanol. These properties show that butanol is more similar to gasoline than ethanol. It has low volatility that makes it easier and safer to handle (Zhang et al. 2016). Moreover, the low vapor pressure, high flash point, and lower autoignition temperature ensures easy start during cold climatic conditions unlike bioethanol which creates cold start problems.

At present, butanol is produced mainly through the oxo process through the use of petrochemical derivatives, which is highly energy intensive. Alternatively, butanol also can be produced using microbial fermentation technology. It is naturally produced by *Clostridium* species, which follows acetone-butanol-ethanol (ABE) fermentation pathway (Qureshi et al. 2010). Currently, various challenges restrict biobutanol production at a commercial scale such as low yields, product inhibition, incomplete sugar use, and high feedstock costs. Furthermore, the separation of butanol from the fermentation broth is costly. This chapter provides a comprehensive review of biobutanol production as a potential biofuel with a key emphasis on the recent advancements.

**TABLE 10.1**
**Fuel Properties of Butanol**

| Parameter | Property |
| --- | --- |
| Molecular formula | $C_4H_9OH$ |
| Molecular mass | 74.12 g mol$^{-1}$ |
| Density | 0.808 kg L$^{-1}$ |
| Boiling temperature | 117.7°C |
| Freezing point | −90°C |
| Flash point | 35°C |
| Vapor pressure | 6 mm Hg |
| Higher heating value | 37300 KJ kg$^{-1}$ |
| Lower heating value | 34400 KJ kg$^{-1}$ |
| Specific heat capacity | 4.18 J g$^{-1}$ K$^{-1}$ |
| Viscosity | 2.63 mPa s |
| Refractive index | 1.39 |
| Octane number | 96 |

## 10.2 BIOBUTANOL PRODUCTION PROCESS: PRINCIPLES, MECHANISMS, AND FEEDSTOCKS

The history of the biochemical route of biobutanol production dates back to the twentieth century when industrial butanol production relied on acetone-butanol-ethanol (ABE) fermentation using the *Clostridium* species (Bharathiraja et al. 2017). However, with the discovery of a more economical petrochemical route for butanol production, the research on ABE declined drastically. Recently, the increasing crude-oil prices with the associated global environmental concerns has renewed interest in biobutanol production using inexpensive, non-food- based energy sources (Zhang et al. 2016). The main butanol-producing bacteria include *Clostridium acetobutylicum, Clostridium beijerinckii, Clostridium saccharobutyli-cum*, and *Clostridium saccharoperbutylacetonicum* (Table 10.2). Most of these bacteria follow the ABE fermentation pathway to produce butanol (Figure 4.6). In this type of fermentation, two separate phases are observed for acid (acidogenesis) and alcohol (solventogenesis) production (Figure 10.1).

Acidogenesis usually happens at the same time as the exponential growth of the microorganism and leads to the generation of volatile fatty acids such as acetate and butyrate along with $CO_2$ and $H_2$. It is observed that butyrate concentration is twice that of acetate after the acidogenic phase (Gottschal and Morris 1981). During this period, the pH of the medium is drastically reduced due to the accumulation of acids, which is detrimental to the growth of bacteria because it destroys the essential proton gradient required for the cells. So, to level off this effect of pH change, the ABE-fermenting microorganisms use the metabolic shift strategy to convert the accumulated acids to solvents such as acetone and butanol (solventogenesis) at the end of the exponential growth phase. Typically, a butanol/acetone ratio of 2:1 is found in most of the *Clostridium* sp. (Ezeji et al. 2007). Although production of

## TABLE 10.2
## Butanol Production by Different Microorganisms Using Various Substrates

| Microbe | Substrate | Butanol Conc. (g L⁻¹) | References |
|---|---|---|---|
| *Clostridium saccharobutylicum* | Cane molasses | 13.4 | Ni et al. (2012) |
| *C. acetobutylicum* GX01 | Cassava starch | 17.1 | Li et al. (2015) |
| *C. acetobutylicum* L7 | Jerusalem artichoke juice | 11.21 | Chen et al. (2010) |
| *Clostridium beijerinckii* P260 | Wheat straw hydrolysate | n.a.[a] | Qureshi et al. (2007) |
| *Clostridium beijerinckii* BA101 | Corn fiber | 13.2 | Qureshi et al. (2008) |
| *Clostridium pasteurianum* | Glycerol | 29.8 | Lin et al. (2015) |
| *Clostridium saccharoperbutylacetonicum* | Sago starch | 9.83 | Al-Shorgani et al. (2012) |
| *Clostridium beijerinckii* BA 101 | DDGS* | 3.62 | Wang et al. (2013) |
| *C. acetobutylicum* | Cassava bagasse | 76.4 | Lu et al. (2012) |
| *C. acetobutylicum* NRRL B-591 | Rice straw hydrolysate | 7.10 | Amiri et al. (2014) |
| *Clostridium sporogenes* BE01 | Rice straw hydrolysate | 3.43 | Gottumukkala et al. (2013) |

[a] n.a. not available; DDGS: distiller's dried grains with solubles.

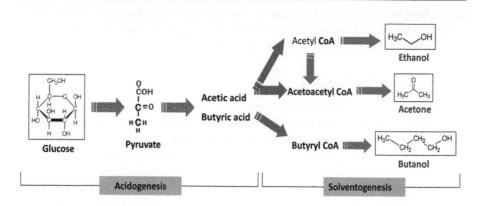

**FIGURE 10.1**   Acetone-butanol-ethanol (ABE) fermentation process.

these solvents causes a rise in pH, their high concentrations in the media is toxic to the growth of bacteria and has damaging effects on the bacteria's cell membrane and some of the membrane proteins (Bharathiraja et al. 2017). To avoid such deleterious effects, bacteria forms endospores to thrive in the unfavorable environmental conditions (Procentese 2015).

Table 10.2 shows that the type of feedstock used for butanol production is dependent upon the strain used during the fermentation process. Most industrial butanol production relies on starch- or sugar-based feedstocks like sugarcane or molasses (Peabody and Kao 2016). Although most reports focus on the use of lignocellulosic feedstocks for butanol production (Nanda et al. 2014; Cao and Sheng 2016), the economic suitability of such process is still too expensive and ineffective. Besides the traditional ABE fermentation process, certain species have been noted to produce butanol through the syngas platform. For example, *Butyribacterium methylotrophicum* is able to produce butanol using syngas as sole substrate (Shen et al. 1999). Similarly, *Clostridium carboxidivorans* can perform anaerobic bioconversion of gaseous C1 compounds (such as carbon monoxide) for butanol production (Fernández-Naveira et al. 2016). Syngas fermentation offers several advantages such as high conversion efficiencies, lower undesirable by-products formation, and simplified recovery process (Worden et al. 1991). Although syngas-based biobutanol production appears promising, the final concentration produced is too low compared to the traditional ABE fermentation process. Thus, for the realization of the commercial biobutanol economy, future investigations should direct towards the improvement of overall butanol yields in ABE and in syngas fermentation processes.

## 10.3   RECENT ADVANCES TO IMPROVE BIOBUTANOL PRODUCTION

Several strategies have been used to improve the biobutanol yields and avoid the inescapable production of alternate by-products such as ethanol, acetone, acetic acid, and butyric acid. Moreover, since high concentrations of butanol are toxic to most ABE-fermenting bacteria, different approaches for obtaining highly robust strains have been considered. In general, the techniques employed for improvement of biobutanol production can be in the form of strain improvement and/or process development as discussed in the following section.

### 10.3.1   STRAIN IMPROVEMENT

The traditional approaches for strain improvement mainly include mutagenesis and recombinant DNA technologies, which manipulate or alter the genes responsible for butanol production in the genome of the microorganism (Ezeji et al. 2007). Various genetic tools (such as gene insertion, overexpression, and protoplast fusion) have been used to introduce active hydrolytic enzymes in the host organism to enhance the carbohydrate use capacity and improve the overall conversion efficiency of the butanol-producing microorganisms. By using such techniques, non-conventional carbon sources such as cellulosic wastes can be used readily as substrates during ABE fermentation. It was reported that by using a chemical mutagen N-methyl-N′-nitro-N-nitrosoguanidine (NTG), a mutagenic strain *Clostridium acetobutylicum* 77 was developed that was able to metabolize glucose faster than its non-mutagenic strain (Matta-el-Ammouri et al. 1986). Similar observation was made in another study where the NTG-mutated strain *Clostridium beijerinckii* BA101 showed elevated levels of butanol due to enhanced capacity for uptake of carbohydrate compared to the parent strain *Clostridium beijerinckii* 8052 (Formanek et al. 1997).

By using rDNA techniques, Qi and others (2018) introduced ferredoxin-NAD(P)$^+$ oxidoreductase (FdNR) and trans-enoyl-coenzyme reductase (TER) genes into the genome of *Clostridium acetobutylicum* that improved NAD(P)H availability and thus in turn enhanced the overall butanol yield.

More recent strategies of strain improvement involve the metabolic engineering approach to redirect the existing metabolic pathways or to introduce *de novo* pathways to generate hyper butanol-producing strains (Zhang et al. 2016). Gaida and others (2016) introduced a CoA-dependent pathway from *C. acetobutylicum* in *C. cellulolyticum* for n-butanol synthesis using crystalline cellulose. Eversloh and Bahl (2011) reviewed the various metabolic strategies used with *C. acetobutylicum* for enhanced butanol production. They concluded that, although achievements have been made to enhance the butanol titers during ABE fermentation, the major disadvantage is the lack of understanding about the metabolic shift from the acidogenic phase to the solventogenic phase of *Clostridium* sp. at the molecular level, which includes inducing signals, regulators, their interaction, and interconnection of regulatory networks. Moreover, for an industrial point of view, using *Clostridium* sp. has several disadvantages. For example, since it is an obligate anaerobe, it is very difficult to maintain strict anaerobic conditions within large-scale systems. In addition, continuous culture conditions, butanol toxicity, and serial sub-culturing are a few other factors that limit the butanol-producing abilities of *Clostridium* sp. (Kharkwal et al. 2009). Due to these inherent disadvantages, attempts have been made to introduce butanol production pathways in other more easily cultivated model organisms such as *E. coli* and *S. cerevisiae* (Atsumi et al. 2008; Shi et al. 2016). Currently, more advanced research is focused on direct $CO_2$ conversion to butanol using synthetic biology approaches with cyanobacteria (Lan and Liao 2011). The combinatorial metabolic and synthetic approaches along with the traditional cloning and recombination techniques are prerequisites to obtain high butanol-yielding strains with desired characteristics. By expanding the opportunities for identification of functional genes and relevant enzymes from various other organisms, butanol titers can be further enhanced.

## 10.3.2 Process Development Strategies

Industrial biobutanol production can be realized not only by the strain improvement techniques but also by improvement in process engineering strategies for bulk production and recovery in an economical and environmentally friendly manner. Recently, the major focus has been on different bioreactor designs and product recovery technologies to improve the efficiency of the ABE fermentation process (Kharkwal et al. 2009). The simplest process used for butanol production is the batch process. However, it has a very low productivity level (Ezeji et al. 2007). To improve the productivity, continuous operation with recycling has been considered by various researchers (Pierrot et al. 1986; Schlote and Gottschalk 1986; Zheng et al. 2013). However, due to the obligatory nature of *Clostridium* sp. and the toxicity of high butanol concentrations, large-scale continuous process is not feasible. To improve the butanol yields and decrease culture toxicity, fed batch operation, immobilization, and *in situ* product recovery by applying methods such as adsorption, gas stripping, pervaporation, perstraction, and membrane separation have been suggested (Nanda et al. 2014). Another strategy that can be used for *in situ* product recovery is vacuum fermentation (Salemme et al. 2016). In this method, the bioreactor is maintained at vacuum conditions and the fermentation product boils off at fermentation temperature. The product is then removed and recovered through condensation. However, this method is effective only when the solvents are more volatile than water. Since the boiling temperature of butanol is higher than that of water, an azeotropic mixture must be formed to lower the boiling point and recover the butanol. Based on the economic assessment by Abdi and others (2016), integrated vacuum fermentation is the only technology that is economically feasible with respect to the current market price of butanol. Another strategy for increasing the butanol yield through process engineering is the manipulation of the redox potential (Zhang et al. 2016). Redox pairs such as $NAD(P)H/NAD(P)^+$ play a key role in directing the synthesis of end metabolite formation during ABE fermentation. In addition, redox potential regulates the acidogenic and solventogenic phases of the microorganism and thus can rechannel the metabolic flux towards more butanol production. Various strategies have been used to manipulate the redox potential of the fermentation media such as the addition of reduced substrates, redox mediators, current supply, and reduced gas sparging (Zhang et al. 2016). However, most of these techniques are limited to small-scale systems and need further investigation for application to large-scale systems.

## 10.4 PROSPECTS OF BIOBUTANOL: TECHNO-ECONOMIC AND ENVIRONMENTAL ASPECTS

The interest for developing biobutanol industries is increasing and multiple companies such as British Petroleum (BP), DuPont, Cobalt, and Chevron Oronite have already started working on commercial-scale butanol production (Tao et al. 2014). Economic assessments on biobutanol production using corn, whey permeate, and molasses have shown that butanol derived through ABE fermentation is not economically competitive compared to the butanol derived through petrochemical processes (Bharathiraja et al. 2017). However, with recent developments in process technologies, the market price of biobutanol can be significantly reduced. Kumar and others (2012) performed a comparative economic assessment of cellulosic and non-cellulosic feedstocks and concluded that for economic feasibility sugarcane and cellulosic materials were suitable substrates in the biobutanol production process with an estimated product cost of 0.59–0.75 $/kg butanol. Similarly, Jang and Choi (2018) conducted a techno-economic analysis of lignocellulosic butanol production and revealed the total production cost and minimum selling price of biobutanol to be $1427/t and $1693/t, respectively, based on the 80,000 t/y capacity plant. They suggested that to secure commercial viability, the production costs need to be reduced by 54%. At present, the various countries that govern the biobutanol market include the United States, Europe, Japan, China, Korea, India, and several Middle East countries (http://www.mynewsdesk.com). It is estimated that by year 2020 the global market for butanol will account for $247 billion due to its advantages over ethanol and biodiesel (Bharathiraja et al. 2017).

Environmentally, it is estimated the vehicles fueled with biobutanol produced using lignocellulosic and waste-based feedstocks could result in a fossil energy saving of 39%–56% and a 32%–48% reduction in GHG emissions compared to conventional gasoline (Niemisto et al. 2013). Brito and Martins (2017) performed a life cycle analysis to assess the environmental impact of the ABE fermentation process using corn and wheat straw as substrates and compared it with the conventional oxo process. They found that the ABE fermentation process using wheat straw as a substrate presented the lowest environmental impact when the allocation method was based on mass. However, when the economic values of the products were considered for allocation, the global impact of ABE fermentation processes was higher than the oxo-process due to the lower economic value of the gases produced during fermentation (Brito and Martins 2017). Pereira and others (2015) also performed a similar life cycle analysis of butanol production in the sugarcane biorefineries of Brazil. They suggested that butanol production using bagasse and straw pentoses with genetically modified organisms can be a suitable alternative to the conventional petrochemical processes and had the lowest environmental impact. In addition, they also showed that inclusion of butanol and acetone as value-added products of the sugarcane biorefinery led to an increase in revenues compared to base scenarios (Pereira et al. 2015). In contrast to these studies, Pfromm and others (2010) pointed out several disadvantages of biobutanol production compared to bioethanol using corn and switchgrass as substrates. They showed that bioethanol production yielded higher economic and environmental benefits compared to biobutanol due to the low yield, low titer, strict sterility requirements, and downstream separation issues of biobutanol. However, with the recent advances in process development and production of robust metabolically engineered strains, these issues can be tackled, thus paving the way for sustainable biobutanol production.

## 10.5  CONCLUSION

The rising environmental concerns over the use of fossil fuels have kindled the search for alternate fuel sources. Butanol has been considered an attractive substitute due to its inherent advantages such as high calorific value and non-corrosive nature. Biochemical production of butanol is more efficient and inexpensive compared to chemical synthesis of butanol. However, the major challenge of ABE fermentation is the cost of the feedstock and the production of toxic by-products that limit the up-scaling of the process. By using inexpensive feedstocks such as lignocellulosics and syngas in place of tradition sugar- and starch-based feedstock and by developing robust industrial strains capable of producing and tolerating high titers of butanol, the economic feasibility of ABE process can be achieved. With the recent advances in synthetic biology applications and process engineering tools along with the development of commercial biobutanol plants, the realization of biobutanol production as a sustainable transportation fuel can be foreseen in the future.

## REFERENCES

Abdi HK, Alanazi KF, Rohani AS, Mehrani P, Thibault J (2016) Economic comparison of a continuous ABE fermentation with and without the integration of an in situ vacuum separation unit. *Can J Chem Eng* 94:833–843.

Al-Shorgani NKN, Kalil MS, Yusoff WMW (2012) Fermentation of sago starch to biobutanol in a batch culture using *Clostridium saccharoperbutylacetonicum* N1-4 (ATCC 13564). *Ann Microbiol* 62:1059–1070.

Amiri H, Karimi K, Zilouei H (2014) Organosolv pretreatment of rice straw for efficient acetone, butanol, and ethanol production. *Bioresour Technol* 152:450–456.

Atsumi S, Cann AF, Connor MR, Shen CR, Smith KM, Brynildsen MP, Chou KJY, Hanai T, Liao JC (2008) Metabolic engineering of *Escherichia coli* for 1-butanol production. *Metab Eng* 10:305–311.

Bharathiraja B, Jayamuthunagai J, Sudharsanaa T, Bharghavi A, Praveenkumar R, Chakravarthy M, Yuvaraj D (2017) Biobutanol—An impending biofuel for future: A review on upstream and downstream processing tecniques. *Renew Sustain Energy Rev* 68:788–807.

Brito M, Martins F (2017) Life cycle assessment of butanol production. *Fuel* 208:476–482.

Cao G, Sheng Y (2016) Biobutanol production from lignocellulosic biomass: Prospective and challenges. *J Bioremediation Biodegrad* 7:1–6.

Chen L, Xin C, Deng P, Ren J, Liang H, Bai F (2010) Butanol production from hydrolysate of Jerusalem artichoke juice by *Clostridium acetobutylicum* L7. *Chin J Biotechnol* 26:991–996.

Ezeji TC, Qureshi N, Blaschek HP (2007) Bioproduction of butanol from biomass: From genes to bioreactors. *Curr Opin Biotechnol* 18:220–227.

Fernández-Naveira Á, Abubackar HN, Veiga MC, Kennes C (2016) Efficient butanol-ethanol (B-E) production from carbon monoxide fermentation by *Clostridium carboxidivorans*. *Appl Microbiol Biotechnol* 100:3361–3370.

Formanek J, Mackie R, Blaschek HP (1997) Enhanced butanol production by *Clostridium beijerinckii* BA101 grown in semidefined P2 medium containing 6% maltodextrin or glucose. *Appl Environ Microbiol* 63:2306–2310.

Gaida SM, Liedtke A, Heinz A, Jentges W, Engels B, Jennewein S (2016) Metabolic engineering of *Clostridium cellulolyticum* for the production of n-butanol from crystalline cellulose. *Microb Cell Fact* 15:1–11.

Gottschal JC, Morris JG (1981) The induction of acetone and butanol production in cultures of *Clostridium acetobutylicum* by elevated concentrations of acetate and butyrate. *FEMS Microbiol Lett* 12:385–389.

Gottumukkala LD, Parameswaran B, Valappil SK, Mathiyazhakan K, Pandey A, Sukumaran RK (2013) Biobutanol production from rice straw by a non acetone producing Clostridium sporogenes BE01. *Bioresour Technol* 145:182–187.

http://www.mynewsdesk.com Global bio-butanol industry market segment up to 2018: Forecast. http://www.mynewsdesk.com/se/pressreleases/global-bio-butanol-industry-market-segment-up-to-2018-forecast-till-2023-2594032. Accessed October 11, 2018.

Jang MO, Choi G (2018) Techno-economic analysis of butanol production from lignocellulosic biomass by concentrated acid pretreatment and hydrolysis plus continuous fermentation. *Biochem Eng J* 134:30–43.

Kaminski W, Tomczak E, Gorak A (2011) Bioutanol—production and purifciaton methods. *Ecol Chem Eng* 18:31–37.

Kharkwal S, Karimi IA, Chang MW, Lee D-Y (2009) Strain improvement and process development for biobutanol production. *Recent Pat Biotechnol* 3:202–210.

Kumar M, Goyal Y, Sarkar A, Gayen K (2012) Comparative economic assessment of ABE fermentation based on cellulosic and non-cellulosic feedstocks. *Appl Energy* 93:193–204.

Lan EI, Liao JC (2011) Metabolic engineering of cyanobacteria for 1-butanol production from carbon dioxide. *Metab Eng* 13:353–363.

Li S, Guo Y, Lu F, Huang J, Pang Z (2015) High-level butanol production from cassava starch by a newly isolated *Clostridium acetobutylicum*. *Appl Biochem Biotechnol* 177:831–841.

Lin DS, Yen HW, Kao WC, Cheng CL, Chen WM, Huang CC, Chang JS (2015) Bio-butanol production from glycerol with *Clostridium pasteurianum* CH4: The effects of butyrate addition and in situ butanol removal via membrane distillation. *Biotechnol Biofuels* 8:1–12.

Lu C, Zhao J, Yang ST, Wei D (2012) Fed-batch fermentation for n-butanol production from cassava bagasse hydrolysate in a fibrous bed bioreactor with continuous gas stripping. *Bioresour Technol* 104:380–387.

Lütke-Eversloh T, Bahl H (2011) Metabolic engineering of *Clostridium acetobutylicum*: Recent advances to improve butanol production. *Curr Opin Biotechnol* 22:634–647.

Matta-el-Ammouri G, Janati-Idrissi R, Rambourg JM, Petitdemange H, Gay R (1986) Acetone butanol fermentation by a *Clostridium acetobutylicum* mutant with high solvent productivity. *Biomass* 10:109–119.

Minteer SD (2011) Biochemical production of other bioalcohols: biomethanol, biopropanol, bioglycerol, and bioethylene glycol. *Handb Biofuels Prod* 258–265.

Nanda S, Dalai AK, Kozinski JA (2014) Butanol and ethanol production from lignocellulosic feedstock: Biomass pretreatment and bioconversion. *Energy Sci Eng* 2:138–148.

Ni Y, Wang Y, Sun Z (2012) Butanol production from cane molasses by *Clostridium saccharobutylicum* DSM 13864: Batch and semicontinuous fermentation. *Appl Biochem Biotechnol* 166:1896–1907.

Niemisto J, Saavalainen P, Pongracz E, Keiski RL (2013) Biobutanol as a potential sustainable biofuel—assessment of lignocellulosic and waste-based feedstocks. *J Sustain Dev Energy, Water Environ Syst* 1:58–77.

Peabody GL, Kao KC (2016) Recent progress in biobutanol tolerance in microbial systems with an emphasis on Clostridium. *FEMS Microbiol Lett* 363:1–6.

Pereira LG, Chagas MF, Dias MOS, Cavalett O, Bonomi A (2015) Life cycle assessment of butanol production in sugarcane biorefineries in Brazil. *J Clean Prod* 96:557–568.

Pfromm PH, Amanor-Boadu V, Nelson R, Vadlani P, Madl R (2010) Bio-butanol vs. bio-ethanol: A technical and economic assessment for corn and switchgrass fermented by yeast or *Clostridium acetobutylicum*. *Biomass Bioenergy* 34:515–524.

Pierrot P, Fick M, Engasser JM (1986) Continuous acetone-butanol fermentation with high productivity by cell ultrafiltration and recycling. *Biotechnol Lett* 8:253–256.

Procentcsc A (2015) Processes for biobutanol production from renewable resources. D. Sc. Dissertation, The University of Naples Federico, Italy.

Qi F, Thakker C, Zhu F, Pena M, San K-Y, Bennett GN (2018) Improvement of butanol production in *Clostridium acetobutylicum* through enhancement of NAD(P)H availability. *J Ind Microbiol Biotechnol* 45(11):993–1002.

Qureshi N, Ezeji TC, Ebener J, Dien BS, Cotta MA, Blaschek HP (2008) Butanol production by *Clostridium beijerinckii*. Part I: Use of acid and enzyme hydrolyzed corn fiber. *Bioresour Technol* 99:5915–5922.

Qureshi N, Saha BC, Cotta MA (2007) Butanol production from wheat straw hydrolysate using *Clostridium beijerinckii*. *Bioprocess Biosyst Eng* 30:419–4270.

Qureshi N, Saha BC, Dien B, Hector RE, Cotta MA (2010) Production of butanol (a biofuel) from agricultural residues: Part I—Use of barley straw hydrolysate. *Biomass Bioenergy* 34:559–565.

Salemme L, Olivieri G, Raganati F, Salatino P, Marzocchella A (2016) Analysis of the energy efficiency of some butanol recovery processes. *Chem Eng Trans* 49:109–114.

Schlote D, Gottschalk G (1986) Effect of cell recycle on continuous butanol-acetone fermentation with *Clostridium acetobutylicum* under phosphate limitation. *Appl Microbiol Biotechnol* 24:1–5.

Shen GJ, Shieh JS, Grethlein AJ, Jain MK, Zeikus JG (1999) Biochemical basis for carbon monoxide tolerance and butanol production by *Butyribacterium methylotrophicum*. *Appl Microbiol Biotechnol* 51:827–832.

Shi S, Si T, Liu Z, Zhang H, Ang EL, Zhao H (2016) Metabolic engineering of a synergistic pathway for n-butanol production in *Saccharomyces cerevisiae*. *Sci Rep* 6:25675.

Tao L, He X, Tan ECD, Zhang M, Aden A (2014) Comparative techno-economic analysis and reviews of n-butanol production from corn grain and corn stover. *Biofuels, Bioprod Biorefining* 8:342–361.

Wang X, Wang Y, Wang B, Blaschek H, Feng H, Li Z (2013) Biobutanol production from fiber-enhanced DDGS pretreated with electrolyzed water. *Renew Energy* 52:16–22.

Worden RM, Grethlein AJ, Jain MK, Datta R (1991) Production of butanol and ethanol from synthesis gas via fermentation. *Fuel* 70:615–619.

Zhang J, Wang S, Wang Y (2016) Chapter one—biobutanol production from renewable resources: Recent advances. *Adv Bioenergy* 1:1–68.

Zheng J, Tashiro Y, Yoshida T, Gao M, Wang Q, Sonomoto K (2013) Continuous butanol fermentation from xylose with high cell density by cell recycling system. *Bioresour Technol* 129:360–365.

# 11 Biodiesel

## 11.1 INTRODUCTION

The unprecedented increase in fossil fuel prices and the associated environmental concerns has intensified the search for alternate renewable fuels. Biomass-based biofuels are considered attractive sources for green energy supply (Demirbas 2008). The concept of using biomass energy dates back to the invention of diesel engine in 1895 by the German scientist, Rudolf Diesel, who originally used vegetable oil as fuel (Ambat et al. 2018). However, use of vegetable oil as fuel was discredited due to the associated disadvantages such as low volatility, high viscosity, low efficiency, and poor atomization upon injection. Recently, however, the use of plant-oil derivatives such as biodiesel has gained momentum owing to its characteristic similarities with petrodiesel. According to the ASTM (American Society for Testing and Materials) standards, biodiesel can be defined as a mixture of fatty acid alkyl esters (mostly methyl esters) produced from lipids through transesterification of acylglycerides (Equation 11.1) or esterification of fatty acids (Equation 11.2) (Suresh et al. 2018).

$$
\begin{array}{llll}
-OOC-R_1 & & CH_3-OOC-R_1 & H_2C-OH \\
-OOC-R_2 + 3CH_3OH \xrightleftharpoons{NaOH} & CH_3-OOC-R_2 + & HC-OH \\
-OOC-R_3 & & CH_3-OOC-R_3 & H_2C-OH
\end{array} \tag{11.1}
$$

Triglycerides    Methanol    Methyl esters    Glycerol

$$R'OH + RCOOH \rightarrow RCOOR' \tag{11.2}$$

Biodiesel has numerous benefits including renewability, biodegradability, high cetane number, high flash point, eco-friendly, and non-toxic nature (Ambat et al. 2018). It can be blended with petrodiesel to reduce greenhouse gas (GHG) emissions. A wide range of feedstocks ranging from edible to non-edible oleaginous crops and residues have been used for extraction of biodiesel (Verma and Sharma 2016). The objective of this chapter is to provide a brief overview of the various biodiesel production strategies. Different feedstocks and process technologies developed so far have been summarized. In addition, the quality standards, the economic aspects, and the environmental impact of biodiesel production processes are emphasized.

## 11.2 FEEDSTOCK FOR BIODIESEL PRODUCTION

The composition and purity of the biodiesel is dependent upon the type of feedstock used for its production (Ambat et al. 2018). In general, it is important that the biodiesel feedstock is rich in lipid content mainly in the form of triglycerides.

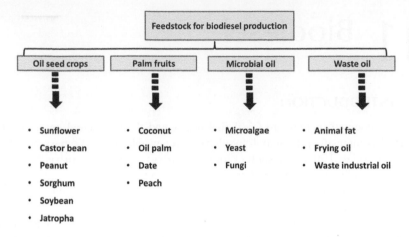

**FIGURE 11.1**   Feedstock for biodiesel production.

Thus, selection of the appropriate feedstock is the key step that influences the yield, productivity, quality, and overall economy of the biodiesel production process. Various sources ranging from edible oil seed crops to non-edible animal fats, algae oils, and microbial oil sources have been used for biodiesel production (Figure 11.1). Currently, major research efforts are exploring new potential biodiesel feedstocks that are independent of regional climate, agricultural conditions, and geographical locations to ensure long-term sustainability of the process. It is estimated that about 350 oil feedstock are available for biodiesel production globally (Kumar and Sharma 2016). However, most present production technologies for biodiesel are dependent upon the conventional first-generation feedstocks (i.e., agricultural and oil seed crops). Brazil, the United States, and the European Union are among the largest producers of biodiesel (Khan et al. 2017). In Europe, almost 70% of biodiesel is made from rapeseed while soybean is the major source for biodiesel in the United States and Brazil. Palm oil corresponds to 21% of the total global production of oil and fats but only 1%–2% of it is used for biodiesel production (André Cremonez et al. 2015).

The use of first-generation feedstocks raises ethical and moral obligations because it directly leads to an increase in food prices and competes with agricultural crops for arable land and water footprint. In India, biodiesel production using edible oil crops is not feasible because more than 68% of edible oils are imported just to meet food requirements (Kumar and Sharma 2016). Thus, much attention has been diverted towards the use of non-edible oil crops such as jatropha oil, castor oil, linseed oil, and karanja and the waste-oriented sources such as waste cooking oils, waste industrial oils, and animal fat waste (Ambat et al. 2018). These non-edible oils crops and waste oil sources are categorized as second-generation feedstocks, which can be obtained at much lower costs and do not directly compete with food crop production. Among these, waste-oriented sources are more economical as they do not depend upon arable land, which is inevitable for the production of non-edible crops. It has been reported that animal waste fats such as those obtained from chicken, tallow, lard, fish processing units, or leather industry fleshing wastes have been used for industrial biodiesel production (Adewale et al. 2015). However, the major disadvantage of using these resources for biodiesel production is their compositional characteristics require complex production techniques. Moreover, the handling and storage issues limit their economic feasibility in the commercial market.

Besides the first and second-generation feedstocks, other precursors for biodiesel production include algal and microbial oils, which are categorized as third-generation feedstock. Algae tend to store a high amount of lipids under nutrient-limited conditions. Compared to other oil crops, they can be cultivated much efficiently and rapidly with low land usage and they do not compete with the production of food and other agricultural crops. Some of the main oil-producing microalgae include *Nanochloropsis oceanica, Chlamydomonas reinhardtii*, and *Prymnesium parvum*, which can accumulate lipid more than 50%w/w oil (dry basis) under nutrient-limited conditions (Adewale et al. 2015). In addition to algae, various yeast and fungal species have been reported to accumulate oils and fats during unfavorable environmental conditions (Ma et al. 2018). The discovery of these oleaginous species provides new opportunities for microbial oil-based biodiesel production. Moreover, with the recent developments in genetic and metabolic engineering tools, the metabolic pathways of plant and microbial species can be manipulated for obtaining high-lipid containing species with the desired fatty acid composition. Nonetheless, oil extraction and its processing, residual management, use of expensive catalysts, and catalyst regeneration still remain the technical challenges that require utmost consideration.

## 11.3 METHODS FOR BIODIESEL PRODUCTION

As stated earlier, fats and oils obtained from biomass cannot be used directly as fuel in diesel engines and need to be converted to a more usable form (i.e., the fatty acid methyl esters (FAME) or biodiesel). The most common method for obtaining industrial biodiesel is transesterification due to its high conversion efficiency (Faried et al. 2017). Other methods available for producing biodiesel include pyrolysis, thermal cracking, microemulsion, and hydrocarbon blending (Baskar and Aiswarya 2016). However, these methods are energy intensive and are seldom used for industrial biodiesel production. The transesterification reaction mainly involves the reaction of triglycerides with short chain alcohol (Equation 11.1). Various factors influence the transesterification process such as the molar ratio (alcohol:oil), type of short chain alcohol used, presence or absence of catalyst, reaction time, and temperature. The type of alcohol directly affects the yield of the FAME obtained. Since the transesterification reaction is an equilibrium reaction, excess alcohol should be added to break the glycerine fatty acid linkages (Meher et al. 2006). However, using a high amount of alcohol makes the ester recovery process complicated and raises the biodiesel cost. Methanol is the most common alcohol used for transesterification as it is inexpensive, polar, and can easily react with triglycerides (Ma et al. 2018). Depending upon the type of reaction (catalytic/non catalytic) and process steps involved, various forms of transesterification processes are categorized as described in the following section.

## 11.3.1 CATALYTIC TRANSESTERIFICATION

The rate of a transesterification reaction is accelerated with the addition of a catalyst. Various types of catalysts have been used for the transesterification reaction as
shown in Figure 11.2. The traditional catalytic transesterification mainly involves
the use of acid and base catalysts that can be homogeneous or heterogeneous in
nature (Baskar and Aiswarya 2016). The homogeneous catalysts usually are added
in a liquid state while the heterogeneous catalysts are in a solid state. Industrially,
homogeneous base catalysts such as sodium hydroxide, potassium hydroxide, or
methoxide are preferred because they are less expensive and generally provide a
faster rate of reaction compared to other catalysts (Aransiola et al. 2014). The major
limitation, however, is their extreme sensitivity to water, which can lead to saponification under high alkaline conditions (Leung et al. 2010). Saponification is detrimental to the transesterication process because it not only consumes the catalyst
but can also form emulsions that make the recovery and purification of biodiesel
obscure. Thus, it is essential that while using homogeneous base catalysts, the starting materials are anhydrous and comprised of lower free fatty acids (FFA). On the
other hand, the homogenous acid catalysts are more suitable for feedstock with high
FFA. Various acids such as sulfuric acid, phosphoric acid, and hydrochloric acid
have been used for homogeneous acid-based transesterification reaction (Avhad and
Marchetti 2015). The major limitation of acid catalyzed process is that it is relatively
a slow process. To improve the process efficiency, many researchers have combined
both acidic and alkaline catalysts in a two-step reaction in which the acid treatment
converts the FFA into esters while the alkaline catalyst is performing the transesterification (Ramadhas et al. 2005; Aransiola et al. 2014). It is estimated that two-step
transesterification can be completed much faster compared to single-step homogeneous acid or base catalysis (Suresh et al. 2018). However, the major disadvantage is
the additional costs incurred for the extra step used to convert FFA to methyl esters.

A possible alternative to a homogeneous acid or base catalyst is to use a solid
heterogeneous catalyst. Unlike homogeneous catalysts, heterogeneous catalysts can

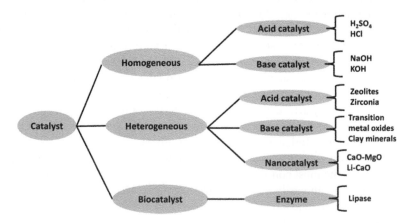

**FIGURE 11.2**   Classification of catalysts used in transesterification.

be regenerated once the transesterification reaction is completed, thus in turn minimizing the material and processing costs. Heterogeneous catalysts lead to a better separation of the final product and can be applied in both batch and continuous modes without the need for additional purification steps (Aransiola et al. 2014). They also can be classified into acid and base heterogeneous catalysts. Nafion-NR50 and sulphated or tungstated zirconia are the most commonly used acidic heterogeneous catalyst while the alkali heterogeneous catalysts include metal oxides such as calcium oxide, zinc oxide, and aluminum oxide. Besides the regenerative capacity, heterogeneous catalysts have several other advantages that include prolonged life time, relatively inexpensive (compared to homogeneous catalysts), less corrosive, less toxic, and environment friendly. However, in spite of several advantages, heterogeneous catalysts decrease the overall rate of reaction due to the diffusional limitation between the three phases (i.e., catalyst, alcohol, and oil). To overcome this limitation, a co-solvent is preferably added that promotes mixing of oil and methanol and accelerates the reaction rate. The most commonly used co-solvents are tetrahydrofuran (THF), dimethyl sulfoxide (DMSO), n-hexane, and ethanol. An alternative strategy to improve diffusion of heterogeneous catalysts is by using structural promoters or catalyst supports. These promoters or supports enhance the surface area or pores of the catalyst, which help in creating more reactive sites for triacylglycerol (TAG) molecules. Recently, heterogeneous nanocatalysts have gained wide attention for application in biodiesel production due to their high catalytic efficiency (Ambat et al. 2018). Compared to conventional catalysts, nanocatalysts have several advantages such as high surface area, high stability, reusability, high resistance to saponification, and high activity. Moreover, nanocatalysts can be modified to obtain desired properties by adjusting synthesis parameters such as concentration, reducing agent, and calcination temperature (Ambat et al. 2018). During transesterification, they are generally used with the help of solid supports such as zirconia, zeolites, and oxides. Table 11.1 shows a comparative analysis of various catalytic transesterification methods used for the production of biodiesel in terms of their conversion efficiencies.

## 11.3.2 Supercritical Transesterification (Non-catalytic Transesterification)

The catalytic transesterification reaction suffers from various disadvantages such as water sensitivity, catalyst consumption and its removal, waste generation, and low purity products (Aransiola et al. 2014). These disadvantages can be addressed through biodiesel production at supercritical conditions (i.e., subjecting the reaction mixture to high temperatures and pressures beyond the critical point). Under such conditions, the reaction mixture becomes homogeneous, thus favoring both esterification and transesterification reactions and does not require the presence of the catalyst. This is a method suitable for any type of feedstock including those containing high FFA and high water content, thus having greater advantage than catalytic transesterification (Baskar and Aiswarya 2016). Moreover, the extra purification steps are not required because no catalyst is involved in the reaction. The typical operating conditions for supercritical transesterification include temperatures in the

**TABLE 11.1**

**Comparison of Various Catalytic Transesterification Methods for the Production of Biodiesel**

| Feedstock | Alcohol | | Catalyst | | Reaction Temperature (°C) | Duration | Biodiesel Conversion (%) | References |
|---|---|---|---|---|---|---|---|---|
| | Type | Molar Ratio (alcohol:oil) | Type | Concentration | | | | |
| Karanja oil (*Pogomia pinnata*) | Methanol | 10:1 | KOH | 1% | 60 | 1.5 h | 92 | Karmee and Chadha (2005) |
| Waste frying oil | Methanol | 245:1 | H₂SO₄ | 3.8 M | 70 | 4 h | 99 | Zheng et al. (2006) |
| Tobacco (*Nicotiana tabacum*) | Methanol | 18:1 | H₂SO₄ | 1% | 60 | 25 min | 91 | Veljković et al. (2006) |
| | Methanol | 6:1 | KOH | 1% | | 30 min | | |
| Soyabean oil (*Glycine max*) | Methanol | 40:1 | TiO₂/ZrO₂ | (11 wt% Ti) | 250 | 2h | >95 | Furuta et al. (2006) |
| | | | Al₂O₃/ZrO₂ | (2.6 wt% Al) | | 2h | >95 | |
| | | | K₂O/ZrO₂ | (3.3 wt% K) | | 2h | >95 | |
| Polanga oil (*Calophyllum inophyllum*) | Methanol | 6:1 | H₂SO₄ | 0.7% | 65 | 2h | >99 | Sahoo et al. (2007) |
| | Methanol | 9:1 | KOH | 1.5% | | 4h | | |
| Trap grease | Methanol | 35:1 | H₂SO₄ | 11.3% | 90 | 4.6h | 90 | Wang et al. (2008) *(Continued)* |

**TABLE 11.1 (Continued)**
**Comparison of Various Catalytic Transesterification Methods for the Production of Biodiesel**

| Feedstock | Alcohol | | Catalyst | | Reaction Temperature (°C) | Duration | Biodiesel Conversion (%) | References |
|---|---|---|---|---|---|---|---|---|
| | Type | Molar Ratio (alcohol:oil) | Type | Concentration | | | | |
| Kusum oil (*Schleichera trigusa*) | Methanol | 10:1 | $H_2SO_4$ | 1% | 50 | 1 h | 97 | Sharma and Singh (2010) |
| | Methanol | 8:1 | KOH | 0.7% | | 1 h | | |
| Palm oil | Methanol | 12:1 | KF/Ca Al Hydrotalcite | 5% | 65 | 5 h | 98 | Gao et al. (2010) |
| Okra oil (*Hibiscus esculentus*) | Methanol | 7:1 | $CH_3ONa$ | 1% | 65 | 2 h | 97 | Anwar et al. (2010) |
| *Scenedesmus obliquus* | Methanol | 75:1 | HCl | 4.5 M | 60 | 7 h | 91.1 | Mandal and Mallick (2012) |
| *Spirulina platensis* | Methanol | 4:1 | $H_2SO_4$ | 5% | 60 | 90 min | 75 | Nautiyal et al. (2014) |
| Microalgae oil (*Chlorella protothecoides*) | Methanol Ethanol | 10:1–42:1 | Supercritical $CO_2$ | — | 270–350 | 10–50 min | 90.8 87.8 | Nan et al. (2015) |

range of 280–400°C, pressures ranging from 10 to 30 MPa, and the reaction time of 2–4 min (Aransiola et al. 2014). Despite its advantages, the major disadvantage of this method is the requirement for high-operating temperatures and pressures along with high alcohol to oil molar ratio which renders its commercial exploitation ineffective.

### 11.3.3  IN SITU TRANSESTERIFICATION

*In situ* transesterification also known as reactive extraction is an alternative to the conventional transesterification process in which the production steps are reduced because it does not involve the oil or fat extraction step prior to the transesterification step (Go et al. 2016). In this method, oil seeds are ground and then reacted directly with alcohol and a catalyst that produces alkyl fatty acid esters (Kasim et al. 2010). By using this method, the amount of alcohol required for the reaction can be significantly reduced. It is noted that by using the *in situ* transesterification process, a high yield and a high purity of biodiesel can be achieved (Baskar and Aiswarya 2016). Moreover, this process lowers the initial capital investments as well as production costs, thereby improving the economic viability of the process. Similar to conventional transesterification, various parameters such as the type of feedstock, catalyst type, particle size, moisture content, molar ratio, and the alcohol type greatly influence the yield and purity of the biodiesel produced (Samuel et al. 2012). Although the technical feasibility of this process has been successfully demonstrated by several researchers, the need for a high volume of alcohol poses major challenge to gaining high marginal profits from the process.

### 11.3.4  ULTRASOUND-ASSISTED TRANSESTERIFICATION

In the ultrasound-assisted transesterification, cavities are created between the oil and alcohol phases by ultrasonication to provide efficient mixing (Baskar and Aiswarya 2016). These cavities improve mixing by enhancing the mass transfer rates, thus improving the overall biodiesel yield. In addition, the ultrasonic waves generate high temperature and pressure during the reaction and provide high catalytic surface area. Furthermore, the high-intensity ultrasonic waves provide the activation energy needed for the reaction initiation. Other advantages of the ultrasound-assisted transesterification process include high reaction rate, short separation time, and simplified post processing (Aransiola et al. 2014). Compared to other conventional mechanical stirrers, this process consumes less energy, thus enhancing the efficiency of the process. It is estimated that biodiesel yield can be increased to about 95%–99% using this technology (Ho et al. 2016).

### 11.3.5  REACTIVE DISTILLATION

In the reactive distillation process, the transesterification reaction and the thermodynamic separation occur simultaneously (Kralova and Sjöblom 2010). Both catalytic and non-catalytic transesterification reactions can be performed using this technology. The use of a high amount of alcohol can be avoided with this process because the equilibrium can be shifted towards methyl ester formation simply by continuous removal of the by-product (Le Chatelier's principle). The easy recycling of product and by-products ensures the purity of the final product (Baskar and Aiswarya 2016). In addition, this process can significantly alleviate the capital investment, operation costs, and provide effective separation (Poddar et al. 2015). The reactive distillation process has been reported as economically acceptable with high selectivity (Aransiola et al. 2014).

### 11.3.6 Enzyme-Catalyzed Transesterification

An environmentally friendly alternative for the chemical catalytic transesterification process is the use of biocatalysts (i.e., enzymes as catalysts for transesterification reaction) (Figure 11.3). Enzymes are proteins produced by the microorganisms that have tendency to catalyze the biochemical reactions. The potential of enzymes for catalyzing the transesterification reaction has been assessed by many researchers (Meher et al. 2006; Marchetti et al. 2007; Avhad and Marchetti 2015). Mainly, biocatalytic transesterification is mediated by lipases (both intra- and extracellular) (Aransiola et al. 2014). These enzymes are used mainly in immobilized forms to improve the enzyme stability and reusability. Due to the high selectivity of enzymes, the overall biodiesel yield can be enhanced with minimal side reactions. Moreover, these reactions can be carried out at ambient temperatures and pressures with lower energy input and require less purification steps compared to conventional chemical catalysts. However, the efficiency of enzymatic transesterification is dependent upon the enzyme source and the prevailing operational conditions. Most enzymes get deactivated in the presence of a high amount of alcohol which further limits the efficiency of the process. Due to the high price of enzymes, their use for biocatalyzed transesterification is restricted to commercial exploitation.

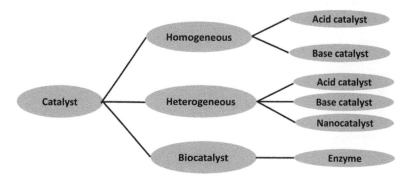

**FIGURE 11.3**   Classification of the catalyst.

## 11.4  STANDARDIZATION OF BIODIESEL

Since the properties of biodiesel are influenced by the type and composition of the feedstock used for its production, it becomes essential to standardize the biodiesel quality prior to its commercialization. The accepted biodiesel standards of various countries are provided in Table 11.2. According to these standards, the biodiesel quality is assessed in terms of different physicochemical properties such as FAME composition, viscosity, density, heating value, flash point, acid value, and iodine value.

An important property of biodiesel is the cetane number, which is related to the ignition delay time and combustion quality (Demirbas 2007). High cetane numbers are preferred because they ensure a good cold start of the engine and minimize the white smoke. It is observed that the esters of saturated fatty acids such as palmitic acid and stearic acids have high cetane numbers (Ma et al. 2018). Therefore, the number of fatty acid carbon chains can indirectly provide the measure of the cetane number. The longer the chain, the more saturated the molecules are, and thus the cetane numbers are higher (Avhad and Marchetti 2015). Although a high degree of saturation favors the cetane numbers, it drastically affects the cloud point of the biodiesel. Cloud point refers to the temperature below which wax or biowax separates from the oils and forms a cloudy appearance. The presence of solidified wax can clog fuel filters and injectors in diesel engines. The higher cloud points affect the practical performance of biodiesel in automotive applications at low temperatures (Kumar and Sharma 2016). Biodiesel obtained from vegetable oils shows high cloud point due to the presence of high saturated fatty acids, which have higher melting points than unsaturated fatty acids. Similarly, viscosity also tends to increase with increase in degree of saturation and chain length. On the contrary, the heat of combustion increases with the increase in fatty acid carbon chain and decreases with unsaturation. Thus, it can be suggested that biodiesel containing saturated and monounsaturated fatty acids would exhibit improved fuel properties (Verma and Sharma 2016).

### TABLE 11.2
### Biodiesel Standards of Various Countries

| Country | Standard Specification |
| --- | --- |
| United States | ASTM D 6751 |
| European Union | EN 14213 (heating fuels) and EN 14213 (automotive fuels) |
| Brazil | ANP 42 |
| India | IS 15607 |
| Japan | JASO M360 |
| South Africa | SANS 1935 |

*Source:* Barabas, I. and Todoru, I.-A., Biodiesel quality, standards and properties, In: *Biodiesel: Quality, Emissions and By-products*, InTech, 2011.

## 11.5   ECONOMY AND ENVIRONMENTAL IMPACT OF BIODIESEL

The economic feasibility of biodiesel production is limited mainly by the cost of raw material, which accounts for almost 80% of the production costs (Hasan and Rahman 2017). At present time, the costs of biodiesel are much higher compared to gasoline or petrodiesel (Meher et al. 2006). In order to minimize this cost, cheaper raw materials and simple production strategies need to be implemented. The major concern for using low-cost non-edible oil crops and waste oils is the presence of higher amount of impurities which require additional pre-treatment, processing, and purification steps. However, these additional steps lead to an increase in overall manufacturing costs, which is undesirable. Thus, for a profitable biodiesel production, a compromise must be made between the use of inexpensive raw materials and the costs incurred for additional pre- and post-processing steps. It is estimated that for the commercial viability of a biodiesel production plant, system variables such as feedstock costs, market price of biodiesel, market price of glycerol, and the plant capacity are crucial (Gebremariam and Marchetti 2018).

Biodiesel has several environmental benefits compared to conventional fossil fuels. It is observed that 60% and 90% of GHG emissions can be reduced by using first-generation and second-generation feedstocks, respectively (Hasan and Rahman 2017). Moreover, using biodiesel in engines reduces sulphur dioxide emissions, which is the primary cause for acid rain (Suresh et al. 2018). Due to the high percentage of oxygen, biodiesel tends to decrease emissions of other harmful compounds such as NOx and PAHs (polycyclic aromatic hydrocarbons) through complete combustion of the fuel. However, excessive land and water usage for biodiesel production can negate the advantages of biodiesel usage. On the other hand, by using degraded land and wastewater for biodiesel production, a positive environmental impact can be achieved. For a sustainable biodiesel production, it is essential to reduce the adverse environmental impacts of biodiesel production. In addition, national polices must be implemented to recognize the global impact of biodiesel production.

## 11.6   CONCLUSION AND PERSPECTIVES

The demand for biodiesel is increasing exponentially because it provides a greener alternative to conventional fossil fuels. Due to the high combustible value, it is much safer to use, store, and transport biodiesel compared to petrol and diesel. Biodiesel is non-toxic in nature and has lubricating properties that improve the engine performance and its longevity. However, due to the high production costs, the market potential of biodiesel is still unsatisfactory. Research in this area must be directed towards the development and use of inexpensive oil-rich feedstocks, selection of low-cost catalysts, and technological improvements with respect to processing and purification steps. Moreover, the quality and capacity of biodiesel must be enhanced to reduce the environmental impact. With the recent developments in process technologies such as *in situ* transesterification, ultrasound-assisted transesterification, and reactive distillation, much of the limitations have been overcome. However, further innovation and supportive government policies are essential for establishing a commercially viable process for biodiesel production.

## REFERENCES

Adewale P, Dumont M, Ngadi M (2015) Recent trends of biodiesel production from animal fat wastes and associated production techniques. *Renew Sustain Energy Rev* 45:574–588.

Ambat I, Srivastava V, Sillanpää M (2018) Recent advancement in biodiesel production methodologies using various feedstock: A review. *Renew Sustain Energy Rev* 90:356–369.

André Cremonez P, Feroldi M, Cézar Nadaleti W, De Rossi E, Feiden A, De Camargo MP, Cremonez FE, Klajn FF (2015) Biodiesel production in Brazil: Current scenario and perspectives. *Renew Sustain Energy Rev* 42:415–428.

Anwar F, Rashid U, Ashraf M, Nadeem M (2010) Okra (Hibiscus esculentus) seed oil for biodiesel production. *Appl Energy* 87:779–785.

Aransiola EF, Ojumu TV, Oyekola OO, Madzimbamuto TF, Ikhu-Omoregbe DIO (2014) A review of current technology for biodiesel production: State of the art. *Biomass and Bioenergy* 61:276–297.

Avhad MR, Marchetti JM (2015) A review on recent advancement in catalytic materials for biodiesel production. *Renew Sustain Energy Rev* 50:696–718.

Barabas I, Todoru I-A (2011) Biodiesel quality, standards and properties. In: *Biodiesel-Quality, Emissions and By-products*. InTech.

Baskar G, Aiswarya R (2016) Trends in catalytic production of biodiesel from various feedstocks. *Renew Sustain Energy Rev* 57:496–504.

Demirbas A (2007) Importance of biodiesel as transportation fuel. *Energy Policy* 35:4661–4670.

Demirbas A (2008) Biofuels sources, biofuel policy, biofuel economy and global biofuel projections. *Energy Convers Manag* 49:2106–2116.

Faried M, Samer M, Abdelsalam E, Yousef RS, Attia YA, Ali AS (2017) Biodiesel production from microalgae: Processes, technologies and recent advancements. *Renew Sustain Energy Rev* 79:893–913.

Furuta S, Matsuhashi H, Arata K (2006) Biodiesel fuel production with solid amorphous-zirconia catalysis in fixed bed bioreactor. *Biomass and Bioenergy* 30:870–873.

Gao L, Teng G, Xiao G, Wei R (2010) Biodiesel from palm oil via loading KF/Ca–Al hydrotalcite catalyst. *Biomass and Bioenergy* 34:1283–1288.

Gebremariam SN, Marchetti JM (2018) Economics of biodiesel production: Review. *Energy Convers Manag* 168:74–84.

Go AW, Sutanto S, Ong LK, Tran-Nguyen PL, Ismadji S, Ju YH (2016) Developments in in-situ (trans) esterification for biodiesel production: A critical review. *Renew Sustain Energy Rev* 60:284–305.

Hasan MM, Rahman MM (2017) Performance and emission characteristics of biodiesel–diesel blend and environmental and economic impacts of biodiesel production: A review. *Renew Sustain Energy Rev* 74:938–948.

Ho WWS, Ng HK, Gan S (2016) Advances in ultrasound-assisted transesterification for biodiesel production. *Appl Therm Eng* 100:553–563.

Karmee SK, Chadha A (2005) Preparation of biodiesel from crude oil of Pongamia pinnata. *Bioresour Technol* 96:1425–1429. doi:10.1016/j.biortech.2004.12.011

Kasim FH, Harvey AP, Zakaria R (2010) Biodiesel production by in situ transesterification. *Biofuels* 1:355–365.

Khan S, Siddique R, Sajjad W, Nabi G, Hayat KM, Duan P, Yao L (2017) Biodiesel production from algae to overcome the energy crisis. *HAYATI Journal of Biosciences* 24:163–167.

Kralova I, Sjöblom J (2010) Biofuels-renewable energy sources: A review. *J Dispers Sci Technol* 31:409–425.

Kumar M, Sharma MP (2016) Selection of potential oils for biodiesel production. *Renew Sustain Energy Rev* 56:1129–1138.

Leung DYC, Wu X, Leung MKH (2010) A review on biodiesel production using catalyzed transesterification. *Appl Energy* 87:1083–1095.

Ma Y, Gao Z, Wang Q, Liu Y (2018) Biodiesels from microbial oils: Opportunity and challenges. *Bioresour Technol* 263:631–641.

Mandal S, Mallick N (2012) Biodiesel production by the green Microalga scenedesmus obliquus in a recirculatory aquaculture system. *Appl Environ Microbiol* 78:5929–5934.

Marchetti JMÃ, Miguel VU, Errazu AF (2007) Possible methods for biodiesel production. *Renew Sustain Energy Rev* 11:1300–1311.

Meher LC, Vidya Sagar D, Naik SN (2006) Technical aspects of biodiesel production by transesterification: A review. *Renew Sustain Energy Rev* 10:248–268.

Nan Y, Liu J, Lin R, Tavlarides LL (2015) Production of biodiesel from microalgae oil (*Chlorella protothecoides*) by non-catalytic transesterification in supercritical methanol and ethanol: Process optimization. *J Supercrit Fluids* 97:174–182.

Nautiyal P, Subramanian KA, Dastidar MG (2014) Kinetic and thermodynamic studies on biodiesel production from Spirulina platensis algae biomass using single stage extraction–transesterification process. *Fuel* 135:228–234.

Poddar T, Jagannath A, Almansoori A (2015) Biodiesel production using reactive distillation: A comparative simulation study. *Energy Procedia* 75:17–22.

Ramadhas AS, Jayaraj S, Muraleedharan C (2005) Biodiesel production from high FFA rubber seed oil. *Fuel* 84:335–340.

Sahoo PK, Das LM, Babu MKG, Naik SN (2007) Biodiesel development from high acid value polanga seed oil and performance evaluation in a CI engine. *Fuel* 86:448–454.

Samuel OD, Eng M, Dairo OU, Ph D (2012) A critical review of in-situ transesterification process for biodiesel production. *Pacific J Sci Technol* 13:72–79.

Sharma YC, Singh B (2010) An ideal feedstock, kusum (*Schleichera triguga*) for preparation of biodiesel: Optimization of parameters. *Fuel* 89:1470–1474.

Suresh M, Jawahar CP, Richard A (2018) A review on biodiesel production, combustion, performance, and emission characteristics of non-edible oils in variable compression ratio diesel engine using biodiesel and its blends. *Renew Sustain Energy Rev* 92:38–49.

Veljković VB, Lakićević SH, Stamenković OS, Todorović ZB, Lazić ML (2006) Biodiesel production from tobacco (*Nicotiana tabacum* L.) seed oil with a high content of free fatty acids. *Fuel* 85:2671–2675.

Verma P, Sharma MP (2016) Review of process parameters for biodiesel production from different feedstocks. *Renew Sustain Energy Rev* 62:1063–1071.

Wang ZM, Lee JS, Park JY, Wu CZ, Yuan ZH (2008) Optimization of biodiesel production from trap grease via acid catalysis. *Korean J Chem Eng* 25:670–674.

Zheng S, Kates M, Dubé MA, McLean DD (2006) Acid-catalyzed production of biodiesel from waste frying oil. *Biomass and Bioenergy* 30:267–272.

# 12 Microbial Electrochemical Technologies and Their Applications

## 12.1 INTRODUCTION

The microbial conversion of biomass to electric current was made possible with the discovery of electricity-producing bacteria in 1911 by M.C. Potter (Potter 1911). Since then, tremendous research efforts are developing various types of microbial electrochemical technologies (METs) to harness the potential of these bacteria for electricity and biochemical production. In general, the METs can be classified into two broad categories depending upon the flow of electrons: (1) electron-producing METs and (2) electron-consuming METs (Figure 12.1). Among these, the most widely studied systems are the electron-producing microbial fuel cells (MFCs), which are used for direct electricity generation using organic substrates (Logan et al. 2006). In conventional MFCs, the anode is the abiotic component where the microbial oxidation of organic substrates occurs while the cathode is the abiotic part where the electrons and protons derived from the anodic oxidation (through the external circuit) recombine with $O_2$ to form water as the sole by-product (Figure 12.1a). Oxygen is the most common reductant used in the cathode chamber of MFCs due to its strong electronegative potential (Oh et al. 2004). However, due to its low solubility in the water, other electron acceptors such as potassium ferricyanide, potassium permanganate, and potassium dichromate have been used seldomly in MFCs (Pandit et al. 2011). Recently, the major focus has shifted towards the use biological cathodes (biocathodes) to reduce the oxygen dependence and improve the overall efficiency of the system (Sun et al. 2012; Jafary et al. 2017).

It was noted that by applying a small amount of voltage to the conventional MFCs, various value-added products such as hydrogen and methane can be generated in the cathode (Figure 12.1b). These electron-consuming METs, known as microbial electrolysis cells (MECs), are comprised of similar anode as MFCs but the cathode is supplied with extra current by using a direct current (DC) power source for hydrogen or methane production (Varanasi et al. 2019). In addition to the MECs, another electron-consuming application of MET is the microbial electrosynthesis (MES) process, wherein the potential of microorganisms to uptake electrons from the cathode is used for bioproduct synthesis and recovery (Figure 12.1c).

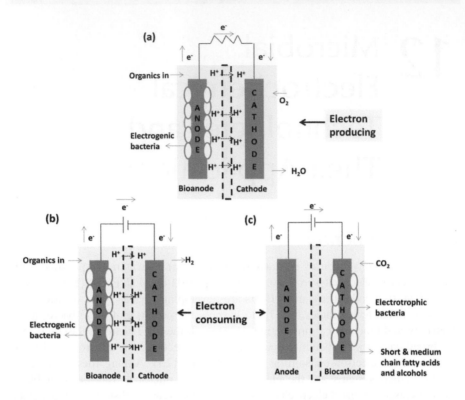

**FIGURE 12.1**  Types of microbial electrochemical technologies: (a) microbial fuel cell (MFC), (b) microbial electrolysis cell (MEC), and (c) microbial electrosynthesis (MES).

Unlike MFCs and MECs, the bacteria in MES tend to perform microbial reduction by harnessing electrons from the cathode to convert inorganic or organic carbon sources to long chain acids, alcohols, or other value-added compounds (Rabaey and Rozendal 2010). In this chapter, the basic principles governing the different types of METs are described with a key emphasis on the various technological advances and their applications. In addition, the current bottlenecks and the possible means to overcome them have been summarized.

## 12.2  PRINCIPLES OF METs

Unlike the traditional chemical fuel cells, METs employ low-cost, sustainable micro-organisms as biocatalysts to oxidize (or reduce) the organic and inorganic substrates. These biocatalysts have inherent metabolic properties. They interact with external solid surfaces or electrodes and perform the desired bioelectrochemical reactions. Irrespective of the type, the major components of METs include an anode, a cathode, electrolyte, membrane (optional), and external circuits (Figure 12.1). The working principle of METs is based on the fundamental thermodynamic principles and microbial physiological processes as described in the following section.

## 12.2.1 Thermodynamic Principles

The thermodynamic evaluation is crucial because it determines the feasibility and extent of the bioelectrochemical reaction. Here, the examples of acetate conversion to electricity in the MFC and acetate conversion to hydrogen in the MEC are described to evaluate the difference in thermodynamic feasibility of electron-producing and electron consuming METs. Considering acetate as the organic substrate ($[CH_3COO^-] = [HCO_3^-] = 10$ mM, pH 7, 298.15 K, $pO_2 = 0.2$ bar), the reactions occurring at the anode and cathode of MFC are given in Equations 12.1 and 12.2:

$$\text{Anode: } CH_3COO^- + 4H_2O \rightarrow 2HCO_3^- + 9H^+ + 8e^- \tag{12.1}$$

$$\text{Cathode: } 2O_2 + 8H^+ + 8e^- \rightarrow 4H_2O \tag{12.2}$$

By combining Equations 12.1 and 12.2, the overall reaction occurring in MFC can be given as:

$$\text{Overall: } CH_3COO^- + 2O_2 \rightarrow 2HCO_3^- + H^+ \; (\Delta G = -847.60 \text{ kJ/mol};$$
$$emf = 1.10 \text{ V}) \tag{12.3}$$

Thus, according to Equation 12.3, the $\Delta G$ is negative and the *emf* is positive implying that electricity generation in MFC is a thermodynamically feasible reaction (i.e., is spontaneous in nature). On the contrary, under similar anodic conditions in the MEC with hydrogen as the final product ($pH_2 = 1$ bar), the reaction occurring at anode is similar to Equation 12.1. However, the hydrogen evolution reaction occurring at cathode is represented as:

$$\text{Cathode: } 8H^+ + 8e^- \rightarrow 4H_2 \tag{12.4}$$

By combining Equations 12.1 and 12.4, the overall reaction occurring in MEC can be given as:

$$\text{Overall: } CH_3COO^- + 4H_2O \rightarrow 2HCO_3^- + H^+ + 4H_2 \; (\Delta G = 93.14 \text{ kJ/mol};$$
$$emf = -0.12 \text{ V}) \tag{12.5}$$

Therefore, according to Equation 12.5, the $\Delta G$ is positive and the *emf* is negative implying that hydrogen generation in MEC is thermodynamically not feasible (i.e., it is non-spontaneous in nature). Hence, electrons must be invested into the MECs to make the reaction feasible at room temperatures. Similar equations can be used for analyzing the thermodynamic feasibility of different bioelectrochemical reactions.

## 12.2.2 Electrogenic Bacteria and Mechanisms
### of Exocellular Electron Transfer

As stated earlier, METs utilize metabolic capabilities of various microorganisms that are capable of interacting with the solid metal surfaces. These microorganisms, known as electrogenic (electron-generating) or electrotrophic (electron-consuming) bacteria, mainly belong to the group α, β, and γ proteobacteria (Varanasi and Das 2018). In addition, a few species belonging to the groups *Firmicutes* and *Acidobacteria* also have been reported (Bretschger et al. 2010). These bacteria are ubiquitous in nature and can be grown in varied environmental conditions (Kiely et al. 2011). Different mechanisms have been elucidated for the e⁻ transfer between the bacteria and the electrode surface as shown in Figure 12.2. They can be broadly classified into direct and indirect electron transfer mechanisms. The indirect e⁻ transfer mostly involves the use of a mediator (endogenous or exogenous) that directs the electron transfer from the bacteria to the electrode surface and vice versa. In the earlier studies of MFCs, an exogenous mediator such as neutral red or methyl viologen was added with bacteria such as *E. coli* to produce appreciable power densities (Logan 2009). However, it was observed that certain bacteria were able generate electricity in the absence of an exogenous mediators, thus depicting a specific trait among certain group of bacteria. These bacteria (i.e., the exoelectrogens) have innate capabilities to secrete endogenous mediators to carry out the exocellular electron transfer. Examples of such bacteria include *Pseudomonas aeruginosa*, which releases phycocyanin and *Shewanella oneidensis* that produces flavins as electron shuttles (Franks et al. 2010). Besides these, certain bacteria were observed to generate power without the need of a mediator. These bacteria showed direct electron transfer either with the help of conductive cellular proteins such as c-type cytochromes of *S. oneidensis* or by the conductive pili or nanowires such as those found in *Geobacter sulfurreducens* (Borole et al. 2011). Over the years, several new electrogenic species capable of direct or indirect exocellular electron transfer have been discovered; however, the versatility of these organisms is inexplicable (Logan 2009). Similar mechanisms

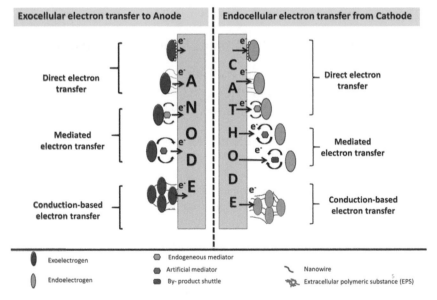

**FIGURE 12.2** Electron transfer mechanisms of electroactive bacteria to (or) from the electrode surface.

(direct and indirect) have been observed for electron uptake by the electrotrophic bacteria where the involved components or mediators operate at different redox potentials (Rosenbaum et al. 2011). However, as compared to the electron transfer process of bioanodes, the electron uptake mechanisms of biocathode are less understood. Some researchers have suggested that these bacteria closely resemble chemolithotrophic bacteria that are dependent upon reduced inorganic compounds as energy sources (Borole et al. 2011). Nevertheless, further studies are needed to completely understand their mechanistic behavior. Screening and exploring the potency of these microbes has brought about the emergence of new applications and designs in the field of wastewater treatment, bioremediation, and $CO_2$ sequestration as described in Section 12.3.

## 12.2.3 Reactor Architecture

The architectural aspects such as the design, configuration, electrode material, electrode spacing, and membrane type play a crucial role in depicting the performance of METs (Pham et al. 2009; Logan 2010; Kondaveeti et al. 2014). They mainly contribute towards the internal resistance of the system and facilitate the electron transfer reactions. Some of the essential features reported for effective performance are described in the following section.

### 12.2.3.1 Design and Configuration

The most conventional design of MFC is the H-type dual chambered system, which consists of a salt bridge that separates both the electrodes (Santoro et al. 2017). However, due to the high internal resistance offered by these systems, they are used mainly for conducting laboratory-scale research. Various other scalable configurations such as membrane-based H-type cell, cuboidal cell, air cathode system, and stack cell configuration have been reported (Logan et al. 2006). The design of MET is dependent mainly upon its type and application. For example, unlike the MFCs where the cathode is kept aerobic for oxygen reduction reaction, the MECs require strict anaerobic conditions in the cathode for hydrogen evolution (Varanasi et al. 2019). Therefore, the design criteria must meet the requirement of the particular application for which MET is intended as detailed in Section 12.3.

### 12.2.3.2 Electrodes

The microbial electrochemistry is confined mainly to the electrode surfaces and thus selection of appropriate material is essential to facilitate the desired oxidation or reduction reaction. The bioanode chosen must be inexpensive, biocompatible, conductive, porous, and stable to favor the bacterial attachment and improve the mass transfer rates (Pant et al. 2011). Similar material attributes are required for the adequate working of the cathode. Most METs rely on carbon-based electrodes as they supply the aforementioned properties (Zhou et al. 2011; Sun et al. 2012). The most commonly used carbon-based electrode materials include activated carbon, graphite, carbon sheet, carbon cloth, carbon felt, and carbon mesh (Table 12.1). Despite their wide range applications, these materials are limited by poor electron transfer rate and often require the addition of a catalyst to aid in the electron transfer reactions.

**TABLE 12.1**
**Common Materials Used in METs**

| Components | Materials | References |
|---|---|---|
| Anode | Stainless steel mesh, graphite plates, graphite fiber, rods, carbon felt, carbon paper, carbon cloth, carbon foam, glassy carbon | Dumas et al. (2007); Logan (2010); Pant et al. (2011); Wei et al. (2011) |
| Cathode | Platinum, graphite, carbon felt, carbon paper, carbon cloth, carbon foam, glassy carbon, stainless steel mesh | Call et al. (2009); Minteer et al. (2012); Ishii et al. (2012); Sharma et al. (2013); Mashkour and Rahimnejad (2015) |
| Ion exchange membrane | Salt bridge, PEM, CEM, AEM, bipolar membranes | Li et al. (2010); Khilari et al. (2013); Kadier et al. (2014) |
| Electrode catalyst | Pt, Pd, carbon nanotubes, $MnO_2$, $WO_3$, polyaniline, graphene, Ni, $MoS_2$, MoC, Fe, nitrogen | Deng et al. (2010); Tokash and Logan (2011); Kuo et al. (2012); Morales-Guio et al. (2014); Varanasi et al. (2016) |

### 12.2.3.3 Catalysts

The poor efficiency of carbon-based electrodes has prompted researchers to modify their surfaces for better catalytic performance. For bioanode, these modifications facilitate bacterial affinity towards the electrode, thereby accelerating the biofilm formation and improving the electron transfer rate (Varanasi et al. 2016). These modifications include pre-treatment techniques such as acid/base treatment, functionalization by heteroatom doping like nitrogen, and nanomaterial impregnation such as graphene and carbon nanotubes (Pham et al. 2009). Cathode materials contribute significantly to determining the performance of METs. For example, due to the sluggish three-phase oxygen reduction reaction (ORR) in the cathode, it is the rate-limiting step. Similarly, the hydrogen evolution reaction (HER) is the rate-limiting step in the MECs. Platinum-based catalysts are most widely used in METs due to their higher kinetic activity. However, their high price and sensitivity have hindered their practical scale application. Various other non-platinum-based ORR catalysts have been used in MFCs to improve the cathode kinetics (Li et al. 2010; Khilari et al. 2015; Rout et al. 2018), but their stability and long-term durability is questionable. In recent times, several other new materials such as three dimensional (3D) foam-based electrodes and polymer-based catalysts have gained much attention owing to their high catalytic activity and stable electronic structure (Cheng et al. 2006; Bhadra et al. 2009).

### 12.2.3.4 Membranes

The anode and cathode of METs usually are separated by a salt bridge or a membrane for the selective transport of ions. Although membraneless systems are more economically efficient, the absence of membrane results in substrate/product diffusivity, which consequently reduces the overall columbic efficiencies (Hamelers et al. 2010). The most commonly used membrane material is Nafion 117 for proton exchange (Cheng and Logan 2008; Christgen et al. 2015). The major constraint of this membrane is the high cost and the pH splitting phenomena, where the pH of the anodic chamber decreases while that in the cathodic chamber increases (Cheng and Logan 2008). This phenomena lowers the system stability and bioelectrochemical performance. Over the past few years, a variety of membrane separators have been extensively explored for METs such as cation exchange membrane, anion exchange membrane, ultrafiltration membranes, bipolar membrane, microfiltration membrane, glass fibers, ceramic membranes, porous fabrics, and other course pore filter materials (Kim et al. 2007; Rozendal et al. 2007; Harnisch et al. 2008; Leong et al. 2013; Rahimnejad et al. 2015; Christgen et al. 2015).

## 12.3   APPLICATIONS OF METs

Although the major application of traditional MFCs was electricity generation, the development of various forms of METs has opened doors for multifaceted applications. Figure 12.3 illustrates the some of the major applications of METs.

### 12.3.1   WASTEWATER TREATMENT

Various organic wastes and wastewaters have been used as substrates in METs to serve as electron donors for the electroactive bacteria (Logan and Rabaey 2012). Almost all biodegradable organic or inorganic matter present in the wastewater can be readily oxidized or reduced in METs. By using these technologies, the energy lost in the operation of aeration and sludge disposal during conventional wastewater treatment processes (such as activated sludge process) can be compensated. In addition, generation of an appreciable amount of electricity or other value-added products while treating targeted wastes such as industrial wastes, domestic wastes, and human excretory wastes or urine shows a promising future for a real-world application of METs.

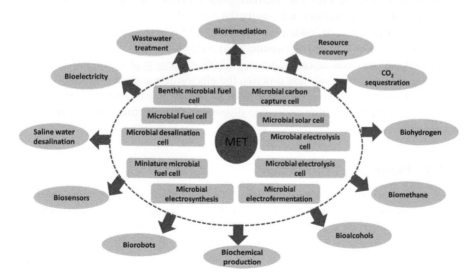

**FIGURE 12.3**   Multifaceted applications of microbial electrochemical technologies.

### 12.3.2  BIOREMEDIATION

Another major application of MET platform is the remediation of underground contaminants by using benthic microbial fuel cells. These systems use the naturally occurring potential difference between the underground anaerobic sediment and the oxygenic seawater for electricity generation (Wang and Ren 2013). By using the indigenous microflora or by bio-augmenting with known electroactive bacteria, the underground contaminants are degraded by concurrent electricity generation. Biodegradation of various xenobiotic compounds including drugs, industrial pollutants, and other naturally occurring environmental pollutants has been demonstrated successfully by using METs (Wu et al. 2014; Kaushik 2015).

### 12.3.3  BIOHYDROGEN PRODUCTION

The primary application of MECs is biohydrogen production. Compared to other biohydrogen production technologies such as fermentation or biophotolysis, MECs can overcome the thermodynamic limitations and improve the substrate conversion efficiencies (Zhang and Angelidaki 2014). Theoretically, an additional 0.3 V is needed by a MEC to produce hydrogen at the cathode, which is much less compared to water electrolysis (Varanasi et al. 2019). However, practically higher voltages are required due to the inherent voltage losses. Various researchers have shown high-rate hydrogen production using these systems in successful pilot scale demonstrations (Logan et al. 2008; Kyazze et al. 2010; Cusick et al. 2011; Gil-Carrera et al. 2013).

### 12.3.4  BIOMETHANE PRODUCTION

It was observed that methane can be produced in MECs by using a biocathode comprised of methanogenic bacteria. Methanogens generate methane in MECs either through direct electron transfer or by using the evolved hydrogen (Siegert et al. 2014). Producing methane has an added advantage of $CO_2$ capture along with methane generation. It is estimated that electrochemical reduction of $CO_2$ using electromethanogenesis has an electron capture efficiency of 96% (Cheng et al. 2009). Compared to the traditional anaerobic digestion process, the electromethanogenesis process offers higher methane productivities (Booth 2009). Moreover, this technology can be integrated easily into the existing anaerobic digestion infrastructure, thereby reducing the capital investments.

### 12.3.5  BIOCHEMICALS PRODUCTION

The discovery of microbial electron uptake from solid electrodes provided a nexus of opportunities for biocathode-based METs such as microbial electrosynthesis (MES) and microbial electrofermentation (MEF). Using these systems, the metabolic and fermentative pathways of microorganisms are directed towards the product of interest (Rabaey et al. 2011). At present, the major application of these systems is focused towards $CO_2$ reduction to multi-carbon compounds such as short or medium chain fatty acids and bioalcohols that have industrial importance. However, compared to the existing industrial processes, no substantial benefit is offered by METs and further technological advancements are needed to make these systems economically competitive.

## 12.3.6 SALINE WATER DESALINATION

By introducing an extra chamber between the anode and cathode of MFCs or MECs along with the addition of an anion exchange membrane (AEM)-facing anode and a cation exchange membrane (CEM)-facing cathode, microbial desalination cells (MDC) can be developed (Kim and Logan 2013). These systems have shown promising application in saline water desalination. In MDCs, the electric potential generated across the anode and cathode is used to drive the desalination process. Different variants of MDCs have been developed over the years such as microbial electrodialysis cells (MEDC), which can perform simultaneous hydrogen production and water desalination (Zhang and Angelidaki 2014). The membranes in microbial saline wastewater electrolysis cells (MSC) involve for treating the saline wastewater present in the anode and cathode chambers (Wang and Ren 2013). Osmotic MDCs use the forward osmosis membrane in place of AEM for wastewater treatment, water extraction and electricity generation (Zhang et al. 2011) and microbial electrolysis, desalination and chemical production cells (MEDCC), which comprises of four chambers with bipolar membrane between anode and AEM (Zhang and Angelidaki 2014). In MEDCCs, the water desalination is accompanied with alkali and acid production for enhanced energy recovery.

## 12.3.7 BIOMASS PRODUCTION AND $CO_2$ SEQUESTRATION

Using the oxygen producing algae and cyanobacteria in the cathode of MFCs can greatly reduce the sluggish kinetics of the ORR with an added benefit of $CO_2$ sequestration. In this method, the modified MFCs known as microbial carbon capture cells (MCC) were developed that can generate appreciable amounts of electricity with simultaneous wastewater treatment at anode and $CO_2$ sequestration at cathode (Pandit et al. 2012). Although promising, these systems must be explored further for practical scale applications. In addition to MCCs, the identification of electroactive photosynthetic organisms expanded further the horizon of METs for direct solar energy conversion to electricity. These METs known as microbial solar cells or bio-solar cells use the photosynthetic efficiencies of certain algae and cyanobacteria to generate biomass for food and feed with simultaneous electricity generation and $CO_2$ sequestration (Strik et al. 2011). The example of an electroactive cyanobacterium is *Synechocystis* PCC-6803, which can perform direct electron transfer to the anode through the nanowires (Santoro et al. 2017). Currently, the research on biosolar cells focuses on the development of artificial photosynthetic system to improve further the photosynthetic efficiencies of the system along with liquid biofuel production.

## 12.3.8 BIOSENSORS

Due to the low power obtained from the MFCs, major efforts have been made to apply these technologies for low power applications (Tender et al. 2008). It is observed that MFCs can readily power small sensor devices such as biosensors (Yang et al. 2015). These biosensors can be used readily for accurate and precise measurement of analytical signals. Various sensors such as the BOD sensor, toxicity sensor, and electrogenic activity-detecting sensor have been tested using MFCs (Veerubhotla and Das 2017). With the development of miniature MFCs and paper-based MFCs in recent years, new novel applications in the field of biosensors and biorobotics can be foreseen in the future.

## 12.4 CHALLENGES AND FUTURE PROSPECTS

Since last decade, the versatility and performance of METs have expanded exponentially. However, there are several technical challenges that must be overcome for successful implementation of this technology at commercial scales. The main disadvantage remains with the reactor architecture using expensive catalysts and membranes, which must be substituted with cheaper alternatives. At present, most of the METs are only demonstrated at laboratory scales and the few pilot-scaled studies performed so far have had problems in reproducing similar current densities (Logan 2010). In addition, the reactor configurations developed so far cannot be integrated with the existing infrastructure and hence require additional capital investments. Compared to the traditional MFCs, recent applications such as bioremediation, resource recovery, heavy metal removal using benthic MFCs, desalination using MDCs, and biochemical production and $CO_2$ reduction using MES are considered more technologically and economically feasible (Wang and Ren 2013). However, these systems are relatively new and only few pilot-scale studies have been reported. Only limited evaluations have been made on the life cycle analysis of these systems and their potential benefits in comparison to the established technologies are uncertain. For establishing the longevity of METs, the energy and environmental footprints of different METs must be determined prior to large-scale applications of these systems. Moreover, the fundamental understanding of the bioelectrochemical basis of electron transfer mechanisms of the electroactive bacteria to and from the solid electrode surfaces is crucial for further development of these technologies. With the recent progress made in the synthesis and application of new materials and the continuing development of innovative technologies, the viability of METs in the long run is inevitable.

## 12.5 SUMMARY AND CONCLUSION

The rapidly developing METs are promising sustainable technologies that have potential applications in the energy and the environment sector. The fundamental basics of METs and their recent advances with respect to targeted oriented applications such as energy generation, wastewater treatment, and biochemical production have been reviewed in details. To further exploit the unique advantages of these systems, a multi-disciplinary approach is essential. Although the feasibility of a few METs has been demonstrated successfully, further research is needed to assert their efficiency, scalability, and durability for commercial applications.

## REFERENCES

Bhadra S, Khastgir D, Singha NK, Lee JH (2009) Progress in preparation, processing and applications of polyaniline. *Prog Polym Sci* 34:783–810.

Booth B (2009) Electromethanogenesis: the direct bioconversion of current to methane. *Environ Sci Technol* 43:4619–4619.

Borole AP, Reguera G, Ringeisen B, Wang ZW, Feng Y, Kim BH, Nevin K et al., (2011) Electroactive biofilms: Current status and future research needs. *Energy Environ Sci* 4:4813.

Bretschger O, Osterstock JB, Pinchak WE, Ishii S, Nelson KE (2010) Microbial fuel cells and microbial ecology: Applications in ruminant health and production research. *Microb Ecol.* 59:415–427.

Call DF, Merrill MD, Logan BE (2009) High surface area stainless steel brushes as cathodes in microbial electrolysis cells. *Environ Sci Technol* 43:2179–2183.

Cheng S, Liu H, Logan BE (2006) Power densities using different cathode catalysts (Pt and CoTMPP) and polymer binders (Nafion and PTFE) in single chamber microbial Fuel Cells. *Environ Sci Technol* 40:364–369.

Cheng S, Logan BE (2008) Evaluation of catalysts and membranes for high yield biohydrogen production via electrohydrogenesis in microbial electrolysis cells (MECs). *Water Sci Technol* 58:853–857.

Cheng S, Xing D, Call DF, Logan BE (2009) Direct biological conversion of electrical current into methane by electromethanogenesis. *Environ Sci Technol* 43:3953–3958.

Christgen B, Scott K, Dolfing J, Head IM, Curtis TP (2015) An evaluation of the performance and economics of membranes and separators in single chamber microbial fuel cells treating domestic wastewater. *PLoS One* 10:1–13.

Cusick RD, Bryan B, Parker DS, Merrill MD, Mehanna M, Kiely PD, Liu G, Logan BE (2011) Performance of a pilot-scale continuous flow microbial electrolysis cell fed winery wastewater. *Appl Microbiol Biotechnol* 89:2053–2063.

Deng L, Zhou M, Liu C, Liu L, Liu C, Dong S (2010) Development of high performance of Co/Fe/N/CNT nanocatalyst for oxygen reduction in microbial fuel cells. *Talanta* 81:444–448.

Dumas C, Mollica A, Féron D, Basséguy R, Etcheverry L, Bergel A (2007) Marine microbial fuel cell: Use of stainless steel electrodes as anode and cathode materials. *Electrochim Acta* 53:468–473.

Franks AE, Malvankar N, Nevin KP (2010) Bacterial biofilms: the powerhouse of a microbial fuel cell. *Biofuels* 1:589–604.

Gil-Carrera L, Escapa A, Mehta P, Santoyo G, Guiot SR, Morán A, Tartakovsky B (2013) Microbial electrolysis cell scale-up for combined wastewater treatment and hydrogen production. *Bioresour Technol* 130:584–591.

Hamelers HV, Ter Heijne A, Sleutels TH, Jeremiasse AW, Strik DP, Buisman CJ (2010) New applications and performance of bioelectrochemical systems. *Appl Microbiol Biotechnol* 85:1673–1685.

Harnisch F, Schröder U, Scholz F (2008) The suitability of monopolar and bipolar ion exchange membranes as separators for biological fuel cells. *Environ Sci Technol* 42:1740–1746.

Ishii S, Logan BE, Sekiguchi Y (2012) Enhanced electrode-reducing rate during the enrichment process in an air-cathode microbial fuel cell. *Appl Microbiol Biotechnol* 94:1087–1094.

Jafary T, Daud WRW, Ghasemi M, Kim BH, Carmona-Martínez AA, Bakar MHA, Jahim JM, Ismail M (2017) A comprehensive study on development of a biocathode for cleaner production of hydrogen in a microbial electrolysis cell. *J Clean Prod* 164:1135–1144.

Kadier A, Simayi Y, Abdeshahian P, Azman NF, Chandrasekhar K, Kalil MS (2014) A comprehensive review of microbial electrolysis cells (MEC) reactor designs and configurations for sustainable hydrogen gas production. *Alexandria Eng J.* 55(1), 427–443.

Kaushik G (2015) Bioelectrochemical systems (BES) for microbial electroremediation: An advanced wastewater treatment technology. *Appl Environ Biotechnol Present Scenar Futur Trends* 1–167.

Khilari S, Pandit S, Ghangrekar MM, Pradhan D, Das D (2013) Graphene oxide-impregnated PVA–STA composite polymer electrolyte membrane separator for power generation in a single-chambered microbial fuel cell. *Ind Eng Chem Res* 52:11597–11606.

Khilari S, Pandit S, Varanasi JL, Das D, Pradhan D (2015) Bifunctional manganese ferrite/polyaniline hybrid as electrode material for enhanced energy recovery in microbial fuel cell. *ACS Appl Mater Interfaces* 7:20657–20666.

Kiely PD, Regan JM, Logan BE (2011) The electric picnic: Synergistic requirements for exoelectrogenic microbial communities. *Curr Opin Biotechnol* 22:378–385.

Kim JR, Cheng S, Oh SE, Logan BE (2007) Power generation using different cation, anion, and ultrafiltration membranes in microbial fuel Cells. *Environ Sci Technol* 41:1004–1009.

Kim Y, Logan BE (2013) Microbial desalination cells for energy production and desalination. *Desalination* 308:122–130.

Kondaveeti S, Lee J, Kakarla R, Kim HS, Min B (2014) Low-cost separators for enhanced power production and field application of microbial fuel cells (MFCs). *Electrochim Acta* 132:434–440.

Kuo CW, Chen BK, Tseng YH, Hsieh TH, Ho KS, Wu TY, Chen HR (2012) A comparative study of poly(acrylic acid) and poly(styrenesulfonic acid) doped into polyaniline as platinum catalyst support for methanol electro-oxidation. *J Taiwan Inst Chem Eng* 43:798–805.

Kyazze G, Popov A, Dinsdale R, Esteves S, Hawkes F, Premier G, Guwy A (2010) Influence of catholyte pH and temperature on hydrogen production from acetate using a two chamber concentric tubular microbial electrolysis cell. *Int J Hydrogen Energy* 35:7716–7722.

Leong JX, Daud WRW, Ghasemi M, Liew K Ben, Ismail M (2013) Ion exchange membranes as separators in microbial fuel cells for bioenergy conversion: A comprehensive review. *Renew Sustain Energy Rev* 28:575–587.

Li WW, Sheng GP, Liu XW, Yu HQ (2011) Recent advances in the separators for microbial fuel cells. *Bioresour Technol* 102:244–252.

Li X, Hu B, Suib S, Lei Y, Li B (2010) Manganese dioxide as a new cathode catalyst in microbial fuel cells. *J Power Sources* 195:2586–2591. doi: 10.1016/j.jpowsour.2009.10.084.

Logan BE (2009) Exoelectrogenic bacteria that power microbial fuel cells. *Nat Rev Microbiol* 7:375–381.

Logan BE (2010) Scaling up microbial fuel cells and other bioelectrochemical systems. *Appl Microbiol Biotechnol* 85:1665–1671.

Logan BE, Call D, Cheng S, Hamelers HVM, Sleutels THJA, Jeremiasse AW, Rozendal RA (2008) Microbial electrolysis cells for high yield hydrogen gas production from organic matter. *Environ Sci Technol* 42:8630–8640.

Logan BE, Hamelers B, Rozendal R, Schröder U, Keller J, Freguia S, Aelterman P, Verstraete W, Rabaey K (2006) Microbial fuel cells: Methodology and technology. *Environ Sci Technol* 40:5181–5192.

Logan BE, Rabaey K (2012) Conversion of wastes into bioelectricity and chemicals by using microbial electrochemical technologies. *Science* 337:686–690.

Mashkour M, Rahimnejad M (2015) Effect of various carbon-based cathode electrodes on the performance of microbial fuel cell. *Biofuel Res J* 2:296–300. doi:10.18331/BRJ2015.2.4.3.

Minteer SD, Atanassov P, Luckarift HR, Johnson GR (2012) New materials for biological fuel cells. *Mater Today* 15:166–173.

Morales-Guio CG, Stern LA, Hu X (2014) Nanostructured hydrotreating catalysts for electro-chemical hydrogen evolution. *Chem Soc Rev* 43:6555.

Oh S, Min B, Logan BE (2004) Cathode performance as a factor in electricity generation in microbial fuel cells. *Environ Sci Technol* 38:4900–4904.

Pandit S, Nayak BK, Das D (2012) Microbial carbon capture cell using cyanobacteria for simultaneous power generation, carbon dioxide sequestration and wastewater treatment. *Bioresour Technol* 107:97–102.

Pandit S, Sengupta A, Kale S, Das D (2011) Performance of electron acceptors in catholyte of a two-chambered microbial fuel cell using anion exchange membrane. *Bioresour Technol* 102:2736–2744.

Pant D, Van Bogaert G, Porto-Carrero C, Diels L, Vanbroekhoven K (2011) Anode and cath-ode materials characterization for a microbial fuel cell in half cell configuration. *Water Sci Technol* 63:2457–2461.

Pham TH, Aelterman P, Verstraete W (2009) Bioanode performance in bioelectrochemical systems: Recent improvements and prospects. *Trends Biotechnol* 27:168–178.

Potter MC (1911) Electrical effects accompanying the decomposition of organic compounds. *Proc R Soc B Biol Sci* 84:260–276.

Rabaey K, Girguis P, Nielsen LK (2011) Metabolic and practical considerations on microbial electrosynthesis. *Curr Opin Biotechnol* 22:371–377.

Rabaey K, Rozendal RA (2010) Microbial electrosynthesis —revisiting the electrical route for microbial production. *Nat Rev Microbiol* 8(10):706.

Rahimnejad M, Bakeri G, Najafpour G, Ghasemi M, Oh SE (2015) A review on the effect of proton exchange membranes in microbial fuel cells. *Biofuel Res J* 1:7–15.

Rosenbaum M, Aulenta F, Villano M, Angenent LT (2011) Cathodes as electron donors for microbial metabolism: Which extracellular electron transfer mechanisms are involved? *Bioresour Technol* 102:324–333.

Rout S, Nayak AK, Varanasi JL, Pradhan D, Das D (2018) Enhanced energy recovery by manganese oxide/reduced graphene oxide nanocomposite as an air-cathode electrode in the single-chambered microbial fuel cell. *J Electroanal Chem* 815, 1–7.

Rozendal RA, Hamelers HVM, Molenkamp RJ, Buisman CJN (2007) Performance of single chamber biocatalyzed electrolysis with different types of ion exchange membranes. *Water Res* 41:1984–1994.

Santoro C, Arbizzani C, Erable B, Ieropoulos I (2017) Microbial fuel cells: From fundamen-tals to applications. A review. *J Power Sources* 356:225–244.

Sharma M, Jain P, Varanasi JL, Lal B, Lema JM, Sarma PM (2013) Enhanced performance of sulphate reducing bacteria based biocathode using stainless steel mesh on activated carbon fabric electrode. *Bioresour Technol* 150, 172–180.

Siegert M, Yates MD, Call DF, Zhu X, Spormann A, Logan BE (2014) Comparison of nonpre-cious metal cathode materials for methane production by electromethanogenesis. *ACS Sustain Chem Eng* 2(4):910–917.

Strik DPBTB, Timmers RA, Helder M, Steinbusch KJJ, Hamelers HVM, Buisman CJN (2011) Microbial solar cells: Applying photosynthetic and electrochemically active organisms. *Trends Biotechnol* 29:41–49.

Sun Y, Wei J, Liang P, Huang X (2012) Microbial community analysis in biocathode micro-bial fuel cells packed with different materials. *AMB Express* 2:21.

Tender LM, Gray SA, Groveman E, Lowy DA, Kauffman P, Melhado J, Tyce RC, Flynn D, Petrecca R, Dobarro J (2008) The first demonstration of a microbial fuel cell as a viable power supply: Powering a meteorological buoy. *J Power Sources* 179:571–575.

Tokash JC, Logan BE (2011) Electrochemical evaluation of molybdenum disulfide as a catalyst for hydrogen evolution in microbial electrolysis cells. *Int J Hydrogen Energy* 36:9439–9445.

Varanasi JL, Das D (2018) Characteristics of microbes involved in microbial fuel cell. In: *Microbial Fuel Cell: A Bioelectrochemical System that Converts Waste to Watts*. Springer International Publishing, Cham, Switzerland. pp. 43–62.

Varanasi JL, Nayak AK, Sohn Y, Pradhan D, Das D (2016) Improvement of power generation of microbial fuel cell by integrating tungsten oxide electrocatalyst with pure or mixed culture biocatalysts. *Electrochim Acta* 199:154–163.

Varanasi JL, Veerubhotla R, Pandit S, Das D (2019) Biohydrogen production using microbial electrolysis cell. In: *Microbial Electrochemical Technology*. Elsevier, Amsterdam, the Netherlands, pp. 843–869.

Veerubhotla R, Das D (2017) Application of microbial fuel cell as a biosensor. In: *Microbial Fuel Cell: A Bioelectrochemical System that Converts Waste to Watts*. Springer International Publishing, Cham, Switzerland, pp. 389–402.

Wang H, Ren ZJ (2013) A comprehensive review of microbial electrochemical systems as a platform technology. *Biotechnol Adv* 31:1796–1807.

Wei J, Liang P, Huang X (2011) Recent progress in electrodes for microbial fuel cells. *Bioresour Technol* 102:9335–9344.

Wu CH, I YP, Chiu YH, Lin CW (2014) Enhancement of power generation by toluene bio-degradation in a microbial fuel cell in the presence of pyocyanin. *J Taiwan Inst Chem Eng* 45:2319–2324.

Yang H, Zhou M, Liu M, Yang W, Gu T (2015) Microbial fuel cells for biosensor applications. *Biotechnol Lett* 37:2357–2364.

Zhang F, Brastad KS, He Z (2011) Integrating forward osmosis into microbial fuel cells for wastewater treatment, water extraction and bioelectricity generation. *Environ Sci Technol* 45:6690–666.

Zhang Y, Angelidaki I (2014) Microbial electrolysis cells turning to be versatile technology: Recent advances and future challenges. *Water Res* 56:11–25.

Zhou M, Chi M, Luo J, He H, Jin T (2011) An overview of electrode materials in microbial fuel cells. *J Power Sources* 196:4427–4435.

# 13 Effect of Reactor Configurations on Gaseous Biofuel Production

The adverse effects of using conventional fossil fuels have triggered an exponential increase in biofuel production from biomass. The process economics of biofuel production is dependent largely upon the type of reactor used during the process. Different configurations of reactors have been developed over the years depending upon the type of gaseous biofuels, conversion technology, and the composition of feedstock used. The simplest design for the production of any biofuel is the batch reactor, which requires less initial capital and infrastructure investment. It can accommodate different types of feedstock and does not require stringent operating conditions. However, it has various disadvantages such as low productivity, large downtime, variation in product quality, and intensive labour and energy requirements. To address these issues continuous flow systems have been developed, which result in consistent product quality and low capital and operating costs per unit of product. Other reactor configurations that have been developed for efficient gaseous biofuels production include packed bed reactors, fluidized bed reactors, and upflow anaerobic sludge blanket reactors. In this chapter, a comprehensive overview of the various reactor configurations used for the biofuels production is presented along with their potential benefits and drawbacks.

The reactor is the heart of any biochemical process. Reactor configurations play a very important role in chemical and biochemical processes. For example, in case of substrate inhibition fed-batch usually is preferred, whereas for product inhibition plug flow reactor is recommended. But the operation of a plug flow reactor poses some problem particularly to maintain no back mixing condition and plug flow, which is also known as piston flow. However, this can be overcome by using cascade reactor or continuous stirred-tank reactor (CSTR) in series. The major problem with the CSTR or chemostat is the cell washout. If the cell mass wasting from the reactor is more than the rate of cell growth, the situation of cell wash out will occur when no cell is present in the reactor. Cell recycle and cell immobilization are the techniques by the virtue of which these problems can be avoided. The major problem with the immobilized whole cell reactor is the gas hold up problem. This problem may be overcome by modifying the reactor configuration such as by use of rhomboidal and tapper (Das et al. 2002; Kumar and

Das 2001). A critical analysis of these reactors is needed for finding out the suitability of the process for gaseous fuels generation such as methane and hydrogen. This chapter deals with the analysis of these reactors.

## 13.1   REACTOR ANALYSIS

### 13.1.1   BATCH REACTOR

The batch reactors are largely in operation in industry. The batch reactor is considered the simplest reactor where the substrates are taken at one time and products are removed at the end of the biochemical process. During the process, there is no feed addition or product withdrawn. This reactor is useful for finding out the cell growth cycle. This growth cycle profile is very important in the case of any fermentation process because the age of the inoculum must be between the mid-log and late-log phase of the microorganism (Das et al. 2014). At that phase of cell growth, microorganisms remain active. The main disadvantage of the batch process is the low productivity and it is not possible to hold one particular phase of growth for an infinite period of time (Baily and Ollis 2010). The batch process is schematically represented in Figure 13.1.

For the analysis of the reactor, material balance and energy balance is done based on the following equation (Levenspiel 1989):

$$\text{Input} + \text{generation} = \text{Output} + \text{disappearance} + \text{accumulation} \qquad (13.1)$$

Steady condition can be achieved in the continuous process when operated the same for an infinite period of time. At steady-state conditions, rate of accumulation = 0. The batch process is an example of the unsteady state operation because substrate

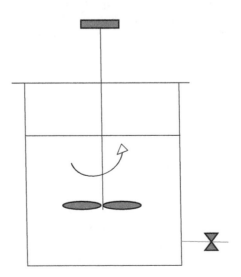

**FIGURE 13.1**   Schematic diagram of a batch process.

concentration changes with respect to time. The rate of the reaction depends on the substrate concentration. So, the rate of reaction in a batch process changes with respect to time. In case of a batch process, the substrate is converted to product.

$$\text{Substrate (S)} \rightarrow \text{Product (P)}$$

Substrate balance in a batch process may be written from Equation 13.1 as follows:

$$0+0 = 0+\left(-r_S\right)V + \frac{ds}{dt}V \tag{13.2}$$

$$\int_0^t dt = -\int_{S_0}^S \frac{ds}{-r_S} \tag{13.3}$$

$$t_{batch} = -\int_{S_0}^S \frac{ds}{-r_S} \tag{13.4}$$

Another disadvantage of the batch process is the down time, which includes the time required for harvesting of the fermentation broth, washing and refilling of the medium for the reactor. So, the total time may be written as

$$t_{total} = t_{batch} + t_{down\ time} \tag{13.5}$$

The volume of the batch reactor is determined based on $t_{total}$.

## 13.1.2 Continuous Stirred-Tank and Plug Flow Reactor

The schematic diagram of the chemostat or CSTR is shown in Figure 13.2.

Under steady-state condition, there is no accumulation of substrate. So, the substrate balance of the reactor may be written based on Equation 13.1 as follows:

$$FS_0 + 0 = FS + \left(-r_S\right)V \tag{13.6}$$

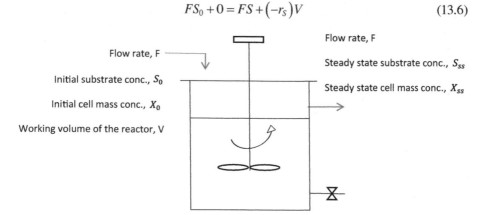

**FIGURE 13.2**   Schematic diagram of a continuous stirred-tank reactor.

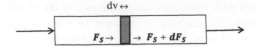

**FIGURE 13.3** Schematic diagram of a continuous plug flow reactor.

$$\frac{V}{F} = \frac{(S_0 - S)}{(-r_S)} = \tau_{CSTR} = \text{Space time of CSTR} \qquad (13.7)$$

The schematic diagram of the plug flow reactor is shown in Figure 13.3. In the plug flow reactor, the flow pattern is such that the velocity gradient across the reactor must be the same and there is no axial mixing. The substrate balance across the differential segment of the reactor of volume $dv$ may be written as Equations 13.8 and 13.9:

$$F_S + 0 = (F_S + dF_S) + (-r_S)dv \qquad (13.8)$$

$$dF_S = -(-r_S)dv$$

$$F\,dS = -(-r_S)dv$$

$$\int_0^V \frac{dv}{F} = -\int_{S_0}^S \frac{dS}{(-r_S)} = \frac{V}{F} = \tau_{PFR} = \text{Space time of PFR} \qquad (13.9)$$

In the case of product inhibition such as the ethanol fermentation process, PFR is found suitable because it takes less time compared to CSTR as shown in Figure 13.4. But PFR is difficult to operate. So, PFR can be replaced by the multiple CSTR in series, which is known as cascade.

### 13.1.3 FED-BATCH REACTOR

A fed-batch reactor is suitable when substrate inhibition limits the bioprocess. There are two strategies: variable volume fed batch and constant volume fed batch. In the

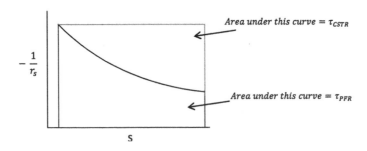

**FIGURE 13.4** Plot of $-\frac{1}{r_S}$ vs. S.

variable volume fed batch, the substrate is added continuously at a constant flow rate $F$, the rate of change in volume ($V$) can be given as:

$$\frac{dV}{dt} = F \tag{13.10}$$

By rearranging and integrating Equation 13.10

$$\int_{V_0}^{V} dV = \int_{0}^{t} F dt \tag{13.11}$$

$$V = V_0 + Ft \tag{13.12}$$

where $V$ is the volume of the reactor at time $t$ and $V_0$ is the initial volume of the reactor (at $t = 0$) (Figure 13.5). Figure 13.6 shows the profiles of cell mass concentration, substrate concentration, specific cell growth rate, and volume of a fed-batch reactor.

At quasi-steady state, $S_{added} \rightarrow S_{consumed}$ and cell mass concentration ($X$) is constant.

The cell mass balance may be written using Equation 13.1

$$FX_0 + \mu XV = 0 + \frac{dXV}{dt} + 0 \quad \left(\text{assuming cell death is negligible}\right) \tag{13.13}$$

$$FX_0 + \mu XV = X\frac{dV}{dt} + V\frac{dX}{dt} \tag{13.14}$$

At steady state, $X_0 = 0$; $\frac{dX}{dt} = 0$

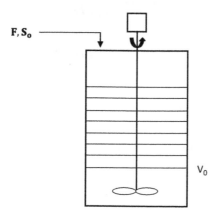

**FIGURE 13.5**  Fed batch process with a variable volume of feed.

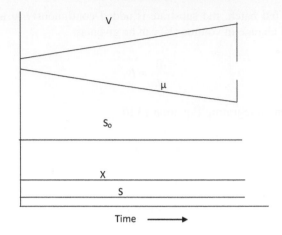

**FIGURE 13.6** Profiles of cell mass concentration, substrate concentration, specific cell growth rate, and volume of a fed-batch reactor.

Therefore,

$$\mu XV = X\frac{dV}{dt} \tag{13.15}$$

$$\mu = \frac{1}{V}\frac{dV}{dt} = \frac{F}{V} \tag{13.16}$$

The dilution rate $(D)$ may be written as:

$$D = \frac{F}{V} \tag{13.17}$$

Thus, from Equation 13.16

$$\mu = D = \frac{F}{V_0 + Ft} \tag{13.18}$$

Applying Monod Kinetics, $\mu = D = \dfrac{\mu_{max}S}{K_s + S} = \dfrac{F}{V_0 + Ft}$ \hfill (13.19)

By rearranging, $S = \dfrac{K_s D}{\mu_{max} - D}$ \hfill (13.20)

The cell mass concentration at time $t$ can be written as:

$$X = \frac{X_t}{V}$$

where $X_t$ is the total biomass concentration

At quasi-steady state, $\dfrac{dX}{dt} = 0$, i.e., $\dfrac{d\left(\dfrac{X_t}{V}\right)}{dt} = 0$

$$\frac{V\left(\dfrac{dX_t}{dt}\right) - X_t\left(\dfrac{dV}{dt}\right)}{V^2} = 0 \tag{13.21}$$

$$\frac{dX_t}{dt} = \frac{X_t}{V}\frac{dV}{dt} = FX \tag{13.22}$$

The total biomass concentration $(X_t)$ can be expressed as

$$X_t = X_0 + Y_{X/S}\left(S_0 - S\right) \tag{13.23}$$

(Since $Y_{X/S} = \dfrac{X_t - X_0}{S_0 - S}$)

When $S = 0$, and $X_0 \ll X_t$, the Equation 13.23 can be written as

$$X_t = Y_{x/S} S_0 \tag{13.24}$$

From Equation 13.22 and Equation 13.24

$$\frac{dX_t}{dt} = FY_{x/S}S_0 \tag{13.25}$$

Integrating Equation 13.25 we get

$$\int_{X_0}^{X_t} dX = FY_{x/S}\,S_0 \int_{0}^{t} dt$$

$$X_t = X_0 + FY_{x/S}\,S_0 t \tag{13.26}$$

In the case of constant volume fed batch, a very concentrated solution of the limiting substrate is added intermittently at a very low flow rate (lower than that of variable fed batch), resulting in insignificant increase in volume (Figure 13.7). Since the limiting substrate is added intermittently, the rate of change in cell mass is dependent on the flow rate such that:

$$\frac{dX}{dt} = A\frac{dX}{dS} = AY_{X/S} \tag{13.27}$$

where $A$ is substrate feed rate in g $L^{-1}$ $h^{-1}$.

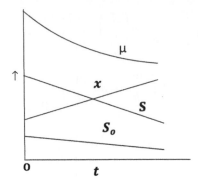

**FIGURE 13.7**   Profiles of different parameters in the case of a constant volume fed batch.

Using Equation 13.1, we can write

$$FX_0 + \mu XV = 0 + \frac{dXV}{dt} + 0 \tag{13.28}$$

Since volume is constant, at $X_0 = 0$; Equation 13.28 can be written as

$$\mu X = \frac{dX}{dt} = AY_{X/S} \tag{13.29}$$

Therefore,

$$\mu = \frac{1}{X} AY_{X/S} \tag{13.30}$$

If $\frac{1}{X} AY_{X/S}$ is less than $\mu_{max}$, the limiting substrate is consumed as soon as it enters the fermenter and thus $\frac{dS}{dt} = 0$.

The biomass concentration changes with time and can be found by rearranging and integrating Equation 13.29 as:

$$\int_{X_0}^{X_t} dX = AY_{X/S} \int_0^t dt \tag{13.31}$$

$$X_t = X_0 + AY_{X/S}t \tag{13.32}$$

where $X_t$ is the total biomass concentration and $X_0$ is the initial biomass concentration.

## 13.2   BIOMETHANATION PROCESSES

Anaerobic digestion processes are used for the stabilization of organic wastes and also for the generation of methane and hydrogen. Methane can be used as a fuel-like compressed natural gas (CNG). Due to its lowest carbon content in the fuel, the air pollution will be minimum with respect to the particulate matters discharged in the air during the combustion of the fuel. This fuel is used in the different sectors such as chemical and biochemical industries, and for automobiles and domestic purposes. On the other hand, hydrogen is considered as zero-carbon fuel because it does not cause any greenhouse effect. Hydrogen when it burns produces only water as by-product (Das and Veziroglu 2001).

## 13.2.1 BATCH/SEMI-CONTINUOUS PROCESSES

The organic wastes may be available in different forms: solid and liquid (Soliva et al. 2004). The anaerobic digestion process for methane production is very old. It can be applied at small scale as well as at commercial scale. There are millions of biogas plants in operation in India and China. These biogas plants mainly operate in batch mode. The anaerobic digesters mainly used for the production of biogas from the organic wastes in India are:

1. Khadi and Village Industries Commission (KVIC) (Floating dome)
2. Janatha (Fixed dome)
3. Deenabandu (Fixed dome, minimizes surface area)

### 13.2.1.1 KVIC Biogas Plant

Floating- and fixed-dome type biogas plants are popular in India. The fixed-dome biogas plants such as Janata and Deenbandhu are optimal in the hilly places in India (Kanwar et al. 1994). The schematic diagram of the KVIC biogas plant is shown in Figure 13.8. The characteristics features of the biogas plant follow (Singh et al. 1997):

- Cylindrical tank made of masonry
- One inlet for the solid organic wastes
- An outlet for digested or spent slurry
- Capacity of the family-size biogas plant is 3 $m^3/d$
- Hydraulic retention time (HRT) varies from 30 to 50 d.

The Department of Agriculture, Government of India, installed about 35,647 of the biogas plants between 1982 and 1997 (Singh et al. 1997; Rastogia et al. 2008). It was observed that 0.332 million family-size (2 $m^3$) biogas plants could produce 0.515 million $m^3$ of biogas per day with an energy equivalent of about 1801.1 tons of fuel wood. The digested materials from these plants could provide nitrogen-rich manure equivalent to 133.0 tons of N, 91.4 tons of $P_2O_5$, and 66.4 tons of $K_2O$ daily. The efficiency of plants depends on the reactor design, raw materials, proper planning, supervision, and adequate monitoring of the operational parameters.

### 13.2.1.2 Janatha Biogas Plant

Janatha biogas plant (Figure 13.9) has a fixed dome and usually remains under the soil to control the change of temperature inside the digester or reactor because the

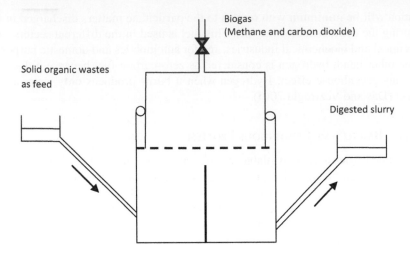

**FIGURE 13.8**    A KVIC biogas plant in India.

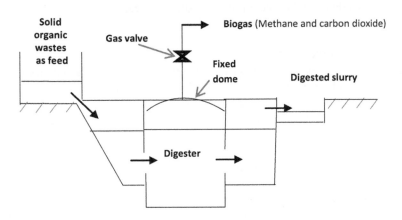

**FIGURE 13.9**    Schematic diagram of the Janatha biogas plant.

ground temperature remains almost constant (Kalia and Kanwar 1998). This process is largely in operation in China in rural places. The characteristics of this plant are:

- Digester located below the ground level
- Fixed dome type anaerobic digester
- Displaced slurry level at the outlet indicates the pressure up to the point of its discharge

Seasonal emission of $CH_4$ ranged between 10 and 178 g $m^{-2}$ $d^{-1}$. The annual average methane emission from the biogas plants in flat areas was 83.1 g $m^{-2}$ $d^{-1}$ compared to 43.1 g $m^{-2}$ $d^{-1}$ in the hilly areas (Khoiyangbam et al. 2004).

## 13.2.1.3 Deenabandu Biogas Plant

The Deenabandu biogas plant (Figure 13.10) also is a batch/semi-continuous process with the following special features (Rastogia et al. 2008):

- Fixed dome
- Requires skilled workers to check quality of materials to ensure no leakage of gas

The Deenabandu biogas plant is approximately 30% cheaper than the Janata plant and is more efficient in terms of gas production per unit volume of digester (Kanwar et al. 1994). The anaerobic digestion process is controlled mainly by two groups of microflora: acidogens and methanogens. The characteristics of these organisms are different with respect to substrate, temperature, and pH. So, the synchronization of the activities of these microorganisms is very important. The acidogens can convert the organic wastes to volatile fatty acids (VFAs) such as acetic, propionic, and butyric acids, whereas methanogens can convert these VFAs to methane and carbon dioxide (Ghose and Das 1982; Das 1985). The pH of the fermentation broth will remain the same if the rate of acid formation is equal to rate of methane generation. So, it is necessary to monitor the pH of the fermentation broth with respect to time. In the case pH increases, the feed rate may be increased to enhance the activity of the acidogenic organisms. On the other hand, if pH decreases then the feed rate is to be reduced to allow the methanogenic organisms to convert the excess VFAs to $CH_4$ and $CO_2$. Temperature plays a very important role in the anaerobic digestion process. In the winter session, the activity of these organisms is reduced significantly. So, the feed rate is adjusted accordingly. The disadvantages of the batch/semi-continuous anaerobic digestion process may be overcome by using a two-stage biomethanation process where both the acidogens and methanogens are allowed to growth separately in two different reactors because the characteristics of these organisms are different from each other with respect to temperature, pH, and substrate.

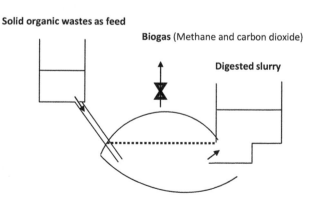

**FIGURE 13.10** Schematic diagram of the Deenabandu biogas plant.

## 13.2.2 Continuous Process

### 13.2.2.1 Chemostat

A continuous stirred-tank reactor (CSTR) is known as chemostat in the biological system (Figure 13.2). CSTR is a very widely used reactor. The microorganisms are active in the log phase of the cell growth cycle. The main advantage of the chemostat is operation of the system in the log phase for an infinite period of time by controlling the dilution rate because biomethanation is a growth associated product. Because of constant mixing pattern, microorganisms are evenly mixed within reactor liquid. Under such conditions, there is a proper contact of substrate and microbe because of that higher mass transfer that can be obtained. However, cell washout is one limitation of CSTR, which inhibits the operation at lower HRT. For that reason, HRT ($1/\mu = 1/D$) must be greater with respect to the maximum specific growth rate of the organisms (Baily and Ollis 2010; Levenspiel 1989; Zhang et al. 2016).

### 13.2.2.2 Anaerobic Contact Process

The main problem with using the chemostat process for biofuel production is the washing out of the cell mass from the reactor. This washing out may take place because the cell mass wasting from the reactor is greater compared to the cell mass growth in the reactor. Another reason might be the HRT of the cell is less than the generation time of the cell. This problem can be overcome either by cell recycling or using an immobilized whole cell system (Sentürk et al. 2010; Hamdi and Garcia 1991). A schematic diagram of cell recycling and anaerobic contact process for the methane production is shown in Figures 13.11 and 13.12. The anaerobic contact process is found most suitable for the treatment of the effluent produced from the citric acid industry because it contents significant amount of $Ca^{+2}$. $Ca^{+2}$ will be precipitated in the form of $CaCO_3$ due to the reaction with the carbon dioxide produced during the anaerobic digestion process. The $CaCO_3$ is an inactive material and present in the sludge. The anaerobic contact process is better because the sludge is removed

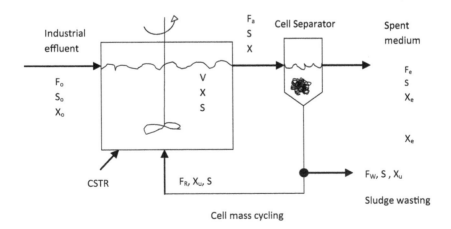

**FIGURE 13.11**   A continuous stirred-tank reactor with cell mass recycling.

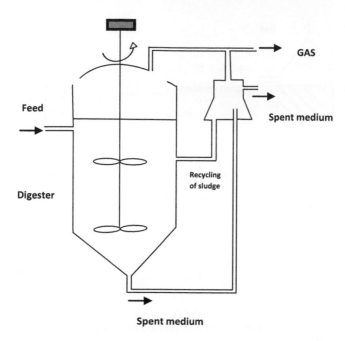

**FIGURE 13.12** Anaerobic contact process for the methane production.

at regular intervals from the process to keep the biomass concentration uniform. Typical features of the anaerobic contact process are:

- Biomass separated in the second tank, recycled to the digester
- Recycle increases solid retention time (SRT) and efficiency
- Requires HRT ≥10 d.

The whole cell immobilized system also is known as a fixed film reactor or anaerobic filter (Figure 13.13). In the immobilized whole cell system, cells are fixed on the solid surface and form a film (Hamdi and Garcia 1991; Aquino et al. 2017; Acharya et al. 2008). There are two types of fixed film reactors: upward and downward (Aquino et al. 2017). The characteristic features of the anaerobic fixed film reactor are:

- Comprised of solid support or packing material for the immobilization of whole cell
- HRT of 0.5–12 d
- Performance of the reactor is better in the bottom part of the reactor due to the present of the suspended cells
- Due to the formation of biomass film on the surface of the solid matrix, the porosity of the packing bed will be reduced drastically which causes channeling

### 13.2.2.3 Two-Stage Biomethation Process

The anaerobic digestion process usually is controlled by two groups of microflora: acidogens and methanogens (Das et al. 1983; Pohland and Ghosh 1971). The characteristics

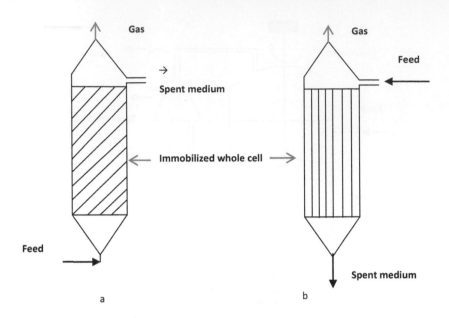

**FIGURE 13.13**    Schematic diagram of (a) an upward and (b) a downward fixed film reactor.

of these organisms differ from each other with respect to substrate, pH, and temperature. The growth rate of acidogens is much higher compared to methanogens. So, it is possible to separate the methanogens from the acidogens by controlling the dilution rate (between 1 and 2 h⁻¹) (Figure 13.14). The methanogen culture can be developed by using volatile fatty acids as the substrate (Das et al. 1983; Das 1988). It has been observed that in case of the two-stage biomethanation process, the time of the fermentation process has been drastically reduced with a higher amount of gas formation (Figure 13.15). Cow dung is a

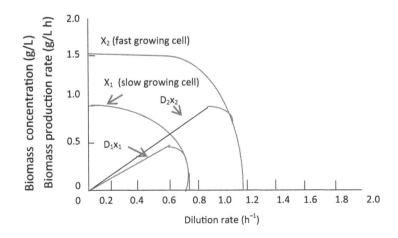

**FIGURE 13.14**    Steady-state substrate and biomass concentration at various dilution rates.

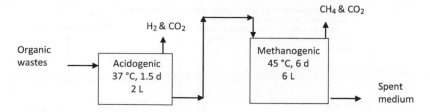

**FIGURE 13.15**    Two-stage biomethanation process.

conventional raw materials for the anaerobic digestion process because it is comprised of not only the substrate but also the microorganisms involved. The degradation efficiency (DE) of the cow dung is low at about 30%. So, attempts have been made to increase DE by manipulating different organic wastes (Pohland and Ghosh 1971; Das et al. 1983; Das 1988; Lindner et al. 2016; Luo et al. 2011; Azbar and Speece 2001). It has been observed that water hyacinth contains a significant amount of hemicellulose, which is considered suitable for the anaerobic digestion process compared to cellulose. Mixed residues comprised of water hyacinth, wastewater grown algae, cow dung, and rice husk are suitable compared to the individual residues. Distillery industries pose severe environmental pollution problem mainly due to its high organic content. Several distillery industries successfully use distillery effluent for stabilization as well as for methane generation. Methane produced from this process is used for the distillation of ethanol. Typical material analysis of the two-stage biomethanation process in the laboratory-scale by using distillery effluent is shown in Figure 13.16. A detailed energy analysis of the process has been reported for the treatment of 100 m³ distillery effluent per day. It has been observed that 65.5% of the energy present in the distillery effluent can be recovered in the form of methane (Das 1985) (Figure 13.17).

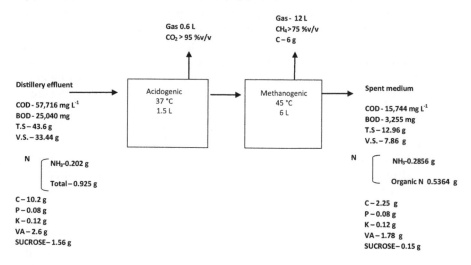

**FIGURE 13.16**    Materials balance of the two-stage biomethation process using distillery effluent.

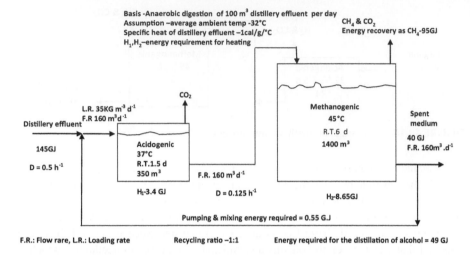

Basis -Anaerobic digestion of 100 m³ distillery effluent per day
Assumption –average ambient temp -32°C
Specific heat of distillery effluent –1cal/g/°C
$H_1, H_2$–energy requirement for heating

CH₄ & CO₂
Energy recovery as CH₄-95GJ

CO₂

L.R. 35KG m³ d⁻¹
F.R 160 m³d⁻¹

Distillery effluent

145GJ

D = 0.5 h⁻¹

Acidogenic
37°C
R.T.1.5 d
350 m³

$H_1$-3.4 GJ

F.R. 160 m³ d⁻¹

D = 0.125 h⁻¹

Methanogenic
45°C
R.T.6 d
1400 m³

Spent
medium

40 GJ
F.R. 160m³.d⁻¹

$H_2$-8.65GJ

Pumping & mixing energy required = 0.55 G.J

F.R.: Flow rare, L.R.: Loading rate          Recycling ratio –1:1          Energy required for the distillation of alcohol = 49 GJ

**FIGURE 13.17**   Energy analysis of the two-stage biomethanation process using distillery effluent.

### 13.2.2.4   Upflow Anaerobic Sludge Blanket Reactor

The upflow anaerobic sludge blanket (UASB) reactor (Figure 13.18) is an efficient reactor for biogas generation from soluble organic wastewater. Microorganisms present in the sludge bed can naturally form granules of 0.2 to 2 mm diameter and have a high sedimentation velocity and thus resist wash out from the system even at high hydraulic load (Barros et al. 2017; Rico et al. 2017). Gatze Lettinga and collaborators developed the UASB reactor concept (Lettinga 1980). The UASB reactor is used mainly for treating mainly soluble wastewaters. The development of granular sludge, which is highly precipitating sludge with high methanogenic activity, is one of the basic conditions for the success of the high-rate anaerobic digestion process using UASB reactors (Hulshoff et al. 1983; Jung et al. 2012). Biogas production takes place under an anaerobic degradation process from the organic wastes. The released gas bubbles moving in upward motion can cause hydraulic turbulence and provide mixing in the reactor without any mechanical parts. The typical features of the UASB reactor are:

- UASB's performance depends on the characteristics of sludge formed
- The influent added from the bottom while gas collected from the top
- Can be operated at the OLR of 10–30 kg chemical oxygen demand (COD)/m³ d
- Localized mixing due to the evolved gas. No external agitation required.
- HRT varies from 0.5 to 7 d
- SRT of 20 d

### 13.2.2.5   Anaerobic Membrane Reactor

Anaerobic membrane reactors (AnMBRs) (Figure 13.19) also can be used for the treatment of wastewater. Recently, attempts have been made to reduce HRT when designing anaerobic reactors. AnMBR deals with combining unit operations of

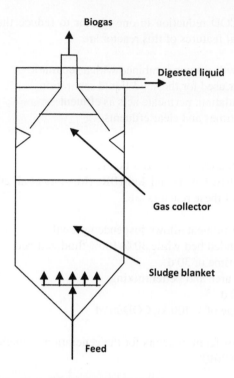

**FIGURE 13.18**   Upflow anaerobic sludge blanket reactor.

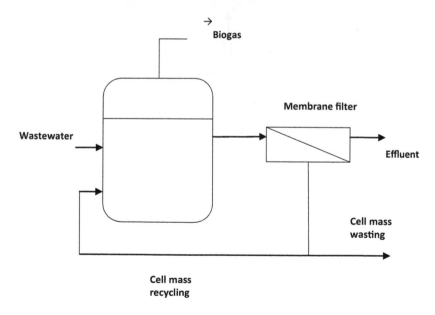

**FIGURE 13.19**   Anaerobic membrane reactor.

solids removal and COD reduction in one reactor to reduce the HRT (David and Stuckey 2012). Typical features of this reactor are:

- Suspended growth reactor combined with a separator
- Membrane filter used for the solid-liquid separation
- Biomass recirculation, permeate acts as effluent
- High retention times and clear effluents

### 13.2.2.6 Anaerobic Expanded/Fluidized Bed Reactor

An anaerobic expanded (AE)/ fluidized bed reactor (FBR) is a high-rate anaerobic digestion process such as UASB and AnMBRs (Mustafa et al. 2014) (Figure 13.20). The special features of these reactors are:

- Upward flow of influent allows suspended growth
- 10%–20% expanded bed while 30%–100% fluidized bed
- Solid retention time of 30 d
- Higher surface area and better mixing
- HRT of 0.2–5.0 d
- OLR in the range of 1–100 kg COD/m³/d

The performance of different reactors for the generation of methane is summarized in Table 13.1 (Gunjan 2010).

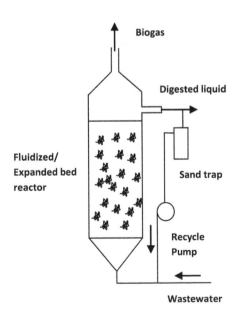

**FIGURE 13.20**　Anaerobic expanded and fluidized bed reactor.

**TABLE 13.1**

**Performance of Various Anaerobic Digesters**

| Anaerobic Digestion Processes | ORL (kg COD/ m³/d) | Substrate Conversion (% COD removal) | Biogas Production (m³/kg) | | HRT (d) | SRT (d) |
|---|---|---|---|---|---|---|
| | | | Minimum | Maximum | | |
| Anaerobic Lagoon | <0.5–2.0 | 65–75 | 0.53 | 0.62 | 30–50 | 50–100 |
| ASBR | 1.2–2.4 | 75–85 | 0.62 | 0.70 | 0.25–0.50 | 50–200 |
| CSTR | 1.0–5.0 | 65–80 | 0.53 | 0.66 | 15–30 | 15–30 |
| Anaerobic Contact | 1.0–8.0 | 85–95 | 0.70 | 0.78 | 0.5–5 | 50–200 |
| Membrane reactor | 2.0–22.0 | 85–95 | 0.70 | 0.78 | 0.5–15 | 30–160 |
| Fixed film | 0.3–20 | 60–90 | 0.49 | 0.74 | 0.5–4 | 14–30 |
| PFR | 2.0–12.0 | 75–85 | 0.62 | 0.70 | n/a | 30 |
| UASB | 15–24 | 85–95 | 0.70 | 0.78 | 0.25–0.50 | 40–100 |

CSTR = Continuous stirred-tank reactor, ASBR = Anaerobic sequential batch reactor, PFR = Plug flow reactor, UASB = Upflow anaerobic sludge blanket reactor, HRT = Hydraulic retention time, SRT = Solid retention time, OLR = Organic loading rate.

## 13.3 BIOHYDROGEN PRODUCTION

Hydrogen production through biological routes is promising because it is a renewable energy source and environmentally friendly. Biohydrogen can be produced by biophotolysis, photofermentation, or dark fermentation. Photo-biological processes require light as a source of energy. However, biohydrogen production by these processes is much less compared to dark fermentation. Several *Enterobacter* and *Clostridium* spp. as well as acidogens present in the anaerobic digester can produce hydrogen from different organic carbon sources like sugar, cane molasses, starchy wastewater, and distillery effluent (Das et al. 2014). Use of a pure microbial strain is unsuitable for the production of hydrogen from the organic wastes because a sterilized condition of process is energy intensive. Reactor configuration is one of the primary deciding factors for high-rate biohydrogen production process. Biohydrogen production is carried out mainly in either a batch or a continuous mode. Batch mode is more suitable for initial optimization studies, but continuous mode is more preferable for the commercial scale. Several studies have reported on the use of CSTRs because of their simple construction, effective homogenous mixing, ease of operation, and the possibility of keeping the system at a certain HRT. Other reactors such as the agitated granular sludge bed reactor (AGSBR), anaerobic cascade reactor (ACR), fluidized-bed reactor, fixed-bed reactor, membrane reactor (MR), and upflow anaerobic sludge blanket reactor (UASB) are suitable for biohydrogen production (Table 13.2).

The main problem with the CSTR is the cell wasting from the reactor. The cell mass concentration is usually varied from 1000 to 4000 mg VSS/L in the CSTR at different HRTs (Show et al. 2010). The CSTR is very easy to operate and to maintain the log phase for an infinite period of time for the maximum growth rate of the cell because hydrogen production is mainly a growth-associated product. Immobilized whole cell reactor as well as cell recycling techniques is effective to overcome the

**TABLE 13.2**

**Performance of Various Bioreactors for the Biohydrogen Production**

| Reactor Type | Volume (L) | Microorganism | Substrate | Hydrogen Yield (mol/mol) | Hydrogen Rate (L/L h) | References |
|---|---|---|---|---|---|---|
| CSTR | – | Mixed culture | Glucose | – | 0.54 | Fang and Liu (2002) |
| CSTR | – | Mixed culture | – | – | 15.09 | Wu et al. (2006) |
| CSTR | – | Mixed culture | – | – | 3.20 | Zhang et al. (2007) |
| PBR | – | E. cloacae IIT-BT 08 | Glucose | – | 1.69 | Das et al. (2002) |
| PBR | 0.48 | E. cloacae BL-21 | Cane molasses | – | 2.17 | Chittibabu et al. (2006) |
| MBR | 1.0 | Fermentative culture | Glucose | 1.3 | 1.5 | Lee et al. (2007) |
| | | | Sucrose | 2.1 | 1.4 | |
| | | | Fructose | 2.7 | 1.4 | |
| **UASB** | | Clostridium sp. | Sucrose | 1.6 | 0.1 | Zhao et al. (2008) |
| CIGSB | 0.88 | Mixed bacteria | Sucrose | 1.9 | 7.7 | Lee et al. (2006) |
| AFBR | – | Mixed bacteria | – | – | 2.4 | Zhang et al. (2007) |
| Column Reactor | – | Anaerobic fermentative bacteria | – | – | 7.49 | Zhang et al. (2008) |
| IBR | 2.5 | C. tyrobutyricum | Food waste | 223 mL/g hexose | 0.3 | Jo et al. (2008) |

cell washout problem in the CSTR (Wu et al. 2006). Mixed acidogenic culture can be developed from digested cow dung and digested anaerobic sludge (Kumari and Das 2017; Mishra et al. 2015). These cultures were suitable for biohydrogen production.

Immobilized solid matrix plays an important role in hydrogen production and retention of cells. The mass transfer resistance is the major limitation of this process (Cheng et al. 2002). Recycling of the effluent will help to improve the substrate conversion and hydrogen production. The performance of several reactor configurations were compared for the production of hydrogen by immobilized whole cell using the environmental friendly solid matrix (Das et al. 2002; Kumar and Das 2001) (Figure 13.21). The rhomboid bioreactor with convergent-divergent configuration produced a maximum rate of hydrogen production (1600 mL/L h) at an HRT of 1.08 h compared to the tapered reactor (1460 mL/L h) and to a tubular reactor (1400 mL/L h). The increased production rate could be attributed to the higher turbulent mixing favoring mass transfer and lower gas hold-up.

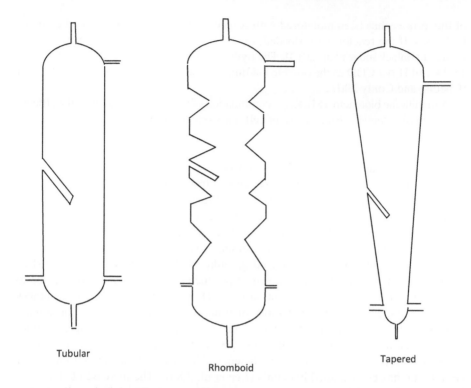

Tubular

Rhomboid

Tapered

**FIGURE 13.21**   Configuration of tubular, rhomboid, and tapered reactors.

A fluidized bed reactor (FBR) is comprised mainly of a stirred-tank and a packed-bed continuous flow reactor. The FBR improves the heat and mass transfer characteristics significantly. Cells are attached to fine sand particles in the form of biofilm in the FBR. These particles remain in suspension mainly due to the dragging force of the upward flow rate of the wastewater. This process increases the microbial activity due the close contact of cells and the organic substrates, which are responsible for the higher degradation rates of organic wastes. The maximum hydrogen production rate and yield of 7.6 L/L h and 0.4–1.7 moles per mole of glucose, respectively, were observed in an anaerobic fluidized bed biofilm reactor using glucose as a substrate (Zhang et al. 2008; Seon et al. 1983). It can be operated at both higher hydraulic loading rates and at low HRTs. High operating energy consumption is the major disadvantage of this process.

Various organic wastes can be treated using various sizes of UASB reactors for the hydrogen production (Chang and Lin 2004). This reactor is used for hydrogen production from sugar industry wastewater, food industry waste, distillery effluent, and beverage industry wastewater. The performance of the UASB reactor depends on the several operating parameters such as retention time, concentration of solids in the feed, organic loading density, pH, and flow recirculation. These parameters were studied to maximize hydrogen production (Yu and Mu 2006).

An anaerobic sequencing batch reactor (ASBR) was suitable for hydrogen production by changing the time of cyclic duration and pH (Chen et al. 2009). Performance of this process has been monitored with respect to COD removal and production of hydrogen. If the reactor is overloaded, the performance of the reactor is drastically reduced (Balachandar et al. 2013). The hydrogen yield was increased from 6.064 to 13.44 mol $H_2$/kg COD as the organic loading rate varied from 6.3 to7.9 kg COD/$m^3$d (Cheong and Conly 2007).

A membrane bioreactor (MBR) is a combination of membrane filtration of the effluent and biomass retention. An increase of cell mass retention time improves the substrate use rate but decreases the rate of hydrogen production (Oh et al. 2004). The biomass concentration increased from 2200 mg/L in a controlled reactor without a membrane to 5800 mg/L in an anaerobic MBR at an HRT of 3.3 h. Under such operating conditions, the rate of production of hydrogen also increased from 0.50 to 0.64 L/L h (Li and Fang 2007). The major disadvantage of MBR is membrane fouling and high operating costs.

There are some reactors that are very effective for biohydrogen production that are proficient with the formation of self-flocculating granular sludge or matrix, immobilized cells, such as the agitated granular sludge bed reactor (AGSBR) (Lin and Lay 2004) and the carrier-induced granular sludge bed reactor (CIGSB) (Lee et al. 2006). These reactors are known to produce a high rate of hydrogen due to enhanced cell retention at high dilution rates. However, their major disadvantages include inefficient mixing and stability of functional granules. Among these reactors, the CIGSB reactor has been shown very efficient in case of hydrogen production (Lee et al. 2006). However, the absence of mechanical agitation may cause a problem with reduced mass transfer efficiency. The effect of the height to diameter (H/D) ratio has been studied in details (Lee et al. 2006). The increase of H/D ratio is responsible for higher rates of hydrogen production (6870 mL/L h) and a higher hydrogen yield (3880 mol $H_2$/mol sucrose). The results indicated that the physical configuration of the reactor and proper upflow velocity can enhance the hydrogen production in a CIGSB reactor (Das and Roy 2017).

## 13.4  BIOHYTHANE PROCESS

The biohythane process is the combination of the biohydrogen production process followed by the bimethanation process. In the biohydrogen production process, organic wastes can be used for hydrogen production. The spent medium of the hydrogen fermentation process is comprised mainly of various volatile fatty acids (VFAs) like acetic, butyric acids, and small amount of ethanol. These VFAs can be used as substrates in the biomethanation process. So, the biohythane process not only generates a maximum amount of gaseous fuels but also increases the stability of the industrial effluent significantly (Das and Roy 2017; Kumari and Das 2018). The advanced reactor configuration as discussed in the preceding section may be implemented for the improvement of the performances of the biohythane processes using different organic wastes.

## 13.5  CONCLUSIONS AND FUTURE PERSPECTIVES

Bioreactor configurations play an important role in the generation of gaseous fuels such as methane and hydrogen. All reactors have advantages and disadvantages, which need to be critically reviewed to determine the most suitable reactor for the biohydrogen and the biomethanation process. Microorganisms involved in the bio-hydrogen production processes are mainly facultative anaerobes whereas those of biomethanation organisms are obligatory anaerobes. These differences must be considered during selection of the reactor configuration. Immobilized whole cell system in the form of fixed film or fluidized bed reactor or UASB may be considered for the high-rate gaseous fuel production. The characteristics of the organic wastewaters differed from each other. So, this parameter must be considered for the selection of the reactor configuration. pH plays a very important role in these gaseous fuel generation process. The biohythane process is most suitable not only for the maximization of gaseous energy recovery from the organic wastes but also for improvement of the conversion efficiency of the process. This process may safeguard our environmental pollution problems to some extent.

## REFERENCES

Acharya BK, Mohana S, Madamwar D (2008) Anaerobic treatment of distillery spent wash—a study on upflow anaerobic fixed film bioreactor. *Bioresour Technol* 99:4621–4626.

Aquino S de, Fuess LT, Pires EC (2017) Media arrangement impacts cell growth in anaerobic fixed-bed reactors treating sugarcane vinasse: Structured vs. randomic biomass immobilization. *Bioresour Technol* 235:219–218.

Azbar N, Speece RE (2001) Two-phase, two-stage, and single-stage anaerobic process comparison. *J Environ Eng* 127:240–248.

Baily JE, Ollis DF (2010) *Biochemical Engineering Fundamentals*. TATA McGraw-Hill Education, New Delhi, India.

Balachandar G, Khanna N, Das D (2013) Biohydrogen production from organic wastes by dark fermentation. In: Pandey A, Chang JS, Hallenbeck P, Larroche C (Eds.), *Biohydrogen*. Elsevier Publisher, Amsterdam, the Netherlands, pp. 103–144.

Barros VG, Duda RM, Vantini JDS, Omori WP, Ferro MIT, Oliveira RA (2017) Improved methane production from sugarcane vinasse with filter cake in thermophilic UASB reactors, with predominance of *Methanothermobacter* and *Methanosarcina* archaea and *Thermotogae* bacteria. *Bioresour Technol* 244:371–381.

Chang FY, Lin CY (2004) Biohydrogen production using an up-flow anaerobic sludge blanket reactor. *Int J Hydrogen Energy* 29:33–39.

Chen SY, Chu CY, Cheng MJ, Lin CY (2009) The autonomous house: a bio-hydrogen based energy self-sufficient approach. *Int J Env Research and Public Health* 6:1515–1529.

Cheng CC, Lin CY, Lin MC (2002) Acid-base enrichment enhances anaerobic hydrogen production process. *Appl Microbiol Biotechnol* 58:224–228.

Cheong DY, Conly LH (2007) Effect of feeding strategy on the stability of anaerobic sequencing batch reactor responses to organic loading conditions. *Bioresour Technol* 42:223–232.

Chittibabu G, Nath K, Das D (2006) Feasibility studies on the fermentative hydrogen production by recombinant *Escherichia coli* BL-21. *Process Biochem* 41:682–688.

Das D (1985) Ph.D. *Dissertation: Optimization of Methane Production From Agricultural Residues*, BERC, IIT New Delhi, India.

Das D (1988) *Scale-up Studies of the Two Stage Biomethanation Process for the Treatment of Cane Molasses Based Distillery Wastes*. IFCON'88 Mysore.

Das D, Badri PK, Kumar N, Bhattacharya P (2002) Simulation and modeling of continuous $H_2$ production process by *Enterobacter cloacae* IIT-BT 08 using different bioreactor configuration. *Enzyme and Microbial Technol* 31:867–875.

Das D, Ghose TK, Gopalakrisnan KS, Joshi AP (1983) *Treatment of Distillery Wastes by a Two Phase Biomethanation Process. Symposium Papers. Energy from Biomass and Wastes VII*, Boca Raton, FL, pp. 601–626.

Das D, Khanna N, Nag Dasgupta C (2014) *Biohydrogen Production: Fundamentals and Technology Advances*. CRC Press, Boca Raton, FL.

Das D, Roy S (2017) *Biohythane: Fuel for The Future*, Pan Stanford Publishing, Singapore.

Das D and Veziroglu TN (2001), Hydrogen production by biological processes: A survey of literature. *Int J Hydrogen Energy* 26:13–28.

David C, Stuckey DC (2012) Recent developments in anaerobic membrane reactors. *Bioresour Technol* 122:137–148.

Fang HHP, Liu H (2002) Effect of pH on hydrogen production from glucose by mixed culture. *Bioresour Technol* 82:87–93.

Ghose TK, Das D (1982) Maximization of energy recovery in biomethanation processes: Part-II use of mixed residue in batch system. *Process Biochem* 17:39–42.

Gunjan A (2010) M.S. dissertation *Commercialization of Anaerobic Contact Process for Anaerobic Digestion of Algae*. Case Western Reserve University, Cleveland, OH.

Hamdi M, Garcia JL (1991) Comparison between anaerobic filter and anaerobic contact process for fermented olive mill wastewaters. *Bioresour Technol* 38:23–29.

Hulshoff Pol LW, de Zeeuw WJ, Velzeboer ZTM, Lettinga G (1983) Granulation in UASB reactors. *Wat Sci Technol* 15:291–304.

Jo J, Lee H, Park DS, Choe D, Park W (2008) Optimization of key process variables for enhanced hydrogen production by *Enterobacter aerogenes* using statistical methods. *Bioresour Technol* 99:2061–2066.

Jung K-W, Kim D-H, Lee M-Y, Shin H-S (2012) Two-stage UASB reactor converting coffee drink manufacturing wastewater to hydrogen and methane. *Int J Hydrog Energy* 37:7473–7481.

Kalia AK, Kanwar SS (1998) Long-term evaluation of a fixed dome Janata biogas plant in hilly conditions. *Bioresour Technol* 65:61–63.

Kanwar SS, Gupta RK, Guleri RL, Singh SP (1994) Performance evaluation of a 1 m³ modified, fixed-dome Deenbandhu biogas plant under hilly conditions. *Bioresour Technol* 50:239–241.

Khoiyangbam RS, Kumar S, Jain MC, Gupta N, Kumar A, Kumar V (2004) Methane emission from fixed dome biogas plants in hilly and plain regions of northern India. *Bioresour Technol* 95:35–39.

Kumar N, Das D (2001) Continuous hydrogen production by immobilized *Enterobacter cloacae* IIT-BT 08 using lignocellulosic materials as solid matrices. *Enzyme and Microbial Technol* 29:280–287.

Kumari S, Das D (2017) Improvement of biohydrogen production using acidogenic culture. *Int J Hydrog Energy* 42:4083–4094.

Kumari S, Das D (2018) Biohythane production from sugarcane bagasse and water hyacinth: A way towards promising green energy production. *J Cleaner Production* 207:689–701.

Lettinga G, van Velsen AFM, Hobma SW, Klapwijk A (1980) Use of the Upflow Sludge Blanket (USB) reactor concept for biological wastewater treatment. *Biotechnol Bioeng* 22:699–734.

Lee KS, Lo YC, Lin PJ, Chang JS (2006) Improving biohydrogen production in a carrier-inducedgranular sludge bed by altering physical configuration and agitation pattern of the bioreactor. *Int J Hydrogen Energy* 31:1648–1657.

Lee KS, Lin PJ, Fangchiang K, Chang JS (2007) Continuous hydrogen production by anaerobic mixed microflora using a hollow-fiber microfiltration membrane bioreactor. *Int J Hydrogen Energy* 32:950–957.

Lee KS, Lin PJ, Chang JS (2006) Temperature effects on biohydrogen production in a granular sludge bed induced by activated carbon carriers. *Int J Hydrogen Energy* 31:465–472.

Levenspiel O (1989) *Chemical Reaction Engineering*. Wiley, New Delhi, India.

Li CL, Fang HHP (2007) Fermentative hydrogen production from wastewater and solid wastes by mixed cultures. *Crit Rev Environ Sci Technol* 37:1–39.

Lin CY, Lay CH (2004) Carbon/nitrogen-ratio effect on fermentative hydrogen production by mixed microfora. *Int J Hydrog Energy* 29:41–45.

Lindner J, Zielonka S, Oechsner H, Lemmer A (2016) Is the continuous two-stage anaerobic digestion process well suited for all substrates? *Bioresour Technol* 200:470–476.

Luo G, Xie L, Zhou Q, Angelidaki I (2011) Enhancement of bioenergy production from organic wastes by two-stage anaerobic hydrogen and methane production process. *Bioresour Technol* 102:8700–8706.

Mishra P, Roy S, Das D (2015) Comparative evaluation of the hydrogen production by mixed consortium, synthetic co-culture and pure culture using distillery effluent. *Bioresour Technol* 198:593–602.

Mustafa N, Elbeshbishy E, Nakhla G, Zhu J (2014) Anaerobic digestion of municipal wastewater sludges using anaerobic fluidized bed bioreactor. *Bioresour Technol* 172:461–466.

Oh SE, Lyer P, Bruns MA, Logan BE (2004) Biological hydrogen production using a membrane bioreactor. *Biotechnol Bioeng* 87:119–127.

Pohland FG, Ghosh S (1971) Developments in anaerobic stabilization of organic wastes-the two-phase concept. *Environ Lett* 1:255–266.

Rastogia G, Tulshiram DR, Milind YY, Yogesh SP, Shouche S (2008) Investigation of methanogen population structure in biogas reactor by molecular characterization of methyl-coenzyme M reductase A (mcrA) genes. *Bioresour Technol* 99:5317–5326.

Rico C, Montes JA, Rico JL (2017) Evaluation of different types of anaerobic seed sludge for the high rate anaerobic digestion of pig slurry in UASB reactors. *Bioresour Technol* 238:147–156.

Seon YH, Lee CG, Park DH, Hwang KY, Joe YI (1983) Hydrogen production by immobilized cells in nozzle loop bioreactor. *Biotechnol Lett* 15:1275–1280.

Sentürk E, Ince M, Onkal Engin G (2010) Kinetic evaluation and performance of a mesophilic anaerobic contact reactor treating medium-strength food-processing wastewater. *Bioresour Technol* 101:3970–3977.

Show KY, Zhang ZP, Tay JH, Liang TD, Lee DJ, Ren N, Wang A (2010) Critical assessment of anaerobic processes for continuous biohydrogen production from organic wastewater. *Int J Hydrog Energy* 35:13350–13355.

Singh SV, Vatsa DK, Verma HN (1997) Problems with biogas plants in Himachal Pradesh. *Bioresour Technol* 59:69–71.

Soliva M, Bernat C, Gil E, Martínez X, Pujol M, Sabate J, Valero J (2004) Organic waste management in education and research in agricultural engineering schools, *International Conference on Engineering Education in Sustainable Development*, Barcelona, Spain.

Wu SY, Hung CH, Lin CN, Chen HW, Lee AS, Chang JS (2006) Fermentative hydrogen production and bacterial community structure in high-rate anaerobic bioreactors containing silicone-immobilized and self-flocculated sludge. *Biotechnol Bioeng* 93:934–946.

Yu HQ, Mu Y (2006) Biological hydrogen production in a UASB reactor with granules. II: Reactor performance in 3-year operation. *Biotechnol Bioeng* 94:988–995.

Zhang ZP, Show KY, Tay JH, Liang TD, Lee DJ, Wang JY (2008) The role of acid incubation in rapid immobilization of hydrogen-producing culture in anaerobic upflow column reactors. *Int J Hydrog Energy* 33:5151–5160.

Zhang F, Yang JH, Dai K, Chen Y, Li QR, Gao FM, Zeng RJ (2016) Characterization of microbial compositions in a thermophilic chemostat of mixed culture fermentation. *Appl Microbiol Biotechnol* 100:1511–1521.

Zhang ZP, Show KY, Tay JH, Liang DT, Lee DJ, Jiang WJ (2007) Rapid formation of hydro-
    gen-producing granules in an anaerobic continuous stirred tank reactor induced by acid
    incubation. *Biotechnol Bioeng* 96:1040–1050.
Zhang ZP, Tay JH, Show KY, Yan R, Liang DT, Lee DJ, Jiang WJ (2007) Biohydrogen pro-
    duction in a granular activated carbon anaerobic fluidized bed reactor. *Int J Hydrog
    Energy* 32:185–191.
Zhang ZP, Show KY, Tay JH, Liang TD, Lee DJ, Wang JY (2008) The role of acid incubation
    in rapid immobilization of hydrogen-producing culture in anaerobic upflow column
    reactors. *Int J Hydrog Energy* 33:5151–5160.
Zhao BH, Yue ZB, Zhao QB, Mu Y, Yu HQ, Harada H, Li Y (2008) Optimization of hydrogen
    production in a granule-based UASB reactor. *Int J Hydrogen Energy* 33:2454–2461.

# 14 Scale-up and Case Studies of Biofuel Production Processes

## 14.1 INTRODUCTION

Energy crisis and the environmental pollution are major concerns in the world. Bioenergy-producing industries are playing a significant role to reduce the net carbon emissions by consumption of biofuels. Biofuels are available in three different forms: solid, liquid, and gas. Solid biofuel mainly include dry lignocellulosic biomass. There are two major problems for the use of solid fuels: transportation and energy conversion efficiency. However, in the case of liquid and gaseous biofuels, these problems can be minimized. Bioenergy sources mainly are renewable in nature. The high production costs, complex conversion technologies, high energy input, and limited resources are among the few concerns associated with other renewables such as wind, solar, and hydrothermal. The main alternative to fossil fuels is biomass (Richard 2010; Larson 2008; Koutinas et al. 2016; Naik et al. 2010, Bauen et al. 2009). It is estimated that a four time increase in present bioenergy production (150 EJ/year) (1 EJ = $10^{18}$ J) can lead to an almost 50% greenhouse gas (GHG) reduction by 2050 (World Energy Resources Bioenergy (2016).

The most common biofuels (Table 14.1) that are produced commercially include ethanol and biodiesel, which require conversion of food-based crops such as grains and oilseeds (first-generation biofuels). Other non-food-based crops such as forest and agricultural wastes, and residues are potent raw materials for biofuel production, but their main disadvantage is high energy input (Uslu et al. 2008; Farrell et al. 2006; Laser et al. 2009). High energy input is the main problem when using lignocellulosic feedstocks. In addition, these feedstocks require pre-treatment for the separation of lignin of which most is not biodegradable. The crystallinity of the cellulose molecules also play an important role in its biodegradation. However, the lignocellulosic feedstock can be converted to fuels by the thermochemical process. Solid food wastes can be easily converted to methane and hydrogen by an anaerobic digestion process. Attempts have been made to convert lignocellulosic feedstock, oil seed, and algae to liquid fuels such as ethanol, biodiesel, and butanol on a large scale.

**TABLE 14.1**
**Energy Densities for Selected Biofuels**

| Fuels | Energy Density (kcal/g) |
|---|---|
| Hydrogen | 34.2 |
| Methane | 12.9 |
| Biodiesel | 11.0 |
| Gasoline | 10.5 |
| Soybean oil | 9.6 |
| Coal | 8.4 |
| Ethanol | 7.1 |
| Methanol | 5.3 |
| Soft wood | 4.9 |
| Hard wood | 4.4 |
| Bagasse | 4.2 |

*Source:* Drapcho, C.M. et al., *Ethanol Production, Biofuels Engineering Process Technology*, The McGraw-Hill Companies McGraw-Hill, New York, 2008.

## 14.2  BIOMETHANATION PROCESS

Most metropolitan cities have been facing environmental pollution problems because of the presence of $CO_2$ and particulate matter in the atmosphere. The energy demand drastically increases mainly due to rapid industrialization and urbanization. This energy demand has been fulfilled by using fossil fuels, which are responsible mainly for the environmental pollution problems. Cities such as New Delhi have been facing a fogging problem mainly due to the presence of particulate matter in the atmosphere. This fogging has been reduced to a great extent by using compressed natural gas (CNG), which is mainly comprised of methane. Natural gas has a limited reserve. So, biomethanation of organic wastes plays an important role not only to produce methane but also to safeguard our environment (Cong et al. 2017). The anaerobic digestion process has been implemented successfully for the treatment of agricultural wastes, food waste, industrial effluent, and wastewater (Figure 14.1) (Das 1985; Das and Roy 2017; Das et al. 1983; Achinasa 2017). Biogas production by using different solid wastes is shown in Table 14.2 (Das 1985). Gaseous energy recovery as methane by using a two-stage biomethanation processes using 100 m³ distillery effluent/day in India is shown in Figure 13.7. Several anaerobic digestion processes were operated successfully in the various countries, which are summarized Table 14.3.

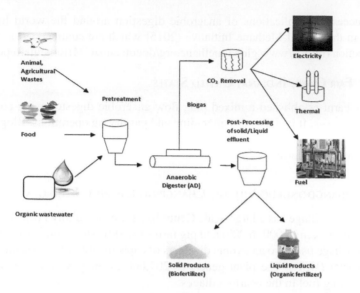

**FIGURE 14.1**   Anaerobic digestion of organic wastes.

---

**TABLE 14.2**

**Comparison of Two-Stage Systems Using Solid Biomass and Liquid Wastes**

| Residues | Parameters | | | | |
|---|---|---|---|---|---|
| | Retention Time (d) | Loading Rate $(Kgm^{-3}d^{-1})$ | COD Reduction (%) | Feed Carbon (%) | Energy Recovery as Methanation (%) |
| Cow dung | 15 | 26.7 | 21 | 20 | 21.4 |
| Mixed biomass (A)[a] | 15 | 26.7 | 40.5 | 39 | 43.5 |
| Distillery wastes | 7.5 | 10.7 | 71 | 70.5 | 79 |

[a]   Mixed biomass (A): water hyacinth, algae, cow dung, rice husk (1:1:1:0.9).

---

**TABLE 14.3**

**Biogas Production in the Various Non-EURO Countries in the World in 2013**

| Sr. No. | Countries | Biogas (TOE*) |
|---|---|---|
| 1. | China | 7,866,844 |
| 2. | United States | 6,348,381 |
| 3. | Russia | 3,029,785 |
| 4. | Thailand | 676,221 |
| 5. | India | 398,969 |
| 6. | South Korea | 236,864 |
| 7. | Turkey | 232,708 |

*Source:*   State of Biogas in the World, Clean Energy Solutions Center, 2017.

*   1 TOE = Energy generated by burning one metric ton (1000 kilograms) of oil ~41.87 GJ.

The successful applications of anaerobic digestion around the world has been available in the Global Methane Initiative (2015) which are summarized in the following section (https://www.globalmethane.org/documents/GMIBenefitsReport.pdf).

### 14.2.1 FAIR OAKS, INDIANA, UNITED STATES

Fair Oaks Farm established a mixed plug flow anaerobic digester using cow dung from 35,000 cows. It includes gas upgrading and gas drying operations. Biogas from the anaerobic digester can produce 865 MMbtu CNG per day, which is used for operating the entire farm.

### 14.2.2 SHANGDONGKUO VILLAGE, CHANGPING DISTRICT, BEIJING, CHINA

Shangdongkuo village has a large-scale Centralized Biogas Supply Project for using wastes from chicken (5,000–6,000) and pig farms (3,000). This project was started in 2008. The village uses two anaerobic digesters of capacity 300 m$^3$ each using Upflow Solids Reactor (USR). The plant generates 207,000 m$^3$ biogas in a year, which is used as cooking fuel in the nearby villages.

### 14.2.3 SANTA ROSILLO VILLAGE, HUIMBAYOC, SAN MARTIN, PERU

BioSynergy uses two digesters of capacity digester 93 m$^3$ each for the treatment of the waste generated from 67 animals (cows and horses) in Santa Rosillo Village. It is estimated that the plant produces 11.65 m$^3$ biogas per day, which is used for combined heat and power (CHP) generation.

### 14.2.4 ELAVAMPADAM MODEL RUBBER PRODUCER'S SOCIETY, INDIA

Over the past few decades, various biogas plants using cattle manure have been established in India. Several chemical and biochemical industries use the anaerobic digester for biogas generation. For example, it is estimated that about 8–12 L of wastewater is generated from rubber-processing industries from the processing of 1 kg of Ribbed Smoked Sheets (RSS), which is used as a feedstock for methane production. The total biogas produced per day is around 10–12 m$^3$, which is used for CHP purposes. Use of biomethane in such industries can reduce firewood consumption by almost 35%.

### 14.2.5 BIOMETHANATION PLANT, PUNJAB, INDIA

The wastes from the dairy complex comprised of 80,000 cattle are used as feedstock in the two high-rate biomethanation plants of capacity 5,000 m$^3$. The plant produces 10,000 m$^3$ of biogas per day, thereby generating 6 million kWh/year electrical power.

### 14.2.6 OREGON, UNITED STATES

Biogas produced from the wastes of three dairy farms in Oregon can generate from 190 kWe to 370 kWe. This energy is used for electricity generation and organic fertilizer.

## 14.3 BIOHYDROGEN PRODUCTION

Several research groups carried out pilot plant studies for the biohydrogen production. Unfortunately, no biohydrogen plant has been operated at the commercial scale. Ren and others (2006) conducted a pilot scale study using a 1.48 $m^3$ bioreactor for hydrogen production from cane molasses. They observed a hydrogen production rate of 8.24 $m^3$ $H_2$/ day and an overall yield of 26.13 mmol/g $COD_{removed}$. Similarly, Jayalakshmi and others (2009) operated a 150 L inclined plug-flow reactor by using kitchen waste as the raw material. They obtained an overall $H_2$ yield of 72 mL $H_2$/g VS added which was much less compared to the previous study. Lin and others (2010) reported a $H_2$ production rate of 15.59 L/L d and a yield of 1.04 mol $H_2$/mol sucrose using an organic loading rate of 240 g COD/L/d. In another biohydrogen pilot-scale study, a 100 $m^3$ bioreactor obtained a $H_2$ yield of 2.76 mol $H_2$/mol of hexose from distillery effluent by using a co-culture of *C. freundii* 01, *E. aerogenes* E10, and *Rhodopseudomonas palustris* P2 (Vatsala et al. 2008). Classen and others (2004) reported a pilot plant-scale hydrogen production at a capacity of 10,200 $Nm^3$ $H_2$ $d^{-1}$ using lignocellulosic feedstock in a 95 $m^3$ thermo bioreactor fermentation followed by photofermentation in a 300 $m^3$ photobioreactor. Other researchers performed a 10 $m^3$ pilot-scale study for biohydrogen production using *Enterobacter cloacae* IIT-BT 08 with cane molasses and groundnut de-oiled cake as co-substrates (Balachandar 2018; Das 2017). They reported a maximum $H_2$ production of 76.2 $m^3$ with a COD removal of 18.1 kg/$m^3$ and an energy conversion efficiency of 26.8%. The conversion of biomass to hydrogen may be improved by using a cheap and abundant renewable substrate like biomass or wastewater along with the development of potential microbial consortia. It is possible to make the process commercially viable. Several pilot plant studies are discussed in the following section.

### 14.3.1 PILOT PLANT STUDIES AT THE INDIAN INSTITUTE OF TECHNOLOGY, KHARAGPUR

It was observed that a medium comprised of cane molasses (CM) 1% (w/v) and groundnut deoiled cake (GDOC) of 2.5% (w/v) is suitable for hydrogen production by using *Enterobacter cloacae* IIT-BT 08. Initially, the production medium of 4.5 $m^3$ consisting of cane molasses 1%w/v and GDOC of 2.5%w/v was added to the 10 $m^3$ bioreactor (Das 2017; Balachandar 2018). Then a 500 L production medium with culture was added as an inoculum to the 10 $m^3$ bioreactor. This addition was followed by the addition of the remaining production medium of 5 $m^3$. The performance of the 10 $m^3$ reactor on cumulative gas production, rate of gas production, and profile of soluble end metabolic products is shown in Figure 14.2. Total cumulative gas ($H_2$ + $CO_2$) was observed to be 158.9 ± 8.8 $m^3$ of which about 45%–50% corresponded to hydrogen (i.e., total cumulative $H_2$ was 76.2 $m^3$). The major end metabolites were ethanol, acetate, and butyrate. A total COD of 18.1 ± 1.2 kg $m^{-3}$ from CM and GDOC was removed during the fermentation process. An increase in the concentration of volatile fatty acids (VFAs) leads to the reduction in pH and buffering capacity of the fermentation medium. It was observed that the biohydrogen production is a growth-associated product (Balachandar 2018).

Evaluation of the amount of substrate conversion for a particular product or biomass can be done with the help of material analysis. Material analysis has been conducted

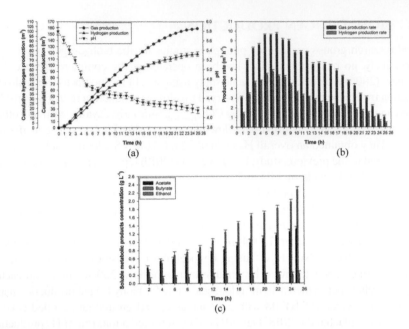

**FIGURE 14.2**  Performance of a 10 m³ reactor on biohydrogen production using organic residues as substrate: (a) cumulative gas production and pH profile, (b) rate of gas and hydrogen production, and (c) soluble end metabolic products.

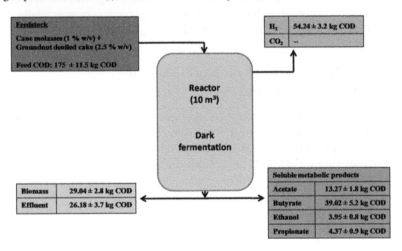

**FIGURE 14.3**  Material analysis in terms of COD of biohydrogen production using groundnut deoiled cake as a co-substrate with cane molasses in a 10 m³ bioreactor.

based on the chemical oxygen demand of the substrate and products (Figure 14.3). COD balance was performed for the 10 m³ reactor based on an input of $175 \pm 10.5$ kg COD. COD balance helps in reflection of experimental data validation and leads to the investigations of underlying mechanism involved (Barker and Dold, 1995). It was observed that most of the substrate was channelized towards the production of hydrogen (30.99%) followed by butyrate, biomass, acetate, and propionate of 22.29, 16.59, 7.58, and 2.49%, respectively. COD distribution analysis helps in the assessment of substrate use for biomass and other metabolite formations. The continuation of the COD distribution analysis included an evaluation of energy recovery to determine the potential of hydrogen production using organic waste for commercial viability. Gaseous energy recovery using CM and GDOC as substrates was estimated and is shown in Table 14.4.

### 14.3.2 FENG CHIA UNIVERSITY, TAIWAN

**Operational parameters of the pilot-scale plant**:
- Seed culture comprised of *Clostridium pasteurianum*, *Bifidobacteria* sp., and *Clostridium tyrobutyricum* (Lin et al. 2011)
- Feedstock storage tanks (2) with carrying capacity of 0.75 m³
- Raw material: synthetic wastewater with simple sugars
- Nutrient storage tank (carrying capacity of 0.75 m³)
- Mixing tank having a carrying capacity of 0.6 m³
- Granular solid bed reactor (SBR) (working volume of 0.4 m³)

**Process conditions:**
- Duration = 67 d
- Temperature = 35°C
- Organic loading rate (OLR) = 40–240 kg COD/m³ d
- Influent sucrose concentration = 20 and 40 kg COD/m³
- Fermenter agitation speed = 10–15 rpm

**Hydrogen production rate and yield:**
- The highest hydrogen yield observed was 1.04 mol of hydrogen per mol of sucrose.
- The highest hydrogen production rate was 15.59 m³/m³ d.

---

**TABLE 14.4**

**Energy Recovery Analysis of Biohydrogen Production in a 10 m³ Pilot-Scale Reactor**

**Values**

| | |
|---|---|
| Heating value of hydrogen | 120 MJ kg⁻¹ |
| Heating value of cane molasses | 8 MJ kg⁻¹ |
| Heating value of groundnut deoiled cake | 12 MJ kg⁻¹ |
| Cane molasses used in 10 m³ of the medium | 100 kg |
| Groundnut deoiled cake used in 10 m³ of the medium | 250 kg |
| Total heating value of the fermentation medium used | 3800 MJ |
| Total cumulative H₂ production from 10 m³ | 6.85 kg |
| Total heating value of H₂ evolved from 10 m³ | 822 MJ |
| Energy recovery = {(Total heating value of H₂ evolved) / (Total heating value of fermentation medium)} × 100 | |
| Total energy recovery (in terms of substrate added) | 21.6% |
| Total energy recovery (in terms of substrate removed) | 37.9% |

### 14.3.3   SHRI AMM MURUGAPPA CHETTIAR RESEARCH CENTRE, CHENNAI, INDIA

**Operational parameters of a pilot-scale plant:**
- Seed culture comprised of *Citrobacter freundii* C01, *Enterobacter aerogenes* E10, and *Rhodopseudomonas palustris* P2 (Vatsala et al., 2008)
- Reactor volume: 100 m$^3$
- Hydrogen production scaled up from 1 L to 100,000 L
- Different reactor capacities considered: 0.125, 1.25, 12.5, and 125 m$^3$
- Height-to-diameter ($H/D$) ratio maintained at 1.28
- Feedstock for biohydrogen production: Distillery effluent

**Process conditions:**
- Time of fermentation: 40 h
- Temperature: 37°C
- Initial COD of the feedstock: 101.2 g/L
- Initial BOD of the feedstock: 58.8 g/L

**Performance of the reactor:**
- 21.38 kg of hydrogen corresponding to 10,692.6 mol obtained through the batch method from reducing sugar (3,862.3 mol) as glucose
- Average yield of hydrogen = 2.76 mol/mol glucose
- Rate of hydrogen production estimated to be 0.53 kg/100 m$^3$h

### 14.3.4   HARBIN INSTITUTE OF TECHNOLOGY, CHINA

**Operational parameters of a pilot-scale plant:**
- Seed culture comprised of *Clostridium* sp., *Enterobacter aerogenes*, and *Ectothiorhodospira vacuolata* (Ren et al., 2006)
- Capacity: 2 m$^3$ (working volume of 1.48 m$^3$)
- Feedstock: Cane molasses (53% w/w reducing sugars)
- OLR varied in the range of 3.11–85.57 kg COD/m$^3$d

**Process conditions:**
- Time of fermentation: 200 days
- Temperature: 35°C
- pH: 7
- Start-up OLR: 6.32 kg COD/m$^3_{reactor}$ d with an HRT of 11.4 h and a substrate concentration of 3000 mg COD/L

**Performance of the reactor:**
- Maximum hydrogen production of 5.57 m$^3$ hydrogen m$^{-3}$
- Specific hydrogen production rate of 0.75 m$^3$ hydrogen kg$^{-1}$ MLVSS d$^{-1}$
- Hydrogen yield 26.13 mol kg$^{-1}$ COD$_{removed}$ within the OLR range (35–55 kg COD m$^{-3}_{reactor}$ d$^{-1}$)
- Hydrogen production rate influenced by accumulation of VFAs
- Hydrogen production rate of 0.75 m$^3$ kg$^{-1}$ MLVSS d$^{-1}$ observed

## 14.4 BIOHYTHANE PRODUCTION

The spent medium of the dark fermentation process is comprised of VFAs such as acetic acid, propionic acid, and butyric acid, which can be readily used as substrates for biomethane production. This two-step hydrogen and methane production by integrating dark fermentation with anaerobic digestion process is termed the biohythane process (Das and Roy 2017). Compared to other process, the biohythane process has enormous potential to maximize the overall energy recovery from a single substrate. Cavinato et al. (2012) used a two-stage biohythane process treating food waste, which consisted of two continuous stirred-tank reactors (CSTRs) of 0.2 m$^3$ and 0.76 m$^3$ capacities for hydrogen and methane, respectively. By maintaining a hydraulic retention time (HRT) of 3.3 d and 12.6 d for hydrogen and methane reactors, respectively, a specific hydrogen yield of 66.7 L kg$^{-1}$ TVS and a methane yield of 0.72 m$^3$ kg$^{-1}$ TVS were obtained.

## 14.5 BIODIESEL PRODUCTION

Biodiesel is a renewable liquid fuel that is obtained from the chemical or enzymatic conversion of oil crops, animal fats, and microbial lipids (Price et al. 2016; Meher et al. 2006; Brask et al. 2011; Ranganathan et al. 2008). The most commonly used process for the conversion is known as transesterification, which involves the reaction of the oil (or fat) and an alcohol (usually methanol) to produce fatty acid methyl ester (biodiesel) in presence of an acid or base catalyst (Gog et al. 2012; Price et al. 2017). The biodiesel has properties similar to petrodiesel and thus can be readily used in diesel engines (Al-Zuhair 2007). It can be used either in pure form or in a blended form without any modification (Price et al. 2016). It is estimated that using blends B20 (20% biodiesel, 80% petroleum diesel) or higher can significantly reduce the GHG emissions.

The world production of biodiesel increased dramatically between 2000 and 2013 from 213 million gallons to 6,289 million gallons. The top production regions were the European Union, Brazil, the United States, Argentina, and China (EIA Monthly Biodiesel Production Report 2014). In 2013, the United States alone produced over 1,300 million gallons of biodiesel from soy bean oil (Atadashi et al. 2011). In 1893, the German engineer Rudolf Diesel became the first person to create a diesel engine. He believed that the use of vegetable oil could fuel machines for agriculture in rural parts of the world. The starting feedstock plays a crucial role for the commercial production of biodiesel. In North America, Australia, and China, major emphasis has been given to converting the waste industrial gases to biodiesel using bio-based gas-to-liquid (GTL) technologies (Wood et al. 2012; Fei et al. 2014). Methanotrophic bacteria tend to convert the methane into microbial lipids, which can be a suitable source for biodiesel production (Helwani et al. 2009; Jin et al. 2014). Although attractive, the GTL-based biodiesel production requires further breakthroughs for use at commercial scales.

Due to the high price of conventional first-generation feedstocks such as oil palm and rapeseed, the focus has been shifted towards the use of non-food oil crops and waste oils such as microalgal lipids, waste cooking oil, and animal fats. Price and others (2017) performed a comparative economic assessment of biodiesel production using the fed batch and the continuous mode of operation. It was observed that the CSTR operation on an average had a reaction time 1.3 times greater than the fed-batch operation. Their cost analysis showed that a profitable margin could be obtained only if higher revenues are obtained by selling biodiesel compared to the initial cost of enzymes. Another major limitation of the transesterification process is the amount of alcohol used. It was estimated that for the lipase catalyzed process, 50% excess methanol was needed to attain 95% biodiesel yield (Fjerbaek et al. 2009); whereas for an alkaline catalyzed process, over 100% excess methanol is needed (Meher et al. 2006). The benefits of biodiesel are ease of use, production simplicity, and cost competitiveness. However, the economical availability of renewable feedstock such as animal fat and vegetable oil will create a promising future for biodiesel.

## 14.6  BIOETHANOL PRODUCTION

Bioethanol is a promising renewable, sustainable, and environmentally friendly biofuel that can reduce and replace the dependence on traditional fossil fuels (Karlsson et al. 2014; Fan et al. 2014; Su et al. 2015; Raman et al. 2015; Ojeda et al. 2011). The United States and Brazil are among world's largest producers of bioethanol (García et al. 2013; Petrobras, 2016) (Table 14.5). There are 285 distillery industries in India producing 2.7 billion liters of alcohol annually. At present, commercial bioethanol is produced mainly from sugar and starch crops. However, use of these crops leads to an increase in food prices and thus alternate feedstocks such as dedicated non-food-based energy crops, agricultural wastes and forestry wastes, and residues are being considered as potential raw materials for bioethanol production. Industrial bioethanol production mainly uses the yeast *Saccharomyces cerevisiae* since it has high productivity, high ethanol tolerance, and can use a wide range of substrates. Its major disadvantage, however, is the inhibition at high temperature conditions and the inability to convert pentoses. To improve the

**TABLE 14.5**
**Ethanol Production in Various Countries**

|  | % of World Production | Million Gallons | % of World Production |
|---|---|---|---|
| United States | 58% | 15,800 | 58% |
| Brazil | 27% | 7,060 | 26% |
| European Union | 5% | 1,415 | 5% |
| China | 3% | 875 | 3% |
| Canada | 2% | 450 | 2% |
| Thailand | 1% | 395 | 2% |
| Argentina | 1% | 310 | 1% |
| India | 1% | 280 | 1% |

*Source:* RFA analysis of public and private data sources.

ethanol yields, various recombinant strains have been used. Depending upon the type of feedstock, the steps in the ethanol production process varies. Simple sugars such as glucose or sucrose can be fermented easily by yeast into ethanol; however, the starchy raw material requires a pre-processing step (i.e., the dry grind process or wet milling process to convert the starch into simple sugars), which can then be fermented to ethanol. For more complex substrates, such as lignocellulosics, several pre-processing steps are needed (such as pre-treatment and hydrolysis) prior to the fermentation step, which increases the cost of production. Various physicochemical parameters influence the ethanol production such as temperature, pH, substrate concentration, time of fermentation, and agitation rate. Immobilizing the yeasts cells can be an alternative strategy to improve the bioethanol yields. The main problem with using lignocellulosic materials as the raw material is the presence of lignin, which is non-biodegradable. Several research groups have carried out intensive research at the pilot scale on bioethanol production (Binod et al. 2010; Petersson et al. 2007; Sanchez et al. 2008). These studies suggest economic feasibility for lignocellulosic ethanol can be realized in the near future.

## 14.7   BIOBUTANOL PRODUCTION

Biobutanol is an alternative liquid biofuel produced by using the *Clostridial*-acetone-butanol-ethanol (ABE) fermentation process. It is advantageous over ethanol due to its high calorific value and high energy density. Moreover, it can be used in the gasoline engines without any modification. The major challenges for industrial butanol production include the low yields, high feedstock costs, and product inhibition (Mitra et al. 2017; Formanek et al. 1997; Xue et al. 2013; Xue et al. 2017). Parekh (1999) assessed the economic viability of the ABE fermentation process in a 200-L pilot-scale bioreactor using glucose/corn steep water (CSW) as the substrate. A comparative analysis was performed on the butanol-producing capacity of *Clostridium beijerinckii* NCIMB 8052 and *Clostridium beijerinckii* BA101. It was observed that *C. beijerinckii* BA101 produced 40% higher butanol compared to *C. beijerinckii* 8052. As in the case of bioethanol, feedstock plays a crucial role in determining the overall production costs. Alternative feedstocks such as lignocellulosics can be cost effective raw materials for butanol production. Intensive studies have been carried out to explore the use of lignocellulosic materials for biobutanol production. Zhang and others (2012) used corn cob as a raw material for butanol production that was pre-treated using $Ca(OH)_2$ to remove the inhibitors. It was observed that the treated corn cob residue yielded two-fold sugars compared to the untreated one. This study suggests that the preliminary pre-treatment is necessary for complete hydrolysis of lignocellulosic substrates. Thus, inexpensive pre-treatment techniques can help in reducing the overall production costs of the process.

## 14.8   CHALLENGES AND CONSIDERATIONS

During the last decade, several efforts have attempted to make the biofuel production processes more feasible. However, major issues remain to be solved. Currently, low substrate conversion efficiency, low product yield, pre-treatment processes, and availability of the raw materials are the biggest bottlenecks in the process. These challenges may be overcome by using efficient bioreactor designs, process modification, suitable feedstock, and suitable microbial strains. Despite a large amount of research in the past and at present, these specific areas must be concentrated upon for further enhancement of biofuel production. Bioethanol and biomethane production on the industrial scale was successful. The biomethanation process will be more attractive at the energy and economic point of view if it is replaced by the biohythane process for the maximization of gaseous energy recovery. The successful use of cellulosic raw materials for the bioethanol and biobutanol production must be explored. Scarcity of the raw materials is the main problem for the biodiesel production. So, an attempt should be made to produce cheaper raw materials.

## REFERENCES

Achinasa S, Achinas V, Euverink GJW (2017) A technological overview of biogas production from biowaste. *Engineering* 3:299–307.

Al-Zuhair S (2007) Production of biodiesel: Possibilities and challenges. *Biofuels Bioprod Biorefining* 1:57–66.

Atadashi IM, Aroua MK, Aziz ARA, Sulaiman NMN (2011). Refining technologies for the purification of crude biodiesel. *Appl Energy* 88:4239–4251.

Balachandar G (2018) *Biohydrogen Production from Organic Wastes and Residues by Dark Fermentation.* PhD dissertation. Indian Institute of Technology Kharagpur, India.

Bauen A, Berndes G, Junginger M (2009) Bioenergy: A sustainable and reliable energy source. A review of status and prospects. *IEA Bioenergy* Main report 1–108.

Binod P, Sindhu R, Singhania RR, Vikram S, Devi L, Nagalakshmi S, Kurien N, Sukumaran RK, Pandey A (2010) Bioethanol production from rice straw: An overview. *Bioresour Technol.* 101:4767–4774.

Brask J, Damstrup ML, Nielsen PM, Holm HC, Maes J, Greyt W (2011) Combining enzymatic esterification with conventional alkaline transesterification in an integrated biodiesel process. *Appl Biochem Biotechnol* 163:918–927.

Cavinato C, Giuliano A, Bolzonella D, Pavan P, Cecchi F (2012) Bio-hythane production from food waste by dark fermentation coupled with anaerobic digestion process: A long-term pilot scale experience. *Int J Hydrog Energy* 37:11549–11555.

Classen PAM, de Vrije T, Budde MAW (2004) Biological hydrogen production from sweet sorghum by thermophilic bacteria. *Proceeding of the World Conference on Biomass for Energy,* Industry and Climate Protection, Rome, Italy.

Cong RG, Caro D, Thomsen M (2017) Is it beneficial to use biogas in the Danish transport sector? An environmental-economic analysis. *J Cleaner Production* 165:1025–1035.

Das D (1985) *Optimization of Methane Production from Agricultural Residues.* PhD dissertation. BERC, IIT Delhi, India.

Das D (2017) A road map on biohydrogen production from organic wastes. *INAE Letts* 2:153–160.

Das D, Ghose TK, Gopalakrisnan KS, Joshi AP (1983) Treatment of Distillery Wastes by a Two Phase Biomethanation Process Symposium papers. *Energy from Biomass and Wastes VII.* Boca Raton, FL, pp. 601–626.

Das D, Roy S (2017) *Biohythane: Fuel for The Future,* Pan Stanford Publishing, Singapore.

Drapcho CM, Nhuan NP, Walker TH (2008) *Ethanol Production, Biofuels Engineering Process Technology.* The McGraw-Hill Companies McGraw-Hill, New York.

EIA Monthly biodiesel production report. Washington, DC: U.S. Energy Information Administration, (2014). http://www.eia.gov/biofuels/biodiesel/production/.

Fan S, Xiao Z, Zhang Y, Tang X, Chen C, Li W, Deng Q, Yao P (2014) Enhanced ethanol fermentation in a pervaporation membrane bioreactor with the convenient permeate vapor recovery. *Bioresour Technol* 155:229–234.

Farrell AE, Plevin RJ, Turner BT, Jones AD, O'Hare M, Kammen DM (2006) Ethanol can contribute to energy and environmental goals. *Science* 311:506–508.

Fei Q, GuarnieriMT, Tao L, Laurens LML, Dowe N, Pienkos PT (2014) Bioconversion of natural gas to liquid fuel: Opportunities and challenges. *Biotechnol Adv* 32:596–614.

Fjerbaek L, Christensen KV, Norddahl B (2009) A review of the current state of biodiesel production using enzymatic transesterification. *Biotechnol Bioeng* 102:1298–1315.

Formanek J, Mackie R, Blaschek HP (1997) Enhanced butanol production by Clostridium beijerinckii BA101 grown in semi defined P2 medium containing 6 percent maltodextrin or glucose. *Appl Environ Microbiol* 63:2306–2310.

García V, Pongrácz E, Phillips PS, Keiski RL (2013) From waste treatment to resource efficiency in the chemical industry: Recovery of organic solvents from waters containing electrolytes by pervaporation. *J Clean Prod* 39:146–153.

Gog A, Roman M, ToSsa M, Paizs C, Irimie FD (2012) Biodiesel production using enzymatic transesterification—Current state and perspectives. *Renew Energy* 39:10–16.

Going Global (2015) *Ethanol Industry Outlook.* Renewable Fuels Association, Washington, DC.

Helwani Z, Othman MR, Aziz N, Fernando WJN, Kim J (2009) Technologies for production of biodiesel focusing on green catalytic techniques: A review. *Fuel Process Technol* 90:1502–1514.

Jayalakshmi S, Joseph K, Sukumaran V (2009) Bio hydrogen generation from kitchen waste in an inclined plug flow reactor. *Int J Hydrog Energy* 34:8854–8858.

Jin M, Slininger PJ, Dien BS, Waghmode S, Moser BR, Orjuela A, Sousa L, da C, Balan V (2014) Microbial lipid-based lignocellulosic biorefinery: Feasibility and challenges. *Trends Biotechnol* 33:43–54.

Karlsson H, Barjesson P, Hansson P, Ahlgren S (2014) Ethanol production in biorefineries using lignocellulosic feedstock-GHG performance, energy balance and implications of life cycle calculation methodology. *J Clean Prod* 83:420–427.

Koutinas A, Kanellaki M, Bekatorou A, Kandylis P, Pissaridi K, Dima A, Boura K et al., (2016) Economic evaluation of technology for a new generation biofuel production using wastes. *Bioresour Technol* 200:178–185.

Larson ED (2008) Biofuel production technologies: Status, prospects and implications for trade and development. *United Nations Conf Trade Dev* 1–41.

Laser M, Larson E, Dale B, Wang M, Greene N, Lynd LR (2009) Comparative analysis of efficiency, environmental impact, and process economics for mature biomass refining scenarios. *Biofuels Bioprod Bioref* 3:247–270.

Lin CY, Wu SY, Lin PJ, Chang JS, Hung CH, Lee KS, Lin YC (2011) A pilot-scale high-rate biohydrogen production system with mixed microflora. *Int J Hydrog Energy* 36:8758–8764.

Meher LC, VidyaSagar D, Naik SN (2006) Technical aspects of biodiesel production by transesterification—A review. *Renew Sustain Energy Rev* 10:248–268.

Mitra R, Balachandar G, Singh V, Sinha P, Das D (2017) Improvement in energy recovery by dark fermentative biohydrogen followed by biobutanol production process using obligate anaerobes. *Int J Hydrog Energy* 42:4880–4992.

Naik SN, Goud V V, Rout PK, Dalai AK (2010) Production of first and second generation biofuels: A comprehensive review. *Renew Sustain Energy Rev* 14:578–597.

Ojeda K, Sánchez E, Kafarov V (2011) Sustainable ethanol production from lignocellulosic biomass -application of exergy analysis. *Energy* 36:2119–2128.

Parekh M, Formanek J, Blaschek HP (1999) Pilot-scale production of butanol by Clostridium beijerinckii BA101 using a low-cost fermentation medium based on corn steep water. *Appl Microbiol Biotechnol* 51:152–157.

Petersson A, Thomsen MH, Hauggaard-Nielsen H, Thomsen AB (2007) Potential bioethanol and biogas production using lignocellulosic biomass from winter rye, oilseed rape and faba bean. *Biomass and Bioenergy* 31:812–819.

Petrobras (2016) Ethanol plants [WWW Document].

Price J, Nordblad M, Martel HH, Chrabas B, Wang H, Nielsen PM, Woodley JM (2016) Scale-up of industrial biodiesel production to 40 m$^3$ using a liquid lipase formulation. *Biotechnol Bioeng* 113:1719–1728.

Raman S, Mohr A, Helliwell R, Ribeiro B, Shortall O, Smith R, Millar K (2015) Integrating social and value dimensions into sustainability assessment of lignocellulosic biofuels. *Biomass and Bioenergy* 82:49–62.

Ranganathan SV, Narasimhan SL, Muthukumar K (2008) An overview of enzymatic production of biodiesel. *Bioresour Technol* 99:3975–3981.

Ren NQ, Li JZ, Li BK, Wang Y, Liu SR (2006) Biohydrogen production from molasses by anaerobic fermentation with a pilot scale bioreactor system. *Int J Hydrog Energy* 31:2147–2157.

Richard TL (2010) Challenges in scaling up biofuels infrastructures. *Science* 329:793–796.

Sanchez OJ, Cardona CA (2008) Trends in biotechnological production of fuel ethanol from different feeds-stocks. *Bioresource Technology* 99:5270–5295.

State of Biogas in the World (2017), *Clean Energy Solutions Center*, United States Department. of Energy, Washington, DC.

Su Y, Zhang P, Su Y (2015) An overview of biofuels policies and industrialization in the major biofuel producing countries. *Renew Sustain Energy Rev* 50:991–1003.

Uslu A, Faaij APC, Bergman PCA (2008) Pre-treatment technologies, and their effect on international bioenergy supply chain logistics. Techno-economic evaluation of torrefaction, fast pyrolysis and pelletisation. *Energy* 33:1206–1223.

Vatsala TM, Mohan Raj S, Manimaran A (2008) A pilot-scale study of biohydrogen production from distillery effluent using defined bacterial co-culture. *Int J Hydrog Energy* 33:5404–5415.

Wood DA, Nwaoha C, Towler BF (2012) Gas-to-liquids (GTL): A review of an industry offering several routes for monetizing natural gas. *J Nat Gas Sci Eng* 9:196–208.

World Energy Resources Bioenergy (2016), *World Energy Council*, UN.

Xue C, Zhao XQ, Liu CG, Chen LJ, Bai FW (2013) Prospective and development of butanol as an advanced biofuel. *Biotechnol Adv* 31:1575–1584.

Xue C, Zhang X, Wang Y, Xiao M, Chen L, Bai F (2017) The advanced strategy for enhancing biobutanol production and high-efficient product recovery with reduced wastewater generation. *Biotechnology for Biofuels* 10:148.

Zhang WL, Liu ZY, Liu Z, Li FL (2012) Butanol production from corncob residue using clostridium beijerinckii NCIMB 8052. *Letts Appl Microbiol* 55:240–246.

# 15 Energy and Economic Analysis of Biofuel Production Processes

## 15.1 INTRODUCTION

The current energy policies are focused mainly upon the development of green and environmentally friendly technologies, which can address the global warming and climate change issues associated with the use of traditional fossil fuels (Demirbas 2009). Among the various renewables, biofuels show immense potential for sustainable and efficient energy generation (Nigam and Singh 2011). There are several benefits of using biofuels over fossil fuels such as greater energy security, reduced carbon emissions, biodegradability, fuel diversity, safer handling, and additional market for agricultural products (Coelho 2012). The most common liquid biofuels include bioethanol obtained using sugarcane or starch and biodiesel obtained using oil crops. Both biofuels require high-quality agricultural land for their production. Although alternate non-food crop-based sources and wastes have been considered as potential feedstock for bioenergy production, several impediments still exist that suppress their full fledge production and commercialization. Energy and economic analysis of any process helps in the assessment of its feasibility in terms of efficiency, sustainability, and commercial viability (Murali et al. 2016). The economics of existing biofuels differ from one another depending upon the type of feedstock, availability, conversion processes, scale of the production, local market, and technological variations (Koutinas et al. 2016). Complete supply chain, upstream, bioprocess, and downstream must be considered for accuracy of the energy and economic analysis (Browne et al. 2011). In addition, the total biofuel costs also must include their impact on the related market such as food. For the realization of a successful economically viable process, the net energy gain should be positive. In this chapter, the various economic aspects essential for establishing and sustaining biofuel markets are discussed along with the energy balance evaluation of certain commonly used biofuel production processes.

## 15.2 BIOMASS AVAILABILITY AND ITS POTENTIAL FOR BIOFUEL PRODUCTION

Biomass contributes to a major portion of the world's total renewable energy supply, but this resource is not evenly spread throughout the world. Several countries produce a relatively less amount of biomass and require imports from other countries to meet the increasing demands (Hu et al. 2017). Therefore, due to the aforementioned reasons, the production and use of bioenergy can play a vital role in developing a country's economy. The recent changes in international climate change policies and mechanisms can have considerable impact on sustainable biomass trade and might influence the agricultural, forestry, and social welfare policies. There still remains a huge amount of biomass that has remained unexploited and unexplored for commercial purposes (IEA Bioenergy 2006). At present, forestry, agriculture, and municipal wastes and residues are used as major sources for heat and electricity generation while a small share of sugar, grain, and vegetable oil crops are used as feedstocks for liquid biofuel production (British Petroleum 2018). There is significant potential to expand the biomass use by using the large volumes of unused lignocellulosic crops, aquatic biomass, organic wastes, and residues for biofuel production. Figure 15.1 depicts the various types of biomass that can be used for biofuel production. However, major portions of these biomass resources are used for other applications that are not part of the energy sector, thereby leading to competition for biofuel markets. To increase the biofuel production, large areas must be cultivated in the future exclusively for dedicated energy crop production. In addition, the improvements in agricultural practices and progress in plant breeding, cultivation, and harvesting techniques can help in increasing the overall biomass availability sufficing for both energy and other applications.

Several factors influence the economics of biomass supply chains. Particularly, the production, collection, pre-processing, and distribution of the raw materials (such as dedicated energy crops, agricultural wastes, and forestry residues) increase the overall costs of the biofuel production processes (Koutinas et al. 2016). Although the logistics involved with the use of conventional food crops and forestry products are well established, these cannot be directly implemented to the second-generation energy crops, wastes, and residues (Naik et al. 2010). To minimize the transportation

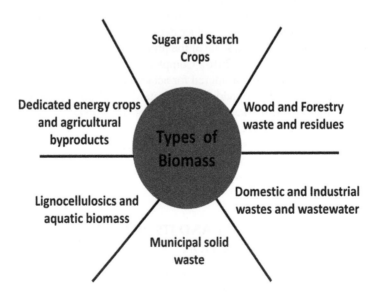

**FIGURE 15.1**   Types of biomass used for biofuel production.

**TABLE 15.1**
**Regional Potential of Biomass Supply**

| Region | Land Use (Mha) | Biomass Types | Primary Biomass Potential (EJ) |
|---|---|---|---|
| Europe | 25 | Mix of conventional and lignocellulosic feedstocks | 6 |
| USA | 20 | Corn, herbaceous crops, forest and agricultural residues, municipal solid wastes | 18.2 |
| China and India | 86 | Sugarcane, corn, soy bean, wheat, palm oil and cellulosic residues | 16.9 |
| Australia | – | Process residues (mainly bagasse and wood) and waste streams (municipal solid wastes and sewage sludge) | 0.07 |

*Source:* IEAWEO-2017 Special Report, Energy Access Outlook, 2017.

costs, increasing the energy density of biomass by reducing the moisture content can be a viable option. Another major factor that impacts the cost of biomass supply is the seasonal variation and storability. For example, the harvesting season of sugarcane typically takes 6–7 months in a year which limits the total operational hours per year (Bauen et al. 2009). In such cases, the biomass needs to be stored for several months prior to its use, which can be problematic. The storage of biomass with high moisture content, specifically, is extremely difficult because it is prone to dry matter loss and self-ignition. The advanced pre-treatment of biomass can reduce these risks, but will lead to increase in the costs and energy requirement of the overall process. Nevertheless, the storage and transport of biomass is much easier compared to the alternate renewables such as solar and wind, which require immediate consumption or a connection to the grid. Table 15.1 presents an overview of global potential of biomass supply for bioenergy production.

## 15.3 ENERGY BALANCE AND EFFICIENCY OF BIOFUELS

The energy analysis of biofuels considers the energy input during the whole life cycle (i.e., from biomass production to biofuel consumption) and the energy output of the final fuel (Kralova and Sjöblom 2010). Figure 15.2 shows the basic steps involved during life cycle of biofuel processing. For almost all types of biofuels, the steps in the life cycle vary, depending upon the type of feedstock, agricultural practices, biomass productivity, seasonal and regional variations, and process technologies. Therefore, the data validation and comparisons in terms of energy flows for each process must be carefully scrutinized. There are two main measures to evaluate the energy performance of biofuel production processes: (1) the energy balance ratio or (2) the net energy ratio (NER). Energy efficiency NER mainly includes the ratio of

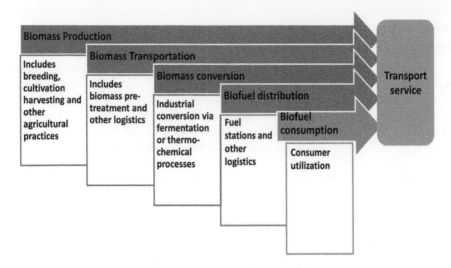

**FIGURE 15.2**   Steps involved in the life cycle of biofuel processing.

energy contained in the final biofuel to the fossil fuel inputs. It does not involve the biomass energy inputs and thus provides an estimate of the substitution efficiency of biofuels (Gnansounou et al. 2009). On the other hand, energy efficiency considers the ratio of biofuel energy to the energy inputs from biomass, fossil fuels, and other renewable energy sources used during the biofuel production. It provides an indication of the energy lost during the biomass conversion to liquid or gaseous biofuel. An NER greater than 1 indicates positive energy impact of the process. However, the energy efficiency of any process is always below 1 because some amount of energy loss from biomass is inevitable.

Several studies have conducted the net energy analysis of various biofuel production processes. Hill and others (2006) evaluated the energy benefits of using biodiesel and bioethanol rather than fossil fuels and estimated that ethanol obtained using corn yielded 25% more energy than the energy invested in its production, while biodiesel yielded 93% more energy. They suggested that the biodiesel production was more energetically favorable than ethanol due to lower agricultural inputs and more efficient conversion of feedstock to fuel. Similarly, Shapouri and others (2003) estimated a positive energy impact of corn ethanol with NER of 1.34. On the contrary, Pimentel (2003) reported that 29% more energy was invested than the energy obtained from a gallon of corn ethanol. Zhang and others (2016) performed the energy balance analysis for various biofuels using crude glycerol (a by-product of biodiesel production) as the primary feedstock. They concluded that the overall NER was 1.16, 0.22, 0.27, and 0.40 for biodiesel, biohydrogen, biogas, and bioethanol production, respectively (Zhang et al. 2016). This result indicates that the production of hydrogen, methane, and ethanol from crude oil are energy intensive processes. Fore and others (2011) estimated the net energy balance of biodiesel production from canola and soybean and estimated NER of 1.78 and 2.05 for canola and soybean biodiesel, respectively.

They suggested that soybean biodiesel was more energetically efficient due to reduced nitrogen fertilizer requirement. A similar observation was made by Pradhan et al. (2008) who estimated NER of 2.55 for soybean biodiesel using unified boundary and bootstrapping technology. Achten and others (2010) performed a life cycle assessment of jatropha biodiesel and found an NER of 1.85, which was significantly improved by the integration of a second stage biogas plant to the system (NER = 3.40). Berglund and Börjesson (2006) assessed the energy performance of a biogas production plant and found that the operation of the biogas plant was the most energy-demanding process accounting for 40%–80% of the total energy input. They concluded that the net energy output of the biogas plant turns negative when transport distance exceeded 200–700 km. Banks and others (2011) performed an energy analysis of an anaerobic digestion process using food waste as the raw material and obtained 90% energy efficiency in terms of energy content of volatile solids added. Manish and Banerjee (2008) performed a comparative energy analysis of various biological hydrogen production processes and obtained an NER of 1.9, 3.0, 3.1, and 1.8, respectively, from dark fermentation, photofermentation, two-stage fermentation, and microbial electrolysis process. All the biological processes were energetically favorable compared to the traditional steam methane-reforming process for hydrogen production (NER = 0.64). Although most studies discussed in this chapter show the positive energy impact of biofuels, there still exists a lack of consistency in defining the system boundaries and energy allocation for the various steps involved during the biofuel processing. The contradictory results obtained among similar processes suggest that a full-scale life cycle assessment, with a consistent estimate of data and assumptions, is crucial for determining the energy, economic, and environmental impact of biofuels.

## 15.4 ECONOMIC ANALYSIS OF DIFFERENT BIOFUELS PRODUCTION PROCESSES

As previously discussed, the market price of biofuels depends upon several factors such as the availability and price of feedstocks, the conversions processes, maintenance, and longevity. Table 15.2 depicts an estimate of production costs for selected biofuels.

The estimates provided in Table 15.2 are based mainly on pilot- or commercial-scale studies, which can be varied with the change in the logistics of the processes. The high production costs remain a critical barrier for the commercialization of second-generation biofuels, although continuing improvements are being achieved. Figure 15.3 illustrates the developmental status of various biofuel technologies.

The production scale of biofuels has significant impact on their costs (de Jong et al. 2017). Moreover, the lack of infrastructure and the bottlenecks along the supply chains affect the economic viability of these processes (Clark et al. 2013). Some of the strategies used for assessing the economy of selected biofuels are discussed in the following section.

## TABLE 15.2
## Production Cost of Various Biofuel Production Processes

| Biofuel | Feedstock (Process) | Cost of Production | References |
|---|---|---|---|
| Ethanol | Corn (dry grind process) | $1.05/gal | Shapouri and Salassi (2006) |
| | Corn (wet milling) | $1.03/gal | |
| | Sugarcane (fermentation) | $2.40/gal | |
| | Sugar beet (fermentation) | $2.35/gal | |
| | Molasses (fermentation) | $1.27/gal | |
| Biodiesel | Vegetable oil (transesterification) | $2.00/gal | Haas et al. (2006) |
| | Soybean oil (transesterification) | $1.14/gal | |
| | Rapeseed oil (transesterification) | $2.62/gal | |
| | Waste cooking oil (transesterification) | $1.05/kg | Mohammadshirazi et al. (2014) |
| Butanol | Corn (ABE fermentation) | $0.55/kg | Qureshi and Blaschek (2000) |
| | Wheat straw (ABE fermentation) | $4.28/gal | PSU (2016) |
| | Corn stover (ABE fermentation) | $1.80/L | Baral and Shah (2016) |
| Hydrogen | Organic waste (dark fermentation) | $1.3/MBTU | Nayak et al. (2013) |
| | Organic waste (pyrolysis) | $7.41/GJ | Das et al. (2014) |
| Methane | Organic waste (anaerobic digestion) | $0.05/kwh | Oreggioni et al. (2017) |
| | Municipal solid waste (anaerobic digestion) | $0.99/m³ | Cucchiella and D'Adamo (2016) |

### 15.4.1 BIOETHANOL

Bioethanol is among one of the most common liquid biofuels that is presently used for commercial purposes. Compared to the traditional fossil fuels, bioethanol is not considered to be economically competitive because most industrial bioethanol plants still rely on first-generation feedstocks. To gain economic acceptance, the cost of biomass conversion to ethanol must be lower than that of current gasoline prices (Chandel et al. 2007). The biomass feedstock costs represent 40% of the overall ethanol production costs and thus is a crucial parameter that needs utmost attention (Sims et al. 2008). Much of the attention has now been diverted towards the use of lignocellulosic feedstock, which represents a low-cost alternative to the traditional energy crops (Mcaloon et al. 2000). Kazi and others (2010) performed a techno-economic analysis of bioethanol production using corn stover and made comparisons between various pre-treatment technologies such as dilute acid, two-stage dilute acid, hot water, and ammonia fiber explosion or AFEX. They found that the dilute acid pre-treatment contributed to the lowest ethanol production

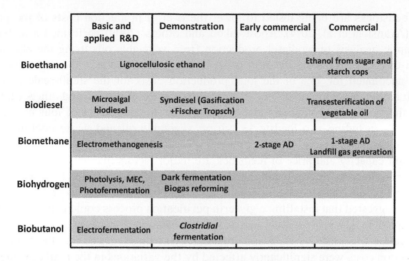

| | Basic and applied R&D | Demonstration | Early commercial | Commercial |
|---|---|---|---|---|
| **Bioethanol** | Lignocellulosic ethanol | | | Ethanol from sugar and starch cops |
| **Biodiesel** | Microalgal biodiesel | Syndiesel (Gasification +Fischer Tropsch) | | Transesterification of vegetable oil |
| **Biomethane** | Electromethanogenesis | | 2-stage AD | 1-stage AD Landfill gas generation |
| **Biohydrogen** | Photolysis, MEC, Photofermentation | Dark fermentation Biogas reforming | | |
| **Biobutanol** | Electrofermentation | *Clostridial* fermentation | | |

**FIGURE 15.3** Development stages of various biofuel technologies. (Adapted from Bauen, A. et al., Bioenergy: A sustainable and reliable energy source. A review of status and prospects. IEA Bioenergy Main report, 1–108, 2009.)

costs ($1.36/L of gasoline equivalent) while the feedstock cost, enzyme cost, and installed equipment costs remained the limiting parameters for the commercial viability of the process. A similar observation was made by Hamelinck and others (2005) in which dilute acid hydrolysis contributed to 35% efficiency in biomass to ethanol conversion. They suggested that with further improvement in hydrolysis-fermentation efficiency, lower capital investments and cheaper feedstock costs, the current bioethanol product costs can be reduced by 60% in the next 20 years. Quintero and others (2013) assessed and compared the economic potential of bio-ethanol using four different lignocellulosic materials (i.e., sugarcane bagasse, cof-fee cut-stems, rice husk, and empty fruit bunches). They used the Aspen Process Economic Analyzer to carry out the economic evaluation and found that the empty fruit bunches corresponded to the lowest ethanol production costs ($0.49/L) with a cogeneration scheme. A similar techno-economic evaluation was performed by Sassner and others (2008) who compared spruce, salix, and corn stover as poten-tial feedstocks for bioethanol production. Their production system was based on $SO_2$ catalyzed steam pre-treatment followed by simultaneous saccharification and fermentation. They concluded that using the pentose fraction for ethanol produc-tion was crucial to obtain good process economy. These studies indicate that use of low-cost feedstocks and enzymes are prerequisites for the cost-effective production of bioethanol.

## 15.4.2 BIODIESEL

Biodiesel refers to the fatty acid alky esters of plant oils or animal fats. They are suit-able alternatives to the conventional petroleum-based diesel fuel. Like bioethanol, the current prices of biodiesel are not economically attractive and new technologies and feedstocks are being tested to lower the overall production costs of the process (Aransiola et al. 2014). Apostolakou and others (2009) performed a techno-economic analysis of biodiesel production from vegetable oils using the classical alkali-catalyzed transesterification process. They observed that the raw material costs accounted for 75% of the total production costs for the small-scale plants, which can increase up to 90% for the large-scale systems. Zhang and others (2003) used the HYSYS® simulations to evaluate the economic impact of four biodiesel processes including two feedstocks (virgin vegetable oil and waste cooking oil) and two catalysts (alkali and acid catalyst). Their sensitivity analysis indicated that the plant capacity and price of feedstock are major factors that affect the biodiesel plant's profitability. Sighabhandhu and Tezuka (2010) estimated the profitability of biodiesel production plant by incorporating the glycerin purification process. Their results suggested that installing a glycerin purification process could benefit the biodiesel plant by glycerin recovery and decrease wastewater treatment costs. A similar observation was made by Haas and others (2006) who suggested that the biodiesel production costs were significantly affected by the variations in the market value of glycerol. Microalgal biodiesel production has gained recent interest due to the various associated advantages (such as limited land usage and high growth rate) over the traditional oil crops (Huang et al. 2010). A comprehensive techno-economic evaluation performed by Xin et al. (2016) suggested that cultivation, harvesting, and downstream conversion processes were critical parameters to obtain overall benefits and high internal rate of returns from an algal biodiesel plant. Further breakthroughs in biodiesel technology by assessing the potential of new and improved feedstocks, catalysts, and transesterification strategies can bring down production costs to acceptable ranges.

### 15.4.3 BIOMETHANATION

Biomethane production using organic wastes and residues has the potential to replace the conventional natural gas obtained from fossil fuels (Cucchiella and D'Adamo 2016). At present, there are three major sectors of commercial biogas plants which include the small-scale household biogas plants, industrial-scale anaerobic digesters, and landfill gas generators.

#### 15.4.3.1 Small Scale

Traditionally, the anaerobic digestion process was used mainly to treat high COD wastes rather that for energy generation purposes. However, during the early nineteenth century, the first small-scale biogas plants emerged, which showed that methane could be produced by decomposing cattle manure (Abbasi et al. 2012). This generated methane could be readily used for domestic heating and electricity generation purposes (Patterson et al. 2011). With the increased recognition for renewable energy sources, there is a renewed interest in biogas plants and in understanding their technical and economic efficiencies. Anaerobic systems capable of recovery and re-use of household and community wastes have prominent applications in developing and industrial countries. The major targets for a small-scale biogas plant include treatment of organic wastes and residues in a broad range of organic loads, energy production and use, improvement in sanitation, and production of high quality fertilizer. The design of a proper digester is crucial for high-rate methane production. Renda and others (2016) performed the economic evaluation of a small-scale biogas plant using the Net Present Value (NPV) analysis for a two-stage biofilm reactor. They concluded that for the two-stage AD prototype of 100 L capacity, and organic municipal waste fed at a rate of 13 kg VS/m$^3$ d, the total payback time would be 8 and 7 years assuming that the biogas was used to power CHP units of 100 kW$_{el}$ and 300 kW$_{el}$, respectively. A similar NPV analysis was performed by Salerno et al. (1995) for a small-scale farm-based biogas plant that was used to power CHP units of <300 kW$_{el}$. They observed a total payback time of 3.5 years, thus suggesting that the investment was economically and financially profitable. These studies suggest that small-scale anaerobic digesters are commercially viable systems that can serve the dual purpose of waste management and energy generation.

### 15.4.3.2 Industrial Scale

The industrial-scale anaerobic digesters are similar to small-scale digesters except that they are bigger in size and allow more storage of gas. They are used to convert the organic fraction of large volume of slurries and sludge into biogas for energy generation (Banks et al. 2011). The biogas is used either directly for heating the reactors or transformed into CHP and fed to the grid. It can be used also to upgrade the quality of natural gas. Typical substrates for industrial-scale anaerobic digesters include excess sludge from wastewater plants or waste slurries from agriculture and dairy industries (Mata-Alvarez et al. 2000). In addition, different energy crops can be added to increase the overall gas yields. The designs of the industrial anaerobic digesters are much more complex compared to small-scale digesters and require efficient construction, operation, and maintenance skills. Therefore, the economic evaluation of the industrial-scale anaerobic digesters must include the costs incurred during aforementioned processes. Carlini and others (2017) performed a comparative economic evaluation of three different anaerobic digester plant sizes (i.e., 100, 500, and 1000 kW), which were fed with organic agro-industrial waste. Their study focused on Italian framework conditions and it was observed that the improvement in break-point was closely related to the increase in plant size. The payback time was found to be 9, 5, and 4 years for the 100, 500, and 1000 kW plants, respectively. Similar studies were conducted by Browne and others (2011) who assessed the cost of gaseous biomethane production in Ireland. They observed that the biomethane produced using an organic fraction of municipal solid waste was cheaper (€ 0.36/L) compared to slaughter house waste (€ 0.65/L) and grass and slurry (€ 1.38/L–€ 1.45/L). These studies suggest that the cost of biomethane production is still in the range of conventional petroleum-derived transportation fuels and requires further advancements to become more competitive. Some of the successful applications of industrial anaerobic digesters across the world include (GMI 2013):

- Mixed plug flow anaerobic digester, Fair Oaks, Indiana, United States (biogas generation: 865 MMbtu/day)
- Upflow solids reactor, Shangdongkuo Village, Changping District, Beijing, China (biogas generation: 207,000 m³/year)

- Trapezoidal lake type anaerobic digester with a PVC-reinforced geo-membrane, Santa Rosillo Village, Huimbayoc, San Martin, Peru (biogas generation: 8.74–11.65 m³/day)
- Below ground cylindrical tank in RCC with hopper bottom digester, fitted with bio-media and a floating dome for gas collection, Elavampadam Model Rubber Producer's Society, India (biogas generation: 10–12 m³/day)
- Intermittently stirred-tank reactors, Biomethanation Plant, Punjab, India (biogas generation: 10,000 m³/day)

### 15.4.3.3 Landfill Gas Generation

Landfill gas is produced during the anaerobic decomposition of organic wastes. It mainly is comprised of 50%–55% methane, which is a high value fuel for gas engines that can be effectively used for power generation (Johari et al. 2012). Use of landfill gas helps in reduction of greenhouse gas (GHG) emissions (Demirbas 2008). Various studies have been conducted to estimate the cost benefits of using landfill gas for power generation. Shin and others (2005) evaluated the economic perspectives of landfill gas electricity generation in Korea using the LEAP (Long-range Energy Alternative Planning) system. They estimated that the annual cost for landfill gas use with gas engine (58 MW), gas turbine (53.5 MW), and steam turbine (54.5 MW) would be 45.1, 34.3, and 24.4 won/kWh, respectively. Moreover, the use of landfill gas for energy purposes accounted for a 75% decrease in global warming potential (Shin et al. 2005). A similar study was performed by Mbav and others (2012) who showed that the operating and maintenance costs for landfill electricity generation at a Western Cape landfill site in South Africa accounted for 210, 130, and 380 $/kW yr, using internal combustion engines, gas turbines, and micro-turbine technology, respectively. Yedla and Parikh (2001) performed an economic evaluation of a landfill system with gas recovery for municipal solid waste management in Mumbai city, India. They concluded that approximately $0.140 billion per annum could be saved with the landfill gas generation system in reference to the existing system of waste disposal. A similar observation was made by Kumar et al. (2004) who showed savings of $0.09 billion per annum with a landfill gas recovery system compared to the existing system of MSW disposal at Port Blair City, Andaman Islands, India. These studies suggest that landfill gas generation technology has the potential to address the global warming issues along with profitable energy generation.

### 15.4.4 BIOHYDROGEN

Biohydrogen is the most promising zero carbon fuel that can provide a solution to the world's energy and climate change issues. Although the technology is still in its developmental stages, a few demonstration plants are already in operation (Das et al. 2014). The economic analysis of this technology can help in assessing the crucial factors that contribute to increasing the costs of the overall process. Han et al. (2016) performed the techno-economic analysis of the dark fermentative hydrogen production from molasses using a continuously mixed immobilized sludge reactor.

They focused mainly on the effect of working volume on the economic performance of a hydrogen-producing plant and concluded that the return on investments increased from 37.2% to 47.3% with scales increasing from 10 m³ to 50 m³, respectively. Moreover, the payback time of 9.7 years and 6.9 years were estimated using the 40 m³ and 50 m³ plants, respectively, suggesting that large-scale biohydrogen plants (>40 m³) would be more economically feasible (Han et al. 2016). Sathyaprakasan and Kannan (2015) evaluated the economics of various biohydrogen production processes and concluded that the direct biophotolysis method was the least favorable in terms of land use and costs while photofermentation and dark fermentation were among the better options available. Balat and Kirtay (2010) reviewed the present and future prospects of hydrogen production from biomass and predicted that steam reforming and gasification from natural gas and biomass, respectively, will be the dominant technologies for hydrogen production by the end of the twenty-first century. On the other hand, Hay and others (2013) suggested that the two-stage biohydrogen production by integrating dark and photofermentation is more economically feasible because it can generate high revenues.

### 15.4.5 BIOHYTHANE

Hydrogen and methane are important gaseous energy carriers that have tremendous industrial applications. Both gases can be produced using biomass through the biohythane process, which is a promising technique for renewable energy generation (Das and Roy 2016). It mainly involves a two-stage fermentation of biomass such that the first stage consists of acidogenic hydrogen production, while the second stage consists of methanogenic methane production. Due to the higher energy efficiency and ease in operation, the biohythane process has gained worldwide attention in recent years. However, several technological hurdles still remain for the commercialization of these processes (Roy and Das 2017). Only a few significant studies have been conducted to evaluate the economic feasibility of biohythane process. Wilquist and others (2012) performed a cost analysis for a biohythane process based on wheat straw, which was assessed using process modelling. They revealed that the capital investments required for the steam-explosion pre-treatment and the gas upgrading equipment were the main contributors to the high production costs of the biohythane process. Moreover, use of expensive nutrients such as yeast extracts also significantly contributed to the overall production costs (Willquist et al. 2012). Therefore, it is essential to replace such expensive nutrients with low-cost organic wastes and residues to improve the economic viability of the process. More recently, the economic aspect of biohythane production was studied in detail by Zuldian et al. (2018) who used palm oil mill effluent as substrate in a 1 m³ bioreactor prototype for the two-stage fermentation process. They showed that the biohythane system was able to generate electricity at 0.735 $KW_{el}$ with a COD input of 35,000 ppm. The total payback time was 7.55 years with an internal rate return of 13% (Zuldian et al. 2018). These studies suggest that with further improvement in research and development (R&D) of biohythane technology, a profitable route for clean energy can be realized.

### 15.4.6  Biobutanol

Butanol production using the classical ABE (acetone-butanol-ethanol) fermentation process has gained renewed interest for renewable energy generation. Biobutanol is considered a promising liquid biofuel that has more beneficial fuel properties than ethanol (Tashiro and Sonomoto 2010). The biobutanol production is limited mainly by the low titres of butanol, which limit the scalability of the process. However, various technological developments over the years has resulted in higher volumetric productivities (Formanek et al. 1997; Tashiro et al. 2004). The reactor designs, mode of operation, solvent recovery, and the type of feedstock are crucial parameters that affect the overall yield and thus in turn the profitability of the process. For biobutanol to be a viable substitute to gasoline, the economics of biobutanol production must be assessed. Van Der Merwe and others (2013) compared the economics of various process designs for biobutanol production from sugarcane molasses using ASPEN Icarus software. According to their studies with respect to the economic conditions of South Africa, the fed-batch fermentation followed by $CO_2$ gas stripping proved to be the only profitable design. Baral and others (2016) assessed the economic viability of ABE fermentation using corn stover as feedstock and suggested that feedstock processing and fermentation process designs were the main hurdles that affected the economic competitiveness of biobutanol. A more comprehensive review of the economics of ABE fermentation performed by Ranjan and Moholkar (2012) concluded that the use of metabolically engineered strains, inexpensive substrates, and superior reactor designs can bring biobutanol close to commercialization.

## 15.5  MAIN CONSTRAINS OF THE BIOFUELS GENERATION PROCESSES

As discussed previously, each biofuel technology has certain technical challenges that are dependent mainly upon their development stage. However, certain common barriers are associated with all the bioenergy systems as discussed in the following sections.

### 15.5.1  Feedstock Availability and Handling

The feedstock quality and the moisture content are the most essential factors that impact the performance and reliability of the bioenergy plant, which in turn affects the economics of the process. The low bulk density biomass can be of varied structures and types, which can be difficult to handle and store (Zheng and Nirmalakhandan 2010). Moreover, as discussed in Section 15.2, most of the energy cops are limited by geographical locations and thus the distribution and supply logistics are limited by the trade policies of various countries.

### 15.5.2  Co-product Contamination

It is usually observed that the biofuel production is associated with generation of multiple by-products. The channelling of the redox equivalents towards the co-product formation can lead to lower yields of product of interest. Moreover, some of these by-products can have detrimental effects on the environment. For example, the left over ash or digestate may contain heavy metals that must be discarded with utmost care.

### 15.5.3 TOXIC EMISSIONS

Although the bioenergy technologies produce fewer toxic emissions compared to the fossil fuels, the increasing concerns over climate change and the stringent environmental regulations require proper life cycle assessment of biofuels prior to their commercialization. Since there are several types of biofuels and blends that originate from a variety of feedstocks, evaluating their effect on environment and human health becomes much more complex.

### 15.5.4 COMPETITION WITH OTHER RENEWABLES

Bioenergy faces competition with alternative renewable sources such as solar, wind, and hydrothermal in almost all its market segments. Although, at present, biomass is much cheaper compared to the other renewables. However, the complex conversion technologies as well as biomass transportation and biofuel infrastructure costs can significantly increase the overall production costs of the process making it economically uncompetitive. In addition, with the technological advances in the different biofuel technologies, an internal competition within the bioenergy sector for the feedstock sources is plausible in the near future.

### 15.5.5 POLICIES AND REGULATIONS

The successful implementation of biofuel technologies in the energy sector will require supportive government policies and subsidies. Similarly, there is a need for clarity in the regulatory aspects of biofuels such as setting the minimal quality standards for the commercialization of targeted product. The bioenergy policies must consider their impact on other sectors such as food and forestry. Any negative impact can affect the public acceptance of biofuels as the alternate energy source.

## 15.6 CONCLUSION AND PERSPECTIVES

Biofuels have the potential to tackle world's major energy and environmental concerns such as fossil fuel depletion, climate change, air and water pollution, and waste management. Therefore, assessing the energy and economic feasibility of the various biofuel production technologies is crucial for the determining their commercial viability. Feedstock selection and its availability are the most critical parameters that determine the scale and economy of a bioenergy system. Most traditional biofuels such as corn ethanol or oil palm biodiesel depend upon agricultural food crops that create land use issues and the food vs. fuel debate. The second-generation feedstocks are more economically favorable but require complex conversion technologies. The advanced biofuels such as biohydrogen and biobutanol are considered potential alternatives to the traditional biofuels. However, various technological barriers and the socio-economic concerns must be overcome to achieve cost parity with fossil fuels.

## REFERENCES

Abbasi T, Tauseef SM, Abbasi SA (2012) A brief history of anaerobic digestion and biogas. In: *Biogas Energy*. Springer, New York, pp. 11–23.

Achten WMJ, Almeida J, Fobelets V, Bolle E, Mathijs E, Singh VP, Tewari DN, Verchot L V., Muys B (2010) Life cycle assessment of Jatropha biodiesel as transportation fuel in rural India. *Appl Energy* 87:3652–3660.

Apostolakou AA, Kookos IK, Marazioti C, Angelopoulos KC (2009) Techno-economic analysis of a biodiesel production process from vegetable oils. *Fuel Process Technol* 90:1023–1031.

Aransiola EF, Ojumu TV, Oyekola OO, Madzimbamuto TF, Ikhu-Omoregbe DIO (2014) A review of current technology for biodiesel production: State of the art. *Biomass and Bioenergy* 61:276–297.

Balat H, Kirtay E (2010) Hydrogen from biomass: Present scenario and future prospects. *Int. J. Hydrog Energy* 35:7416–7426.

Banks CJ, Chesshire M, Heaven S, Arnold R (2011) Anaerobic digestion of source-segregated domestic food waste: Performance assessment by mass and energy balance. *Bioresour Technol* 102:612–620.

Baral NR, Shah A (2016) Techno-economic analysis of cellulosic butanol production from corn stover through acetone-butanol-ethanol fermentation. *Energy Fuels* 30:5779–5790.

Baral NR, Slutzky L, Shah A, Ezeji TC, Cornish K, Christy A (2016) Acetone-butanol-ethanol fermentation of corn stover: Current production methods, economic viability and commercial use. *FEMS Microbiol Lett* 363:33

Bauen A, Berndes G, Junginger M (2009) Bioenergy: A sustainable and reliable energy source. A review of status and prospects. *IEA Bioenergy* Main report 1–108.

Berglund M, Börjesson P (2006) Assessment of energy performance in the life-cycle of biogas production. *Biomass and Bioenergy* 30:254–266.

British Petroleum (2018) *BP Statistical Review of World Energy*. British Petroleum Company.

Browne J, Nizami AS, Thamsiriroj T, Murphy JD (2011) Assessing the cost of biofuel production with increasing penetration of the transport fuel market: A case study of gaseous biomethane in Ireland. *Renew Sustain Energy Rev* 15:4537–4547.

Carlini M, Mosconi EM, Castellucci S, Villarini M, Colantoni A (2017) An economical evaluation of anaerobic digestion plants fed with organic agro-industrial waste. *Energies* 10:1165.

Chandel K, Chan E, Rudravaram R, Narasu L, Rao V, Ravindra P (2007) Economic and environmental impact of bioethnaol production technologies: An appraisal. *Biotechnol Mol Biol Rev* 2:014–032.

Clark CM, Lin Y, Bierwagen BG, Eaton LM, Langholtz MH, Morefield PE, Ridley CE, Vimmerstedt L, Peterson S, Bush BW (2013) Growing a sustainable biofuels industry: Economics, environmental considerations, and the role of the conservation reserve program. *Environ Res Lett* 8:025016.

Coelho ST (2012) Traditional biomass energy: Improving its use and moving to modern energy use. In: *Renewable Energy: A Global Review of Technologies, Policies and Markets*. Taylor & Francis Group, Hoboken, NJ, pp. 230–261.

Cucchiella F, D'Adamo I (2016) Technical and economic analysis of biomethane: A focus on the role of subsidies. *Energy Convers Manag* 119:338–351.

Das D, Khanna N, Dasgupta CN (2014) Biohydrogen production process economics, policy, and environmental impact. In: *Biohydrogen Production: Fundamentals and Technology Advances*. CRC Press, Boca Raton, FL, pp 343–370.

Das D, Roy S (2016) *Biohythane: Fuel For the Future*. CRC Press, Boca Raton, FL.

de Jong S, Hoefnagels R, Wetterlund E, Pettersson K, Faaij A, Junginger M (2017) Cost optimization of biofuel production—The impact of scale, integration, transport and supply chain configurations. *Appl Energy* 195:1055–1070.

Demirbas A (2009) Political, economic and environmental impacts of biofuels: A review. *Appl Energy* 86:S108–S117.

Demirbas A (2008) Biofuels sources, biofuel policy, biofuel economy and global biofuel projections. *Energy Convers Manag* 49:2106–2116.

Fore SR, Porter P, Lazarus W (2011) Net energy balance of small-scale on-farm biodiesel production from canola and soybean. *Biomass Bioenergy* 35:2234–2244.

Formanek J, Mackie R, Blaschek HP (1997) Enhanced butanol production by Clostridium beijerinckii BA101 grown in semidefined P2 medium containing 6 percent maltodextrin or glucose. *Appl Environ Microbiol* 63:2306–2310.

GMI (2013) Successful Applications of Anaerobic Digestion From Across the World by Global Methane Initiative (GMI). https://www.globalmethane.org/documents/GMI Benefits Report.pdf. Accessed November 18, 2018.

Gnansounou E, Dauriat A, Villegas J, Panichelli L (2009) Life cycle assessment of biofuels: Energy and greenhouse gas balances. *Bioresour Technol* 100:4919–4930.

Haas MJ, McAloon AJ, Yee WC, Foglia TA (2006) A process model to estimate biodiesel production costs. *Bioresour Technol* 97:671–678.

Hamelinck CN, Van Hooijdonk G, Faaij APC (2005) Ethanol from lignocellulosic biomass: Echno-economic performance in short-, middle- and long-term. *Biomass and Bioenergy* 28:384–410.

Han W, Liu Z, Fang J, Huang J, Zhao H, Li Y (2016) Techno-economic analysis of dark fermentative hydrogen production from molasses in a continuous mixed immobilized sludge reactor. *J Clean Prod* 127:567–572.

Hay JXW, Wu TY, Juan JC, Md. Jahim J (2013) Biohydrogen production through photo fermentation or dark fermentation using waste as a substrate: Overview, economics, and future prospects of hydrogen usage. *Biofuel Bioprod Bior* 7:334–352.

Hill J, Nelson E, Tilman D, Polasky S, Tiffany D (2006) Environmental, economic, and energetic costs and benefits of biodiesel and ethanol biofuels. *Proc Natl Acad Sci* 103:11206–11210.

Hu J, Cadham WJ, Vandyk S, Saddler JN (2017) Will biomass be used for bioenergy or transportationbiofuels? What drivers will influence biomass allocation. *Front Agric Sci Eng* 4:473.

Huang G, Chen F, Wei D, Zhang X, Chen G (2010) Biodiesel production by microalgal biotechnology. *Applied Energy* 87:38–46.

IEA (2017) WEO-2017 *Special Report: Energy Access Outlook*. IEA, Paris, France.

IEA Bioenergy (2006) *The Availability of Biomass Resources for Energy*. 2006.

Johari A, Ahmed SI, Hashim H, Alkali H, Ramli M (2012) Economic and environmental benefits of landfill gas from municipal solid waste in Malaysia. *Renew Sustain Energy Rev* 16:2907–2912.

Kazi FK, Fortman JA, Anex RP, Hsu DD, Aden A, Dutta A, Kothandaraman G (2010) Techno-economic comparison of process technologies for biochemical ethanol production from corn stover. *Fuel* 89:S20–S28.

Koutinas A, Kanellaki M, Bekatorou A, Kandylis P, Pissaridi K, Dima A, Boura K et al., (2016) Economic evaluation of technology for a new generation biofuel production using wastes. *Bioresour Technol* 200:178–185.

Kralova I, Sjöblom J (2010) Biofuels-renewable energy sources: A review. *J Dispers SciTechnol* 31:409–425.

Kumar S, Gawaikar V, Gaikwad SA, Mukherjee S (2004) Cost-benefit analysis of landfill system with gas recovery for municipal solid waste management: A case study. *Int J Environ Stud* 61:637–650.

Manish S, Banerjee R (2008) Comparison of biohydrogen production processes. *Int J Hydrogen Energy* 33:279–286.

Mata-Alvarez J, Macé S, Llabrés P (2000) Anaerobic digestion of organic solid wastes. An overview of research achievements and perspectives. *Bioresour Technol*. 74:3–16.

Mbav WN, Chowdhury S, Chowdhury SP (2012) Feasibility and cost optimization study of landfill gas to energy projects based on a Western Cape landfill site in South Africa. In: *Proceedings of the Universities Power Engineering Conference*. IEEE, pp 1–6.

Mcaloon A, Taylor F, Yee W, Ibsen K, Wooley R (2000) Determining the Cost of Producing Ethanol from Corn Starch and Lignocellulosic Feedstocks. *Agriculture* 44.

Mohammadshirazi A, Akram A, Rafiee S, Bagheri Kalhor E (2014) Energy and cost analyses of biodiesel production from waste cooking oil. *Renew Sustain Energy Rev* 33:44–49.

Murali P, Hari K, Puthira Prathap D (2016) An economic analysis of biofuel production and food security in India. *Sugar Tech* 18:447–456.

Naik SN, Goud V V., Rout PK, Dalai AK (2010) Production of first and second generation biofuels: A comprehensive review. *Renew Sustain Energy Rev* 14:578–597.

Nayak BK, Pandit S, Das D (2013) Biohydrogen. In: *Air Pollution Prevention and Control*. John Wiley & Sons, Chichester, UK, pp 345–381.

Nigam PS, Singh A (2011) Production of liquid biofuels from renewable resources. *Prog Energy Combust Sci* 37:52–68.

Oreggioni GD, Luberti M, Reilly M, Kirby ME, Toop T, Theodorou M, Tassou SA (2017) Techno-economic analysis of bio-methane production from agriculture and food industry waste. *Energy Procedia* 123:81–88.

Patterson T, Esteves S, Dinsdale R, Guwy A (2011) An evaluation of the policy and techno-economic factors affecting the potential for biogas upgrading for transport fuel use in the UK. *Energy Policy* 39:1806–1816.

Pimentel D (2003) Ethanol fuels: Energy balance, economics, and environmental impacts are negative. *Nat Resour Res* 12:127–134.

Pradhan A, Shrestha DS, Gerpen J Van, Duffield J (2008) The energy balance of soybean oil biodiesel production: A review of past studies. *Trans ASABE* 51:185–194.

PSU (2016) 11.3 Economics of butanol production. In: E-education.psu.edu. https://www.e-education.psu.edu/egee439/node/721. Accessed November 15, 2018.

Quintero JA, Moncada J, Cardona CA (2013) Techno-economic analysis of bioethanol production from lignocellulosic residues in Colombia: A process simulation approach. *Bioresour Technol* 139:300–307.

Qureshi N, Blaschek HP (2000) Economics of butanol fermentation using hyper-butanol producing Clostridium beijerinckii BA101. *Food Bioprod Process Trans Inst Chem Eng Part C* 78:139–144.

Ranjan A, Moholkar VS (2012) Biobutanol: Science, engineering, and economics. *Int J Energy Res* 36:277–323.

Renda R, Gigli E, Cappelli A, Simoni S, Guerriero E, Romagnoli F (2016) Economic feasibility study of a small-scale biogas plant using a two-stage process and a fixed bio-film reactor for a cost-efficient production. *Energy Procedia*. 95:385–392.

Roy S, Das D (2016) Biohythane production from organic wastes: Present state of art. *Environ Sci Pollut Res* 23:9391–9410.

Salerno M, Gallucci F, Pari L, Zambon I, Sarri D, Colantoni A (1995) *Costs-benefits Analysis of A Small-Scale Biogas Plant and Electric Energy Production*. Agricultural Academy, Bulgaria.

Sassner P, Galbe M, Zacchi G (2008) Techno-economic evaluation of bioethanol production from three different lignocellulosic materials. *Biomass and Bioenergy* 32:422–430.

Sathyaprakasan P, Kannan G (2015) Economics of bio-hydrogen production. *Int J Environ Sci Dev* 6:352–356.

Shapouri H, Duffield JA, Wang MQ (2003) The energy balance of corn ethanol revisited. *Trans ASAE* 46(4):959.

Shapouri H, Salassi M (2006) *The Economic Feasibility of Ethanol Production from Sugar in the United States*. USDA Rep 78.

Shin HC, Park JW, Kim HS, Shin ES (2005) Environmental and economic assessment of landfill gas electricity generation in Korea using LEAP model. *Energy Policy* 33:1261–1270.

Sims R, Taylor M, Jack S, Mabee W (2008) From 1st to 2nd Generation Bio Fuel Technologies: An overview of current industry and R&D activities. *IEA Bioenergy* 1–124.

Singhabhandhu A, Tezuka T (2010) A perspective on incorporation of glycerin purification process in biodiesel plants using waste cooking oil as feedstock. *Energy* 35:2493–2504.

Tashiro Y, Sonomoto K (2010) Advances in butanol production by clostridia. *Current Research, Technology and Education Topics in Applied Microbiology and Microbial Biotechnology*, 2:1383–1394.

Tashiro Y, Takeda K, Kobayashi G, Sonomoto K, Ishizaki A, Yoshino S (2004) High butanol production by *Clostridium saccharoperbutylacetonicum* N1-4 in fed-batch culture with pH-Stat continuous butyric acid and glucose feeding method. *J Biosci Bioeng* 98:263–268.

Van Der Merwe AB, Cheng H, Görgens JF, Knoetze JH (2013) Comparison of energy efficiency and economics of process designs for biobutanol production from sugarcane molasses. *Fuel* 105:451–458.

Willquist K, Nkemka VN, Svensson H, Pawar S, Ljunggren M, Karlsson H, Murto M, Hulteberg C, Van Niel EWJ, Liden G (2012) Design of a novel biohythane process with high H2and CH4production rates. *Int J Hydrog Energy* 37:17749–17762.

Xin C, Addy MM, Zhao J, Cheng Y, Cheng S, Mu D, Liu Y, Ding R, Chen P, Ruan R (2016) Comprehensive techno-economic analysis of wastewater-based algal biofuel production: A case study. *Bioresour Technol* 211:584–593.

Yedla S, Parikh JK (2001) Economic evaluation of a landfill system with gas recovery for municipal solid waste management: A case study. *Int J Environ Pollut* 15:433.

Zhang X, Yan S, Tyagi RD, Surampalli RY, Valéro JR (2016) Energy balance of biofuel production from biological conversion of crude glycerol. *J Environ Manage* 170:169–176.

Zhang Y, Dubé MA, McLean DD, Kates M (2003) Biodiesel production from waste cooking oil: 1. Process design and technological assessment. *Bioresour Technol* 89:1–16.

Zheng X, Nirmalakhandan N (2010) Cattle wastes as substrates for bioelectricity production via microbial fuel cells. *Biotechnol Lett* 32:1809–1814.

Zuldian P, Hastuti ZD, Murti SDS, Adiarso A (2018) Biohythane system using two steps of POME fermentation process for supplying electrical energi: Economic evaluation. In: *IOP Conference Series: Materials Science and Engineering*. IOP Publishing, p. 012010.

# Index

Printed and bound by CPI Group (UK) Ltd, Croydon, CR0 4YY

24/10/2024

01778307-0005